New Advances in Indoor Acoustics and Thermal Comfort for Sustainable Buildings

New Advances in Indoor Acoustics and Thermal Comfort for Sustainable Buildings

Guest Editors

Davide Borelli
Umberto Berardi

MDPI

Basel • Beijing • Wuhan • Barcelona • Belgrade • Novi Sad • Cluj • Manchester

Guest Editors
Davide Borelli
Dipartimento di Ingegneria Meccanica, Energetica, Gestionale e dei Trasporti
Università degli Studi di Genova
Genoa
Italy

Umberto Berardi
ArCoD Department
Polytechnic University of Bari
Bari
Italy

Editorial Office
MDPI AG
Grosspeteranlage 5
4052 Basel, Switzerland

This is a reprint of the Special Issue, published open access by the journal *Energies* (ISSN 1996-1073), freely accessible at: https://www.mdpi.com/journal/energies/special_issues/Indoor_Acoustics_Thermal_Comfort_Building.

For citation purposes, cite each article independently as indicated on the article page online and as indicated below:

Lastname, A.A.; Lastname, B.B. Article Title. *Journal Name* **Year**, *Volume Number*, Page Range.

ISBN 978-3-7258-5159-1 (Hbk)
ISBN 978-3-7258-5160-7 (PDF)
https://doi.org/10.3390/books978-3-7258-5160-7

Cover image courtesy of Umberto Berardi

Contents

energies

MDPI

Review

Energizing Sustainable Agriculture: Advances in Greenhouse Heating through Microwave-Based Technologies

Francesco Cepolina [1], Federico Silenzi [2], Leonardo Cirillo [1], Corrado Schenone [1,*] and Matteo Zoppi [1]

1 Department of Mechanical, Energy, Management and Transportation Engineering, University of Genoa, Via all'Opera Pia 15a, 16100 Genoa, Italy; francesco.cepolina@edu.unige.it (F.C.); leonardo.cirillo@edu.unige.it (L.C.); matteo.zoppi@unige.it (M.Z.)

2 National Institute for Nuclear Physics, Genoa Section, Via Dodecanneso 33, 16145 Genoa, Italy; federico.silenzi@ge.infn.it

* Correspondence: corrado.schenone@unige.it

Abstract: For the agricultural sector to develop sustainably in the future, progress toward more environmentally friendly technologies and methods is crucial. It is necessary to increase output while reducing the demand for energy, agrochemicals, and water resources. Although greenhouses can be utilized successfully for this purpose, significant technical advancements are required, especially when it comes to heating, to lower the use of fossil fuels and boost energy efficiency. Microwaves can warm plants without heating the entire greenhouse volume, which takes a significant amount of energy to compensate for heat loss in the outdoor environment. In this paper, through a thorough examination of the state of the art, a general overview of novel greenhouse heating systems based on radiation is reported. First, the strengths and weaknesses of microwave heating are discussed, and finally, the use of microwaves for soil sterilization is examined. All outcomes suggest these irradiation-based technologies can contribute significantly to energetically sustainable agriculture; moreover, they can be used to increase plant comfort.

Keywords: microwaves heating; greenhouse heating; pest control; soil sterilization

1. Introduction

The greenhouse is essential to agriculture because it offers a regulated environment that optimizes plant growth by controlling variables like temperature and humidity. Prolonged growing seasons, higher food yields, and defense against erratic weather patterns are some benefits of greenhouses. The high upfront expenses of greenhouse setup, worries about energy use, and possible environmental effects are some of the challenges they present. Greenhouses are essential for meeting the expanding food needs of the world's population despite these obstacles. In doing so, they enhance overall food security and meet the needs of a growing population. They also promote sustainable agriculture by facilitating year-round farming, increasing resource efficiency, and reducing the impact of climate change on crop output.

The goal of agricultural advances has been to boost cultivar yields and quicken the growth of crops. To support the workers, agricultural exoskeletons can be used for lifting heavy weights in the greenhouse in confined spaces and for harvesting [1]. The most important factor in agriculture is the environmental conditions, which can greatly affect crop results and yield. Greenhouses introduce the concept of "protected agriculture" by physically shielding crops from harsh environmental conditions while trying to control the temperature and humidity inside the greenhouse itself. The spectrum is divided into different regions based on wavelength [2].

Ultraviolet (UV) radiation is followed by visible light, which approximately corresponds to photosynthetically active radiation (PAR). Beyond visible light, there is near-infrared (NIR) radiation, and finally, far-infrared (FIR) radiation, which encompasses longer wavelengths associated with heat and thermal imaging. Greenhouse solar heating exploits the segregation of solar radiation into the greenhouse volume, increasing its inner temperature. A greenhouse is a structure with a roof and external surfaces made of transparent materials, i.e., glass, plastic, or nylon sheets. The cover material must be transparent to the short infrared waves coming from the Sun but not to radiation in the IR spectrum. The greenhouse effect happens because the ground is heated by the incoming short-wave radiation. It then reemits a portion of the energy acquired with a different wavelength in the infrared spectrum. Long-wave infrared radiation cannot cross the covering material, which absorbs it and eventually reemits the absorbed energy inside the greenhouse volume. While this effect benefits heat crops during colder seasons, the phenomenon is dependent on external atmospheric conditions, where sun irradiation is fundamental but relatively scarce compared to mild seasons. This explains why, over the years, the need for control of the greenhouse climatic conditions has increased. Therefore, the energy efficiency and consumption of a modern structure design are crucial factors. Often, there is a need to adopt measures to heat or cool the greenhouse. For instance, in the case of heating, depending on the installation area and local climatic conditions, the fuel consumption varies greatly: the energy consumption of greenhouse systems ranges from 60 to 80 kWh/m^2 per year in the Mediterranean region to 460–930 kWh/m^2 annually in central and northern Europe [3]. To give a contextualized scenario, the cost of heating energy accounts for up to 30 to 40% of the entire cost of greenhouse crop agricultural production in Italy [4]. As a result, one of the primary goals of a modern agricultural enterprise is to reduce production costs and boost production efficiency by making environmentally friendly decisions that will lower fuel consumption and CO_2 emissions while increasing crop cycle energy efficiency, defined as the proportion between plant production and primary energy consumed (either carbon fossil or renewable). The other two key parameters are the economic value of crop production and the carbon ratio.

Greenhouse heating systems are usually powered by electricity or fossil fuels, with a significant environmental and economic footprint. Hence, there is a need to optimize the energy performance of greenhouses using new technologies that allow one to maintain the advantages of this type of cultivation while reaching a more economically and environmentally sustainable equilibrium. Several types of technological enhancements have been introduced for this purpose, ranging from solar panels (thermals and photovoltaics) to geothermal [5], from biomass re-use [6] to cogeneration, infrared waves, and microwaves (frequencies mid ranges 3000 GHz and 3 GHz).

Among the various technologies aimed at increasing the energy efficiency of greenhouses, the application of microwaves appears very promising. This is thanks to the great flexibility of use and the possibility of combining different functions, such as heating or soil sterilization. Besides, microwave heating would permit, at least at a theoretical level, a tuning of energy based on the growth needs of the plants by modulating the supply capable of simultaneously achieving rapid vegetation and contained consumption.

The goal of this review is to demonstrate how far microwave technology has come in agriculture. Managing all the factors at once is challenging: farmers want to decrease their carbon footprint while minimizing running costs and keeping cheap investment costs and a dependable heating system. Research in the field of greenhouse microwave heating is still under development, making it difficult to forecast the industrial cost of the equipment. The carbon footprint can be minimized using renewable energy that powers efficient equipment. The high efficiency of microwave heating may be a smart green option (compared to burning fossil fuels) if the microwave generators are powered using "green energy" (from renewable sources). The efficiency of the energy production plant is another key parameter: for example, a fuel cell plant using cogeneration is one of the most efficient options for producing heat and electricity.

The overall saving that can be achieved using microwave technology also depends on how the electric energy is generated. A fuel cell plant is more efficient than a combustion plant [7]. Cogeneration enhances the efficiency of the combustion and fuel cell plants (Table 1).

Table 1. Comparison among combustion plant, fuel cell, and cogeneration plant.

	Combustion Plant	Fuel Cell Plant	Cogeneration Plant
Efficiency	40% to 45% A combustion power plant's efficiency is influenced by the fuel used, the plant layout, and the generation technology.	40% to 60% Fuel cells skip the combustion step and use an electrochemical method to convert fuel directly into electricity.	80% to 90% Combined heat and power (CHP) are very efficient. They concurrently produce energy and use the waste heat for other uses, like heating or cooling.
Environmental Impact	Due to the release of greenhouse gases such as carbon dioxide (CO_2), sulfur dioxide (SO_2), nitrogen oxides (NO_x), and particulate matter, combustion-based power plants have a significant negative influence on the environment.	Since fuel cells generate power through an electrochemical reaction rather than combustion, they have a moderate environmental impact. However, depending on the fuel used to power the fuel cell, the environmental impact may change.	Cogeneration facilities typically have a low to moderate impact on the environment. They increase overall efficiency and use less fuel to produce heat and electricity independently by using waste heat.
Fuel Flexibility	Combustion power plants rely on fossil fuels, such as coal, oil, or natural gas.	Fuel cells can run on hydrogen, natural gas, methanol, or biomass-based fuels.	Cogeneration plants often rely on conventional fuels such as fossil fuels or biomass.

A cogenerator or CHP, which can produce at least two power flows—thermal power recovered from the cooling oil through specific exchangers and electrical power from the generator connected to the engine—is the optimal choice when considering a greenhouse, as it requires heat and electricity. Therefore, the proposed system enables the recovery of secondary thermal energy that would otherwise be wasted, increasing the overall efficiency of the system [8–10]. The heat recovered can be used to produce hot water, which in turn can transfer heat to the root portion of the crops, which does not require high thermal gradients but benefits from the increase in temperature. The electrical energy produced by the CHP can be converted into microwaves capable of heating the vegetables. For example, depending on the cultivar, the leaf temperature should preferably be set between 10 °C and 20 °C, considering basil (essential for the agricultural economy in the Ligurian Region), using microwaves with a wavelength of 5.8 GHz. With this frequency, a shallower leaf penetration depth is guaranteed, and less energy is released into the environment [8]. Moreover, the control process and the design of the heating system are essential and can be optimized using the latest cutting-edge technologies, including artificial intelligence, which may further improve process optimization [11].

In this framework, the present work aims to gather most information available in the literature, organizing it in a review that shows the benefits associated with using microwave-based technologies in protected agriculture, from heating to soil disinfection and pest control, and invasive species reduction. Many available sources focus on specific topics, giving deep insights that can be too specific. The present work tries to broaden the analysis while preserving the most useful information.

To summarize, the intense progress that has recently accompanied the development of technology for microwave application in the agricultural field to greenhouse systems deserves to be described and analyzed in terms of technological effectiveness and possible energy consumption reduction. This is precisely the purpose of this paper. In this regard, after Section 1 provides the framework in which this work is developed, microwave application for greenhouse heating is analyzed in Section 2. Then, Section 3 examines the use of microwaves for soil sterilization in terms of energy efficiency and pollutant reduction. Section 4 briefly discusses the application of microwave thermal treatments for seed drying. Section 5 is dedicated to discussing the influence of the main parameters (irradiation uniformity, frequency, power, and time) to offer readers a reasoned analysis of the strengths and weaknesses of this technology. Section 6 presents a commentated summary of the presented work.

2. Microwave Heating

Heating is a crucial factor for plant development, given its direct and indirect roles. Maintaining the correct temperature avoids the growth of molds, fungi, and other pathogens, leading to increased growth and overall health of the crops. Among the available heating technologies, microwave heating generates an electric field with a magnetic field component [12]. While polarization leads to dielectric heating, conduction generates magnetic loss heating (Table 2). For greenhouse heating, waves between 1 GHz and 300 GHz are relevant.

Table 2. Microwave dielectric and magnetic heating.

	Electric Field	Magnetic Field
Phenomena	Polarization	Conduction
Heating	Dielectric	Magnetic loss
Parameters	• Dipolar polarization • Ionic conduction	• Eddy current, hysteresis • Resonance, residual losses

This range falls within the Industrial, Scientific, and Medical (ISM) band, a collection of electromagnetic spectrum areas dedicated to industrial, scientific, and medical use. Dielectric heating refers to the heating of the dielectric itself caused by an electromagnetic field. The polarized molecules in the dielectric, which continuously rotate by the magnetic field's fluctuating orientation, cause this phenomenon. The material's dielectric properties are also crucial; the propensity to reduce the strength of the internal electric field is measured by the dielectric permittivity. The magnetic permeability of a substance describes its propensity to become magnetized in the presence of a magnetic field. The amount of water and free charge carriers, such as salts, in an organic material significantly impacts its dielectric characteristics. When added to pure water, salts reduce a material's dielectric constant and raise the dissipated power density. Conversely, organic compounds with low water content have less solvent available for the salts to dissolve, lowering ionic conduction by decreasing ion mobility. This is primarily a loss process, especially at low frequencies.

Water quantity and its physical presence in the medium state directly affect its relaxation frequency, i.e., the frequency at which the dielectric loss factor achieves a maximum:

- If the water is in a solid state, the relaxation frequency is of the kHz order of magnitude.
- If the water is bound to the molecules of the (plant) biomass inside which it is located, the relaxation frequency is of the MHz order of magnitude [13–15].
- If the water is free, the relaxation frequency is of the GHz order of magnitude.

According to Henry et al. [16], the difference in power dissipation in organisms containing free water compared to those with only bound water can be up to 10^4 times. For instance, free water is found in fruits rich in juice (e.g., oranges, lemons, watermelon, and tomatoes), while bound water is primarily found in vegetables like cacti and pine tree needles.

Heredia et al. [17] investigated the feasibility of exploiting a combination of osmotic dehydration and microwave drying as a potential new process that could improve the quality of dried tomatoes. The authors investigated the interaction between microwaves and vegetable tissues, reporting that the dielectric properties of materials depend, amongst other factors, on the frequency of the electric field applied. To support this claim, Sun et al. [18] show that the microwave frequency directly impacts heating and its homogeneity: at low RF frequencies, the power loss emitted by the waves is greater, and it diminishes in the microwave regime, while at a higher frequency, the losses increase roughly linearly. For these reasons, the ideal power loss density and power penetration depth must be balanced. As the frequency increases, so too does the power loss density. The ISM band availability and the impact of the wavelength on the homogeneity of the field patterns inside the installation restrict the choice of frequency. Hence, the ISM frequencies of 915 MHz, 2450 MHz, and 5800 MHz offer the best compromise between power loss density and penetration depth.

Since air is transparent to microwaves, atmospheric temperature is unaffected because only the water inside the plant is heated. This allows the heating energy to be directed toward the vegetables rather than being dispersed in large volumes of air. The associated experimental results are generally consistent.

Teitel et al. [19] investigated the use of microwave radiation as a heating source to ripen tomatoes and peppers in a greenhouse cavity, focusing on eliminating grey mold. A shielding metal cavity that measured 0.78 m × 0.78 m × 1.7 m was used to house the microwave devices. A 2 m × 2 m × 2 m polyethylene greenhouse contained the cavity. The control group consisted of crops grown in an identical greenhouse heated conventionally. The plants were heated with a 500-watt microwave source and thermostatically controlled with thermocouples, resulting in roughly four on/off cycles every minute. The plants were heated without causing major damage or increasing their vulnerability to grey mold. The experiment duration was two weeks; considering the lifespan of the developing plants, the short test period was highly inadequate. The temperature range for the target crops was not indicated. From an energy point of view, the authors determined that the microwave system is ~75% more efficient than a hot air heater. A few years later, the experiment proposed by Teitel et al. [19] was replicated and enhanced by Guess et al. [20].

The authors set up an experiment to compare the growth of tomato plants heated by microwave (MW) with two control groups: a cold control group (CC) and hot control group (HC). The HC group received conventional hot air heating, while the CC group received no heating. The MW group underwent additional differentiation to assess how the spatial and temporal distribution of microwaves impact the growth process. A linearly polarized antenna (Antenna 1) with a 9.14 dB gain and beamwidths of 40.6° and 98.4° in the horizontal and vertical planes, respectively, was placed 35 cm away from a tomato plant on a rotating dish. A circularly polarized antenna (Antenna 2) with a 10.45 dB gain and a horizontal and vertical beamwidth of 35° was placed in front of two additional plants. To further test the influence of spatial distribution, one of the plants was placed on a rotating dish positioned 55 cm in front of the antenna's center line, while the other was positioned 55 cm away from the antenna, without any motion-inducing devices. Both Antenna 1 and Antenna 2 received a 60 W power feed.

The plant heated using linearly polarized microwaves showed an average temperature of 17.5 °C. The other two plants, exposed to the circularly polarized field, were maintained at an average temperature of 21 °C. The former displayed noticeably more burn damage despite the latter being exposed to higher power and temperature. This is explainable, considering that none of the plants were precisely aligned with the electric field, and circularly polarized fields provide superior heating uniformity. This also clearly shows that using a rotating dish does not significantly improve the thermal homogeneity between the considered samples. According to the study, tomato crops can be heated and grown successfully in greenhouses using microwaves, and the growth rate is solely based on plant temperature. The experiments revealed that the hot control plants outperformed the microwave group during the growth period (Figure 1); however, it must be considered that the microwave-heated plant group was, on average, cooler than the hot control group, likely due to insufficient MW power output [20].

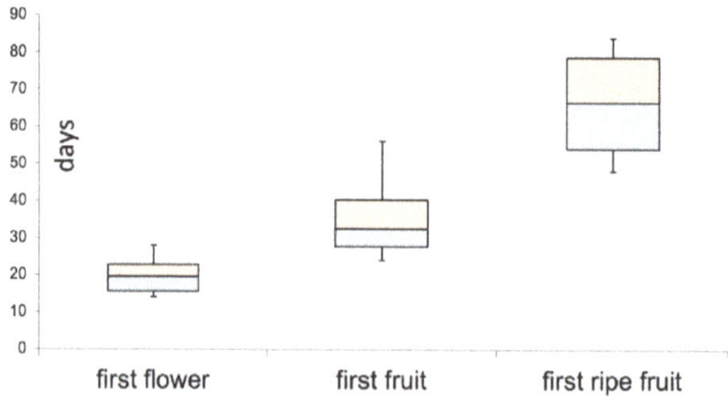

Figure 1. Growth comparison between experimental groups [20].

The output power of the amplifiers and the antenna gain can be controlled so that the power increases with the plant's growth to achieve maximum efficiency. As shown in Figure 2, circular polarization of the MW field can prevent non-uniform heating-related damage to vegetable tissues.

Figure 2. Leaf damage induced by a linearly polarized MW field.

When one considers the power requirements of microwave heating, it is necessary to keep in mind that the vast surface area and relatively thin geometries of the various morphological parts of vegetables cause the thermal energy to dissipate quickly into the environment.

Simulations and experiments revealed that a 200–300 W/m^2 power density was needed to heat plants. Plant death is caused by excessive cold (low power densities of $\leq$50 W/m^2) or excessive heat (power densities $\geq$ 500 W/m^2) [20]. To assess the potential damage induced by prolonged exposure to microwaves over a long period, Schmutz et al. [21] investigated the impact of prolonged microwave exposure on fir and beech trees. For three and a half years, four groups of trees were subjected to microwave radiation at 2.45 GHz and various power densities. The group exposed to the highest power density had an average temperature that was 4 °C higher than the other groups. There were no noticeable microwave effects on the plants, and plant height and canopy transparency—a characteristic used to measure forest conditions related to leaf density on different trees—were similar among the groups. It is unknown how much microwaves may have contributed to the calcium shortage in the high-power density group during the experiment, which was not addressed in the study conclusions.

Another interesting study investigated heat transmission in leaves [22] or peels [23] exposed to microwave radiation. In a closed-circuit wind tunnel, the thermal behavior of microwave-exposed leaves from several species with constant and intermittent wind conditions was investigated. Microwave energy from a transmitter (Model EMS, Microtron 200) operating at a frequency of 2450 MHz was used to heat a single leaf. The authors investigated the heat transfer coefficient of different varieties, finding that it is dependent on the species. For instance, water hyacinth leaves displayed a two- to four-times higher transient heat transfer coefficient with respect to tobacco leaves, and leaf temperatures were in good agreement with the mathematical model predictions analyzed in the paper. The research also showed a linear relationship between wind intermittency and equilibrium leaf temperatures for dry and wet leaves. Leaves with more water were heated more by the same amount of microwaves. The Huber [24] greenhouse, heated by microwaves, is now analyzed (Figure 3)

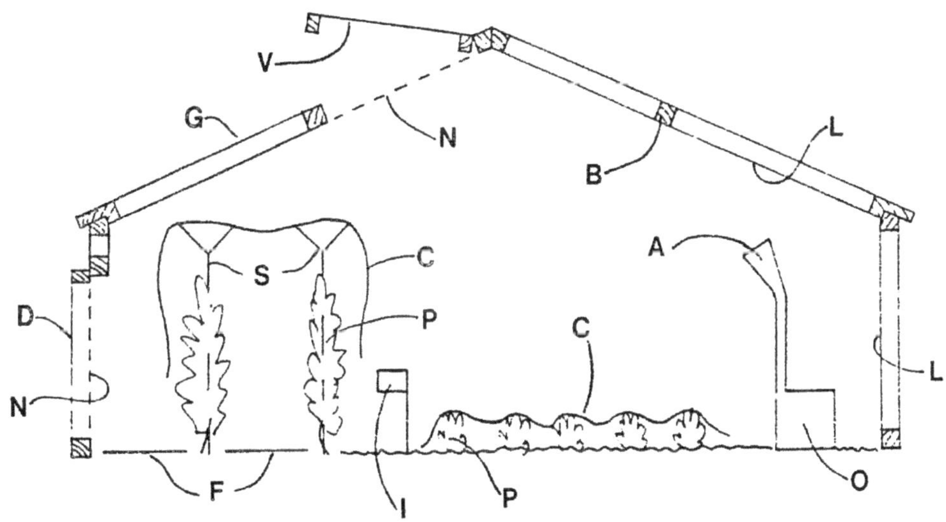

Figure 3. General system setup as proposed by Huber's patent n° GB2120065A. Meaning of letters: "D" door, "O" magnetron oscillator, "A" horn antenna, "L" reflecting liner, "G" glazing, "B" glazing bars, "V" ventilators, "N" metal mesh, "F" metal foil, "P" plants to be heated, "C" film of thin, clear, polythene sheet, "S" plant supports, "I" infra-red detector.

Crops are particularly sensitive to temperature changes. Thus, the fundamental obstacle to heating plants using microwaves has been recognized as the non-uniformity of the electric field. For this reason, Huber patented a system that aims to increase the uniformity of the microwave field, by irradiating a space lined with special reflective material, causing multiple reflections to combine and form a complex pattern of waves that fill the area.

The components adopted to create the glasshouse heating control systems are potentially subject to microwaves, heat/cold stress cycles and high humidity [25]. The major downfall of the author's proposal is that, given the difficulty of creating a homogeneous wave pattern within the cavity, it is quite difficult to set up the system for a larger greenhouse (this may be a limiting factor), and the reflected barriers proposed in the patent must be positioned above the plant causing unpleasant physical barriers to natural light.

The greenhouse's inner surfaces, coated in reflective material, can be thought of as a huge microwave cavity. To achieve a homogeneous radiation density, the microwaves are emitted using a horn antenna (1 kW horn antenna at 2.45 GHz), and after being reflected from the ceiling and the walls, the radiation reaches the crops.

An infrared sensor is used in a closed-loop feedback system to regulate the power. Huber claims that carefully designed reflecting bodies in the cavity improve the uniformity of the electromagnetic field radiating the plants; moreover, neighboring leaves exchange heat to make up for heating inequalities.

Like Huber [24], the University of Genoa proposed a microwave-heated greenhouse for growing basil in a patent titled "System and method for heating greenhouses" [8]. A sketch of the proposed system is shown in Figure 4; note that for the sake of brevity, the components listed are not extensively described.

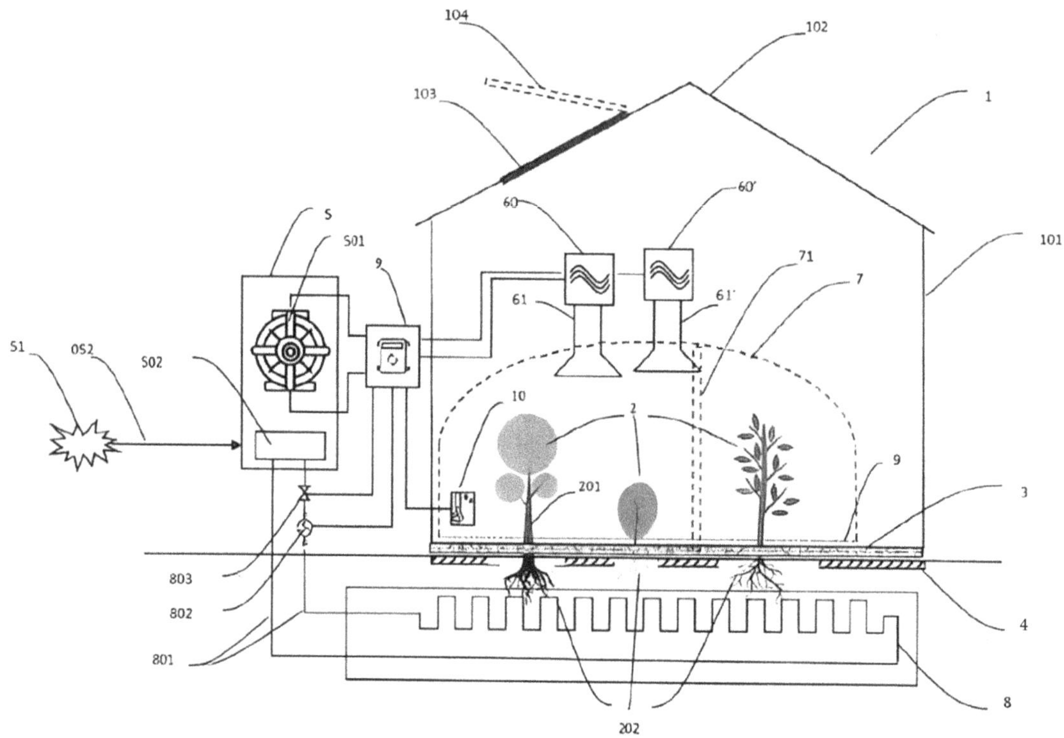

Figure 4. General system setup as proposed by Carusi et al. [8] patent n° WO2021079263A1. Meaning of symbols: "1" greenhouse, "3" soil, "4" subsoil, "101" protection structure, "102" roof, "103" opening, "104" open opening, "2" containment of plants, "201" aerial region, "202" root region, "S1" natural gas energy from anaerobic digesters, "0S2" conduits, "S" generator, "S01" electrical energy production device from natural gas, "S02" heat exchanger, "801" circulating fluid, "8" heat exchangers, "202" root regions, "2" plants. "802" hydraulic pump, "803" block valve, "60" microwave generators, "61" antenna, "10" climatic sensors, "7" membrane, "71" supports, "9" conductive sheet, "802" valve, "803" pump.

A volume above the ground is defined by a membrane that can transmit light and reflect microwaves. It is advantageous to combine the usage of membranes or nets with ground-covering technologies that perform microwave shielding. It is important to verify that the equipment inside the greenhouse is not damaged by the microwave duty cycle [25]. By carefully selecting the system control parameters, it is feasible to transport heat to plants in a homogeneous way, improving the efficiency of the transmission mechanisms in comparison with conventional hot-air systems and reducing heat loss.

Another technology that can be fruitfully employed is microwave lamps, which are deemed an effective approach to promote the growth of plants with high lighting requirements. The efficiency of microwave-powered lamps that provide an intense light source for rice production has been tested in a comparative study [26], the set-up of which is sketched in Figure 5. The combined 3.4 kW output of the lights was used to illuminate a growth chamber that was ~3 m^3. Rice was cultivated using four distinct cultivars in the growing chamber and the open field with a 16-h photoperiod comparing the growth traits and crop yields. According to the findings, microwave-illuminated plants have a dry weight that is 2–3 times higher than those only exposed to sunlight.

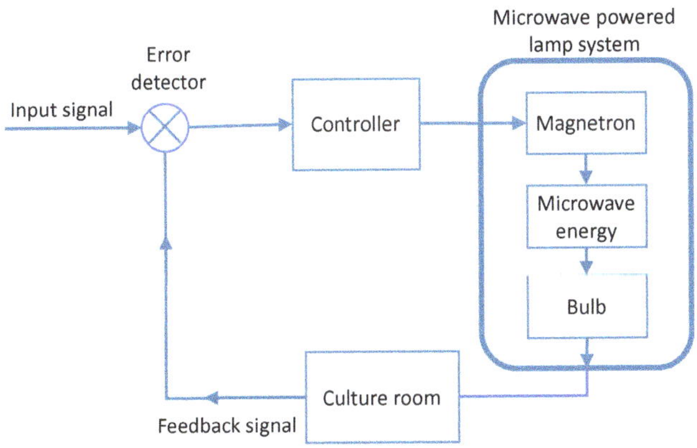

Figure 5. Experimental set-up and microwave-lamp schematic.

During photosynthesis, the plants in agricultural greenhouses solely use the visible spectrum. Plant development is impacted by the increase in greenhouse temperature brought on by the remaining solar spectrum. By absorbing the near-infrared radiation in the solar spectrum and transmitting the visible spectrum, spectrum-selective nanofluids offer an effective and long-lasting solution to this issue [27,28].

The use of microwaves in greenhouses offers benefits and limitations. Microwaves facilitate regulated settings and longer growing seasons by providing accurate dielectric heating for plant development. In comparison to traditional approaches, the experiments demonstrate potential energy savings through efficient heating with minimal damage. However, the dispersion of the electric field is not uniform, which limits scalability. Although sophisticated wave patterns and reflective surfaces are examples of innovations that attempt to address this, larger greenhouses may find it difficult to apply them. In addition, more research is necessary before widespread adoption can be achieved due to concerns about the effects of prolonged microwave exposure on plant health and possible equipment interference.

3. Use of Microwave for Soil Sterilization and Pest Control

Sterilization of soil and seeds promotes good crop growth. Microwave special molecular-level heating capabilities and their resulting enhancement on agri-food sciences have been thoroughly demonstrated. One of the most promising uses of microwaves in agriculture is pest control, which has the potential to slow down or even prevent the germination of invasive specimens [29].

At very small scales, due to the widespread use of commercial microwave ovens, thermal disinfection with microwaves is a frequent procedure used in home gardening and horticulture. The soil is put into plastic bags, sealed, and placed into the oven. When the oven is turned on, the MW radiation acts on the water in the soil and induces a temperature increment, eventually vaporizing the water. The soil may become contaminated if the temperature increases too much due to the thermal dissociation of nutrients, which can lead to the formation of toxic chemical compounds. The recommended wattage is between 600 and 650 W, and the recommended exposure time is between 90 and 120 s. The same process can also be used to sterilize compost [30]. Since microwaves heat water to ~100 °C to sterilize the soil, most of this homemade treatment consists of steaming.

When bigger scales are considered, the power and energy requests become more of a concern: large amounts of power are required to deliver sufficient energy for efficient pest management in open-field activities [31]. For instance, if a hypothetical device traveling at 1 km/h required 1500 J/cm^2 of energy to treat a 1-m broad band, the microwave power delivered to the soil would be ~4100 kW. Hence, the microwave disinfection of soil would be much more suitable for greenhouse applications because there is less material to sterilize, and the application is simpler. The alleged "nonthermal" impacts of microwaves on living things have not yet been fully shown. The fatal mechanisms appear to be thermal, and, in many cases, selective or differential dielectric heating can explain the reported outcomes attributed to "nonthermal biological consequences."

Microwave soil heating may also impact abiotic soil characteristics and soil-living organisms. Such an alteration may have an indirect effect on seedling emergence. Soil microwave treatment can reduce the appearance of invasive species seedlings by directly heating seeds and, in a subsequent phase, by altering the characteristics and functionality of the soil. There are significant inter- and intra-species differences in seed survival and seedling emergence in response to soil warmth. As a result, this approach may not be as effective in the long run for managing invasive species with high dispersal rates (i.e., specimens whose seeds can travel long distances before arriving at the germination site, like dandelions). The authors suggest that before using this approach in open-field crops, research must be conducted on how soil microwave heating affects ecosystem elements serving crucial functional functions, such as soil abiotic factors (e.g., pH, nutrient availability, and organic matter content).

Nevertheless, there is ample literature showing how pest control using microwaves is very effective, as shown by Nelson [31]. The author reported that soil microwave treatment can destroy seeds of several species due to the thermal effects induced in seeds. The seedling emergence of various species is prevented by soil microwave treatment. The effectiveness of microwave treatment depends on soil characteristics, such as soil texture and moisture, as well as treatment intensity and duration. As an example, as Nelson indicated, the soil temperature may be raised to 85 °C after 8 min of treatment with a 2-kW power source, causing up to 98% fewer seedlings to emerge than in the control group. The author suggests that the treatment is effective considering the thermophysical characteristics of the soil sample, which allows it to reach a highly homogeneous temperature profile in the treated sample.

Maynaud et al. carried out an experimental analysis aimed at investigating the effects on biological components as well as the chemical and physical characteristics of the soil. The results of a batch microwave system demonstrated the immediate impact of 915 MHz microwave treatments [32]. The microwave equipment comprised an 840 mm × 620 mm rotating chamber, with two 5 kW generators with power ratings between 1 and 10 kW used to create microwaves at a frequency of 915 MHz while cooling the magnetron using chilled water circulation. The impact of microwaves on seventeen physical, chemical, and biological soil characteristics was examined. Table 3 shows the effectiveness of the proposed treatment.

Table 3. Microwave treatment effects on microbial density and biomass compared to a control group [32].

Microwave Treatments		Microbial Biomass	Density of Total Heterotrophic Bacteria	Density of Fluorescent Pseudomonas
Power	**Time**	**mg OC/kg DS**	**log_{10} CFU/g DS**	**log10 CFU/g DS**
2 kW	4 min	223.8	7	6.5
4 kW	2 min	269.7	7.1	6.6
2 kW	8 min	114.2	6.7	4.4
4 kW	4 min	135.3	6.6	4.7
Control		342.7	7.3	6.5

Comparison between the treatment samples and controls shows that the best results were obtained with a combination of the lowest radiation power (2 kW) applied for the longest time (8 min). Overall, it is possible to infer that after the sample underwent microwave treatments, the biological indicators decreased compared to the control, up to 50% when considering the microbial biomass.

The authors also tested the effect of microwave disinfection treatment on the germination of seeds. On alluvial soil from a grassland, four 915 MHz microwave treatments combining power and exposure time were applied using Festuca seeds as an internal reference. Festuca seed germination was entirely suppressed by two treatments, 2 kW/8 min and 4 kW/4 min, raising the soil temperature to at least 80 °C. After increasing the temperature to 50 60 °C, the two treatments (2 kW/4 min and 4 kW/2 min) did not affect seed germination. Inorganic phosphorus and dissolved organic carbon levels were likewise increased by these treatments. Only soil samples receiving the 2 kW/4 min, or 4 kW/2 min treatments saw increased nitrate concentration.

Mahdi et al. [33] performed several experiments to assess the use of microwave radiation in soil sterilization. They found that lower energy emission levels are beneficial to seed germination. High emission levels have the potential to harm seeds, which would slow down or stop germination. The radiation dosage was increased over time, demonstrating that various effects are induced by longer exposures to higher power rates.

Specific research has been developed to assess the microwave irradiation (MWI) effects on Chinese cabbage growth [34]. The effects of MWI on vermicast potency, plant growth, and biochemical activity in Biko seedlings were examined. Different microwave power output levels (0, 100, 200, 300, 400, and 800 W) were applied to fresh, moist vermicasts. The amount of total aerobic plate content, nutrients, and water loss were evaluated. Although the total aerobic microbial plate count was highest for 200 W and 300 W MWI treatments, the optimum overall environment for growth was provided by the 400 W treatment, followed by the 800 W treatment. This result indicated an inhibiting factor with excess nutrients present and its detrimental effect on the overall growth of the cabbage plant. Plants cultivated in the fresh, moist vermicast medium showed comparatively lower growth and yield. As a result, the MWI treatment at 400 W increased nutrient bioavailability and other elements for better plant development and yield.

By creating heat effects that stop germination, microwaves provide an efficient way to sterilize soil and seeds, especially when used for pest management. On the one hand, microwave ovens are used in home gardening on a small scale to provide effective disinfection. On the other hand, problems with energy and power requirements arise at larger sizes, making soil disinfection more appropriate for greenhouse environments. Although microwave treatment has proven effective in killing seeds and delaying the emergence of seedlings, further research is needed to determine the long-term effects on soil properties, particularly in open-field crops. Evidence indicates that thermal mechanisms predominate, notwithstanding the possibility of nonthermal effects. Microwave treatments have been shown to lower biomass and microbial density, which can impact soil health. Furthermore, studies on the growth of Chinese cabbage have revealed the ideal microwave power settings for nutrient availability and plant development.

4. Other Uses for Microwaves in Agriculture

Another possible application of microwaves in agriculture strongly correlated with energy consumption is production and seed drying. Trabelsi & Nelson [35] summarize investigations on radio frequency and microwave dielectric heating in agriculture. The cost of lower frequency radio frequency (RF) and microwave dielectric heating equipment with capacities large enough for the practical drying scale of most agricultural products and the energy requirements for operation have proven too high for practical applications. Dielectric heating has not been developed to control insects that infest agricultural products for the same reasons.

Because seeds are high-value products and typically far lower quantities are handled, the application of microwave dielectric heating for seed treatment is more economical and likely practical than other agricultural applications. That is, lower RF power levels would need to be delivered, which might lower equipment prices. The advantages of new procedures must be assessed considering costs and marketing factors, as with any commercial development. If the competitive benefits are adequate, improved quality attributes of items due to dielectric heating procedures might be sufficient to justify the costs of the RF and microwave power equipment and its operation.

Microwaves have been successfully applied for manure treatment [36–38], especially for energy harvesting processing, meaning that the MW could potentially be used for biogas production and bioenergy recovery processes to boost the sustainability of greenhouse heating further [39].

It may be possible to recover energy from animal waste to produce electricity, which in turn could be used to power MW generators or CHE generators.

Microwaves show potential in agriculture for treating manure and drying seeds. Microwave dielectric heating equipment is cost-effective for high-value seed treatment; however, it is expensive for large-scale drying. Improved qualitative features and competitive advantages are prerequisites for economic viability.

5. Discussion

There are several possible uses of microwaves in agriculture, all with important energy implications. Achieving the most consistent heating of exposed bodies is the key challenge when using radiative heating in greenhouses. This issue involves many different aspects and technologies.

The anisotropy of the material's dielectric characteristics is a fundamental subject that contributes to non-uniformity in the dielectric heating of organic materials. As a result, the spread of heat in various directions is intrinsically inhomogeneous. Further inhomogeneity is caused by the electromagnetic field that the plant radiates, which is not quite uniform throughout the body.

A second key topic is the effect of frequency on dielectric heating. How are different species of vegetables affected by the wavelength of irradiated energy? How is the frequency spectrum relevant to achieving an effective temperature distribution? The state of the art of this subject is discussed in this section.

5.1. Uniformity of Heating Supply

The essential variables to obtain good results from the dielectric heating of plants are the type and frequency of polarization used and, subsequently, the wavelength. The experimental results evaluating the best selection of these factors are consistent.

Electromagnetic wave polarization affects how the heat is dissipated within the dielectric. The relative location of the radiated item and the antenna affects the field created by linearly polarized waves. Strong directionality heat is produced by linearly polarized waves, which are unsuitable for growing plants. Waves that are elliptically polarized improve heat diffusion.

Finally, the best heat conformity is provided by circularly polarized waves [8]. As circular polarization offers excellent thermal uniformity, it is appropriate for heating irregular bodies that are noticeably large. The relative positioning of the body and the antenna is almost entirely decoupled.

Circular polarization guarantees greater heating uniformity of the body and a smaller temperature range within the body and between different samples exposed to the same field. In the case of linear polarization, the volume loss density is much higher in one direction than in the others. In contrast, in the case of circular polarization, the power dissipation is more similar considering different axes.

The research performed by Guess [9] demonstrates that the various tomato plant components heat differently. The stem approaches burning while the leaves receive relatively little heat, with hotspots where the stem and leaves are joined. Microwave power can provide the leaves with the energy they need to grow properly, however, doing so will cause the stem to burn. Several tactics can increase thermal uniformity, affecting the plant's physical structure and the way the applied electromagnetic field is controlled.

Choosing an appropriate wavelength and, in the case of irregularly shaped items, circular polarization of the wave itself are necessary to provide good heating uniformity. According to Geedipalli et al. [40], rotating the plant and using rotating blades capable of obstructing the incident electromagnetic field results in more uniform outcomes. From an industrial point of view, the authors posit that the rotation of each plant is impractical.

Gunasekaran et al. [41,42] reported how the application of a duty cycle improves the depth to which heat reaches the same wavelength. However, this places a constraint on the maximum average temperature that the body can achieve. The two proposed methods, while providing improvement from the view of heating uniformity, do not seem to provide enough benefit to solve the problem. Improper dielectric heating may lead to the burning of stems. Guess [20] suggests that to obtain better results, it is advisable not to focus on the physical level of the plant but rather on the level of control of the incident wave. It is possible to:

- control and vary the power during the different growth phases of the plant, possibly also with a feedback control system capable of controlling the latter according to the temperature.
- control and vary the wavelength of the field according to the plant's size and state of growth.

In particular, with the onset and growth of even large fruits, it could be useful to lower the frequency of the incident field to increase the depth of heat penetration, while in the early stages, better results could be obtained by using a higher frequency. As there are no large bodies, it is possible to irradiate the bodies with two antennas positioned on opposite sides, producing electric fields that are out of phase with each other.

5.2. *Effect of Frequency on Dielectric Heating*

Kundu et al. have studied the biological effects and potential hazards of coconuts exposed to radio frequency (RF) electromagnetic fields; a 3D model has been created to simulate the coconut behavior [43]. The selection of an appropriate operating frequency is paramount to achieve good heating of the dielectric in question. Three fundamental quantities can be defined which are useful in understanding the effects that the frequency variation causes on how the power dissipates in the dielectric [8]:

- the attenuation factor of the wave p;
- depth at which the electric field is attenuated by a factor of $1/e$;
- depth at which the power is attenuated by a factor of $1/e$;

Table 4 summarizes the dielectric constants and the dielectric loss factors of different vegetables and fruits as proposed by Venkatesh [28]. These values greatly influence how the microwave radiation interacts with the vegetable tissue and thus must be considered when the microwave antenna is designed. Moreover, the values are clearly dependent on the selected frequency.

Table 4. Dielectric constant and loss factor for different vegetables and fruits [15].

Permittivity of Fresh Fruits and Vegetables at 23.8 °C						
Fruit/Vegetable	**Moisture Content**	**Tissue Density**	**Dielectric Constant Frequency ε′**		**Dielectric Loss Factor Frequency ε″**	
name	%	g/cm^3	915 MHz	2.45 MHz	915 MHz	2.45 MHz
Apple	88	0.76	57	54	8	10
Avocado	71	0.9	47	45	16	12
Banana	78	0.4	64	60	19	18
Cantaloupe	92	0.93	68	66	14	13
Carrot	87	0.99	59	56	18	15
Cucumber	97	0.85	71	69	11	12
Grape	82	1.1	69	65	15	17
Grapefruit	91	0.83	75	73	14	15
Honeydew	89	0.95	72	69	18	17
Kiwifruit	87	0.99	70	66	18	17
Lemon	91	0.88	73	71	15	14
Lime	90	0.97	72	70	18	15
Mango	86	0.96	64	61	13	14
Onion	92	0.97	61	64	12	14
Orange	87	0.92	73	69	14	16
Papaya	88	0.96	69	67	10	14
Peach	90	0.92	70	67	12	14
Pear	84	0.94	67	64	11	13
Potato	79	1.03	62	57	22	17
Radish	96	0.76	68	67	20	15
Squash	95	0.7	63	62	15	13
Strawberry	92	0.76	73	71	14	14
Sweet potato	80	0.95	55	52	16	14
Turnip	92	0.89	63	61	13	12

The temperature distribution resulting from heating is strongly influenced by these closely related parameters. If the penetration depth is very shallow, the power is dissipated at the surface, causing a strong inhomogeneity in the temperature distribution. A good power penetration depth value for a spherical element equals its radius, thus guaranteeing an almost homogeneous power distribution [16].

The dielectric properties of materials, power penetration depth, power loss density, wave frequency, and wavelength are the key parameters for evaluating dielectric heating. As frequency increases, there is a higher power loss density and, consequently, more superficial heating of the exposed material [9]. The presence of dissolved salts increases the power dissipated at the surface and decreases the heat penetration depth. Considering, for example, a frequency of 2.45 GHz (belonging to the ISM band) in pure water, where no dissolved salts are present, the power penetration depth is ~20 mm, which is approximately twice that for tomatoes.

Based on these observations, it can be deduced that using waves with frequencies in the infrared spectrum leads to an almost total power dissipation at the surface level. The inside of irradiated bodies would be heated by conduction of the heat present on the surface. However, this process in organic materials is significantly slower than heating by electromagnetic waves. During this process, there would be a rapid dispersion of heat into the environment by convection or radiation, making the heating system less efficient.

Using electromagnetic waves with frequencies belonging to the radio-frequency spectrum, the power penetration depth would be significantly higher than with infrared. This would ensure that the irradiated bodies are heated to an adequate depth, guaranteeing a more homogeneous temperature distribution. However, the power penetration depth is almost always greater than the largest plant to be irradiated, causing energy dissipation in the air or in objects that the wave will later penetrate, lowering the system's overall efficiency. The use of an intermediate frequency may solve the problem.

In addition, the radio frequency spectrum is used extensively in communication and sensor technology, and the usable ISM band is extremely limited. This leads to high design costs due to the stringent design constraints that must be met.

Therefore, the optimal choice for agricultural applications falls within the microwave spectrum, particularly the lower frequency spectrum, where the power penetration depth is approximately equal to the largest element in a plant. This ensures optimal results in terms of heating uniformity and minimal energy waste [16]. Microwave heating powered by cogeneration minimizes energy, reduces the carbon footprint, and lowers operating costs. The three potentially useful ISM bands are the 896 MHz band (available in the UK, while the 915 MHz band is available in the USA), 2.45 GHz band, and 5.8 GHz band. Excluding the lower frequency band because it is not universally available, the optimal frequency seems to be the 2.45 GHz band [43]. This frequency would guarantee the best compromise between the depth of power penetration and the density of power dispersed in the plant.

5.3. *Effect of Irradiation Power and Time*

Other aspects that must be considered include comparisons of the total energy requirements as well as the value of the time saved by faster operations. In this sense, power and time are key parameters when applying microwaves in agriculture for different uses (e.g., heating of plants, pest control, and soil sterilization).

The use of microwaves for sterilization and pest management relies on electromagnetic waves interacting with targeted microorganisms or insects [28]. The procedure efficacy is influenced by exposure duration and power. Overall, obtaining successful sterilization and pest control while avoiding adverse environmental impacts requires selecting the best power and time combination. According to the specific typology of organic material being sterilized and the degree of contamination, different amounts of power and time may be needed. In Table 3, a summary of the most relevant experimental results present in the literature is provided.

Table 5 shows the power and irradiation time used for different applications. The applications considered are soil sterilization, pest and invasive species control, and heating. The results show that it is necessary to adequately select the time and power intensity to achieve the envisaged result, limiting the side effects. Regarding the power, it greatly varies depending on the beneficial effect that is being pursued. For example, thermal sterilization requires a higher power (up to 4000 kW if the sterilization involves open-field terrain, which is not the best application for microwave sterilization), thus representing a critical factor when considering the economics of the treatment. The duration of the treatment is another essential factor that must correspond with the power applied. Even a low power level can be dangerous if applied for an incorrect amount of time, leading to economic issues. Table 3 indicates that, if heating is required, the power output must be significantly lower than in the sterilization process: microwave plant heating is generally performed using a circularly polarized field with a power request that typically ranges from 200 to < 500 W/m^2.

Table 5. Use of microwave in agriculture, key parameter values as proposed by the literature.

Application	Microwave Power	Time	Results	Reference
Soil Sterilization	2 kW	8 min	thermal sterilization	[31,44]
	4 kW	4 min	thermal sterilization	[31,44]
Pest Control	4000 kW	device speed: 1 km/h	thermal sterilization (open field)	[30]
Invasive Plant Species Control	2 kW	8 min	thermal sterilization	[44]
	2 kW	8 min	thermal sterilization	[44]
	4 kW	4 min	thermal sterilization	[44]
Plant Heating	200–300 W/m^2	indefinite	heating	[19]
	500 W	4 on/off cycles/min	heating	[18]

6. Conclusions

The subject covered by this review is broad; due to length constraints, only a limited number of publications were selected as references. Furthermore, this field of study is expanding quickly, so additional findings should be available soon. The literature reveals advancements in using microwaves for agriculture, including seed drying and manure treatment. However, challenges persist, such as high large-scale drying costs and limited practicality. Economic viability depends on competitive benefits, quality improvements, and thorough cost-benefit analyses. The impacts of microwave treatment on soil characteristics and living organisms warrant further investigation for open-field crops.

Over the past fifty years, research on the possible use of microwave dielectric heating for various agricultural applications has answered many of the questions that arise when considering such uses. In particular, the employment of these technologies in heating greenhouses has been studied, where energy consumption plays a leading role in economic terms and crop selection. The identification of several of these informational sources will benefit those who wish to research new dielectric heating applications for agriculture, especially for greenhouse cultivation.

Analysis of the state of the art allows us to cast the following conclusions, which may help organize upcoming experiments:

- Considering power plant efficiency, it is possible to save between 25 and 60% of energy compared with a conventional heating system. Estimates of such savings reported in the literature vary and fluctuate depending on the market price of energy. Thus, the economic aspect must be evaluated with great care and is partially outside the scope of the present review.
- Adverse non-thermal effects in the short or long term have not been observed following the use of microwaves for heating crops. Thermal effects induce any damage to vegetable tissues, especially when poor microwave field spatial homogeneity is considered.
- The inhomogeneity of heating between different parts of the plants limits the temperature attainable by the plants; some parts may be damaged while others are under-heated. Hence, the primary challenge is to make the field irradiating the plants and their heating homogeneous.
- Thermal dishomogeneity is also intrinsic due to the anisotropy of the dielectric properties of organic materials. This effect can be mitigated, but no solutions can eliminate it.
- It is advisable to use circularly polarized antennas to obtain the most uniform heating. The quality of polarization is a factor to be closely considered. The heating generated by linearly polarized antennas is strongly directional.

The following specific recommendations are proposed:

- The frequency at which the waves are propagated is a fundamental parameter for good heating uniformity. A frequency of the ISM band that provides good results is 2450 MHz. Alternatively, 915 MHz and 5800 MHz should also be evaluated.
- It is estimated that the power required to achieve a temperature difference of 10 °C between room and plants is approximately 200–300 W/m^2. The power per unit area must be > 50 W/m^2 and < 500 W/m^2.
- Operating antennas on an on-off duty cycle does not provide tangible results in improving heating uniformity.
- The rotation of crops does not give tangible results in improving heating uniformity, even if it is a common practice when food heating is considered.

The state-of-the-art analysis demonstrates that it is possible to consider microwave generators as an alternative solution to traditional heating systems for greenhouse crop production. Moreover, microwaves can be exploited for disinfection and adverse species control (vegetal, insect, or microbial). From the economic point of view, there is the potential to decisively lower the energy requirements of greenhouses, given the capacity of microwaves to deliver energy directly to the leaf surface, without the need to heat all the air volume inside the greenhouse. Moreover, since microwave generators are powered by electric energy, it could be possible to integrate them with alternative and sustainable energy sources (e.g., photovoltaics) or more sustainable technologies (e.g., CHP generators). The economic analysis must be carefully carried out, given the volatility of the energy market and the instability of the world political situation. For these reasons, an accurate economic analysis is out of the scope of the present review.

As technology rapidly advances, microwave utilization in agriculture, particularly for greenhouse heating, is expected to change significantly in the years to come. Subsequent research endeavors may concentrate on optimizing economically viable large-scale drying methods. Integrating microwaves with alternative energy sources, such as photovoltaics or sustainable technologies, has the potential to improve overall efficiency and sustainability in agricultural practices.

Author Contributions: Conceptualization, F.C., F.S. and L.C.; methodology, C.S. and M.Z.; formal analysis, C.S. and M.Z.; investigation, M.Z.; data curation, F.C. and F.S., writing—original draft preparation, All; writing—review and editing, F.C., F.S. and C.S.; visualization, F.C.; supervision, M.Z.; funding acquisition, C.S. All authors have read and agreed to the published version of the manuscript.

Funding: The research has been funded by the Italian research program: "Rural Development Program 2014–2020 of the Liguria Region" M16.02 "Support for pilot projects and the development of new products, practices, processes and technologies" Agricultural sector.

Conflicts of Interest: The authors declare no conflict of interest.

References

1. Reverberi, G.; Cepolina, F.; Pietronave, G.; Testa, M.; Ramadoss, V.; Zoppi, M. The minimal exoskeleton, a passive exoskeleton to simplify pruning and fruit collection. *Eng. Agric. Environ. Food* **2023**, *16*, 37–42. [CrossRef]
2. Adesanya, M.A.; Na, W.-H.; Rabiu, A.; Ogunlowo, Q.O.; Akpenpuun, T.D.; Rasheed, A.; Yoon, Y.-C.; Lee, H.-W. TRNSYS Simulation and Experimental Validation of Internal Temperature and Heating Demand in a Glass Greenhouse. *Sustainability* **2022**, *14*, 8283. [CrossRef]
3. Rapetti, M.; Sacile, R.; Fossa, M.; Minuto, G. *White Book: Innovative Methodologies and Technologies for Monitoring, Controlling and Increasing Energy Efficiency in Protected Crops*; Smart Agro-Manufacturing Laboratory (SAM-LAB): Rome, Italy, 2013; pp. 34–36.
4. Cirillo, L. Use of Dielectric Heating for Consumption Reduction in Modern Greenhouse. Master's Thesis, University of Genoa, Genoa, Italy, 2022.
5. Bakos, G.C.; Fidanidis, D.; Tsagas, N.F. Greenhouse heating using geothermal energy. *Geothermics* **1999**, *28*, 759–765. [CrossRef]
6. Krajnc, N.; Jemec, T. *Promotion of Residual Forestry Biomass in the Mediterranean Basin: Situation Report on Forest Biomass Use in Mediterranean Region*; Gozdarski Inštitut Slovenije: Ljubljana, Slovenija, 2012.
7. Ellis, M.W.; Von Spakovsky, M.R.; Nelson, D.J. Fuel cell systems: Efficient, flexible energy conversion for the 21st century. *Proc. IEEE* **2001**, *89*, 1808–1818. [CrossRef]

8. Carusi, E.; Giordano, R.; Lo Cascio, E.; Minuto, G.; Pietronave, G.; Schenone, C.; Zoppi, M. Greenhouse Heating System and method. Patent No. WO2021079263A1, 29 April 2021.
9. Guess, M.J. Heating of Greenhouse Crops with Microwave Energy. Ph.D. Thesis, University of Leeds, Leeds, UK, 2011.
10. Yeo, U.-H.; Lee, S.-Y.; Park, S.-J.; Kim, J.-G.; Cho, J.-H.; Decano-Valentin, C.; Kim, R.-W.; Lee, I.-B. Rooftop Greenhouse: (2) Analysis of Thermal Energy Loads of a Building-Integrated Rooftop Greenhouse (BiRTG) for Urban Agriculture. *Agriculture* **2022**, *12*, 787. [CrossRef]
11. Cepolina, E.M.; Cepolina, F.; Ferla, G. Brainstorm on artificial intelligence applications and evaluation of their commercial impact. *IAES Int. J. Artif. Intell.* **2022**, *11*, 799. [CrossRef]
12. El Khaled, D.; Novas, N.; Gazquez, J.A.; Manzano-Agugliaro, F. Microwave dielectric heating: Applications on metals processing. *Renew. Sustain. Energy Rev.* **2018**, *82*, 2880–2892. [CrossRef]
13. Li, H.; Qu, Y.; Yang, Y.; Chang, S.; Xu, J. Microwave irradiation—A green and efficient way to pretreat biomass. *Bioresour. Technol.* **2016**, *199*, 34–41. [CrossRef]
14. Galema, S. Microwave chemistry. *Chem. Soc. Rev.* **1997**, *26*, 233–238. [CrossRef]
15. Speir, T.W.; Cowling, J.C.; Sparling, G.P.; West, A.W.; Corderoy, D.M. Effects of microwave radiation on the microbial biomass, phosphatase activity and levels of extractable N and P in a low fertility soil under pasture. *Soil Biol. Biochem.* **1986**, *18*, 377–382. [CrossRef]
16. Henry, M.; Gaudillat, L.; Costa, C.; Lakkis, F. Free and/or bound water by dielectric measurements. *Food Chem.* **2003**, *82*, 29–34. [CrossRef]
17. Heredia, A.; Barrera, C.; Andrés, A. Drying of cherry tomato by a combination of different dehydration techniques: Comparison of kinetics and other related properties. *J. Food Eng.* **2007**, *80*, 111–118. [CrossRef]
18. Sun, J.; Wang, W.; Yue, Q. Review on Microwave-Matter Interaction Fundamentals and Efficient Microwave-Associated Heating Strategies. *Materials* **2016**, *9*, 231. [CrossRef] [PubMed]
19. Teitel, M.; Shklyar, A.; Dikhtyar, V.; Jerby, E.; Elad, Y. Development of a Microwave System for Greenhouse Heating. In *Proceedings of the International Conference and British-Israeli Workshop on Greenhouse Techniques Towards the 3rd Millennium*; Haifa, Israel, 5–8 September 1999, International Society for Horticultural Science (ISHS): Leuven, Belgium, 2000; pp. 189–195. [CrossRef]
20. Guess, M.J.; Hunter, I.C.; Abunjaileh, A.I. Improving energy efficiency by heating greenhouse crops with microwaves. In Proceedings of the IEEE MTT-S International Microwave Symposium, Baltimore, MD, USA, 5–10 June 2011; pp. 1–4.
21. Schmutz, P.; Siegenthaler, J.; Stäger, C.; Tarjan, D.; Bucher, J.B. Long-term exposure of young spruce and beech trees to 2450-MHz microwave radiation. *Sci. Total Environ.* **1996**, *180*, 43–48. [CrossRef]
22. Barelli, E. Dielectric Relaxation in Biological Materials. Master's Thesis, University of Bologna, Bologna, Italy, 2015.
23. Kratchanova, M.; Pavlova, E.; Panchev, I. The effect of microwave heating of fresh orange peels on the fruit tissue and quality of extracted pectin. *Carbohydr. Polym.* **2004**, *56*, 181–185. [CrossRef]
24. Huber, M.F. Heating Plants. UK Patent Application No. GB2120065A, 22 April 1982.
25. Cepolina, F.; Gabrielli, S.; Ferla, G.; Marcantoni, E.; Spinelli, L. Failure analysis of fiberglass cover used for photovoltaic plants. *J. Appl. Polym. Sci.* **2023**, *140*, e53961. [CrossRef]
26. Kozai, T.; Kitaya, Y.; Oh, Y.S. Microwave-Powered Lamps as a High Intensity Light Source for plant growth. *Greenh. Environ. Control. Autom.* **1994**, *399*, 107–112.
27. Sajid, M.U.; Bicer, Y. Experimental investigation of spectrum selective nanofluids for thermal management of greenhouses in hot arid climates. *Appl. Therm. Eng.* **2023**, *223*, 120019. [CrossRef]
28. Sajid, M.U.; Saghir, M.Z.; Dincer, I.; Bicer, Y. Spectrum splitting through CuS–ZnO/water hybrid nanofluid for agricultural greenhouse cooling applications: An experimental study. *J. Therm. Anal. Calorim.* **2023**, *148*, 8387–8401. [CrossRef]
29. Venkatesh, M.S.; Raghavan, G.S.V. An overview of microwave processing and dielectric properties of agri-food materials. *Biosyst. Eng.* **2004**, *88*, 1–18. [CrossRef]
30. Song, Z.; Wang, Z.; Wu, L.; Fan, Y.; Hou, H. Effect of microwave irradiation pretreatment of cow dung compost on bio-hydrogen process from corn stalk by dark fermentation. *Int. J. Hydrogen Energy* **2012**, *37*, 6554–6561. [CrossRef]
31. Nelson, S.O. A review and assessment of microwave energy for soil treatment to control pests. *Trans. ASAE* **1996**, *39*, 281–289. [CrossRef]
32. Maynaud, G.; Baudoin, E.; Bourillon, J.; Duponnois, R.; Cleyet-Marel, J.C.; Brunel, B. Short-termeffect of 915-MHz microwave treatments on soil physicochemical and biological properties. *Eur. J. Soil Sci.* **2019**, *70*, 443–453. [CrossRef]
33. Mahdi, W.M.; Al-Badri, K.S.L.; Al-Samarrai, G.F. Use of microwave radiation in soil sterilization and effects on the bacteria, fungi, and growth characteristics of chickpea plant (*Cicer arietinum* L.). *Plant Arch* **2019**, *19*, 2064–2069.
34. Abbey, L.; Udenigwe, C.; Mohan, A.; Anom, E. Microwave irradiation effects on vermicasts potency, and plant growth and antioxidant activity in seedlings of Chinese cabbage (*Brassica rapa subsp. pekinensis*). *J. Radiat. Res. Appl. Sci.* **2017**, *10*, 110–116. [CrossRef]
35. Nelson, S.O.; Trabelsi, S. Models for the Microwave Dielectric Properties of Grain and Seed. *Trans. ASABE* **2011**, *54*, 549–553. [CrossRef]
36. Mong, G.R.; Chong, C.T.; Ng, J.-H.; Lam, S.S.; Chong, W.W.F.; Ani, F.N. Pyrolysing Horse Manure via Microwave-Induced Heating for Bioenergy Recovery. *Chem. Eng. Trans.* **2020**, *78*, 139–144.

37. Tunile, I.; Srinivasan, A.; Liao, P.H.; Lo, K.V. Microwave enhanced advanced oxidation treatment of dairy manure. *J. Environ. Sci. Health* **2018**, *53*, 824–830. [CrossRef]
38. Yu, T.; Deng, Y.; Liu, H.; Yang, C.; Wu, B.; Zeng, G.; Lu, L.; Nishimura, F. Effect of alkaline microwaving pretreatment on anaerobic digestion and biogas production of swine manure. *Sci. Rep.* **2017**, *7*, 4765. [CrossRef]
39. Marttila, M.P.; Uusitalo, V.; Linnanen, L.; Mikkilä, M.H. Agro-Industrial Symbiosis and Alternative Heating Systems for Decreasing the Global Warming Potential of Greenhouse Production. *Sustainability* **2021**, *13*, 9040. [CrossRef]
40. Geedipalli, S.S.R.; Rakesh, V.; Datta, A.K. Modeling the heating uniformity contributed by a rotating turntable in microwave ovens. *J. Food Eng.* **2007**, *82*, 359–368. [CrossRef]
41. Gunasekaran, S.; Yang, H.-W. Effect of experimental parameters on temperature distribution during continuous and pulsed microwave heating. *J. Food Eng.* **2007**, *78*, 1452–1456. [CrossRef]
42. Gunasekaran, S.; Yang, H.-W. Optimization of pulsed microwave heating. *J. Food Eng.* **2007**, *78*, 1457–1462. [CrossRef]
43. Kundu, A.; Gupta, B.; Ray, S. A Study of Specific Absorption Rate in Coconut Exposed to RF Radiation. *Microw. Rev.* **2014**, *20*, 2–11.
44. Castellini, A.; Farinelli, A.; Minuto, G.; Quaglia, D.; Secco, I.; Tinivella, F. EXPO-AGRI: Smart automatic greenhouse control. In Proceedings of the 2017 IEEE Biomedical Circuits and Systems Conference (BioCAS), Turin, Italy, 19–21 October 2017; pp. 1–4. [CrossRef]

energies

MDPI

Review

A Review of Earth-Air Heat Exchangers: From Fundamental Principles to Hybrid Systems with Renewable Energy Integration

Hanna Koshlak

Department of Sanitary Engineering, Kielce University of Technology, Aleja Tysiąclecia Państwa Polskiego, 7, 25-314 Kielce, Poland; hkoshlak@tu.kielce.pl

Abstract: Earth-Air Heat Exchangers (EAHEs) provide a compelling solution for improving building energy efficiency by harnessing the stable subterranean temperature to pre-treat ventilation air. This comprehensive review delves into the foundational principles of EAHE operation, meticulously examining heat and mass transfer phenomena at the ground-air interface. This study meticulously investigates the impact of key factors, including soil characteristics, climatic conditions, and crucial system design parameters, on overall system performance. Beyond independent applications, this review explores the integration of EAHEs with a diverse array of renewable energy technologies, such as air-source heat pumps, photovoltaic thermal (PVT) panels, wind turbines, fogging systems, water spray channels, solar chimneys, and photovoltaic systems. This exploration aims to clarify the potential of hybrid systems in achieving enhanced energy efficiency, minimizing environmental impact, and improving the overall robustness of the system.

Keywords: earth-air heat exchanger; renewable energy; sustainable building; HVAC systems

1. Introduction

The escalating global energy crisis exacerbated by climate change has intensified the search for sustainable and energy-efficient building solutions. The building sector is a significant contributor to global energy consumption. As urbanization and population growth continue, the demand for energy in buildings is steadily increasing. Heating and cooling systems, in particular, play a pivotal role in energy consumption, accounting for a substantial portion of the total energy demand in buildings. Recent studies [1–4] have highlighted the escalating trend in global building energy consumption and the growing share of heating and cooling. Furthermore, according to the World Energy Outlook 2023 [5], global energy consumption in buildings has been steadily increasing over the past decade. In 2022, the sector consumed approximately 1% more energy than in the previous year, maintaining an average annual growth rate of just over 1%. Electricity has become the primary energy source for buildings, accounting for around 35% of the total energy use in 2022, up from 30% in 2010. This shift toward electricity is driven by the increasing electrification of heating and cooling systems, as well as the growing adoption of electric appliances. Despite the increasing use of renewable energy sources, fossil fuels continue to play a significant role in building energy consumption. In fact, the use of fossil fuels in buildings has increased at an average annual rate of 0.5% since 2010. This trend is concerning, as it contributes to greenhouse gas emissions and climate change.

Global energy consumption in buildings is a complex issue driven by several interconnected factors. Primarily, urbanization and population growth have led to a significant increase in the number of buildings, thus increasing the demand for energy services. For example, the largest increase was observed in regions with rapidly developing urbanization, such as Southeast Asia, India, and China, and in some African countries, such as Nigeria [6,7].

Furthermore, changing climate patterns have exacerbated the need for heating and cooling. More frequent and intense heatwaves have led to increased reliance on air conditioning systems, while colder winters have increased the demand for heating [8–11]. The impact of climate change is particularly pronounced in regions with extreme weather conditions, such as those experiencing prolonged heat waves or severe cold spells [12].

In addition, increasing living standards have contributed to an increase in energy consumption in buildings [4,13]. As people's expectations of indoor comfort rise, there is a growing demand for energy-intensive appliances, such as air conditioners, refrigerators, and electronic devices [14,15]. Moreover, the trend toward larger homes and more complex building designs further exacerbates energy consumption.

These factors have collectively led to a substantial increase in energy demand for building heating and cooling, particularly in developing countries where urbanization and economic growth are rapidly accelerating.

As the demand for energy continues to grow, there is an urgent need for innovative technologies that can reduce energy consumption and greenhouse gas emissions. As a result, there is a growing demand for innovative technologies that can reduce energy consumption and improve indoor environmental quality. Among the promising solutions, earth-air heat exchangers (EAHEs) have emerged as a viable option to improve the energy performance of buildings. Earth-air heat exchangers (EAHEs) have emerged as a promising technology for design and operation. Their roots can be traced back to ancient civilizations, who recognized the thermal properties of Earth and used them for various purposes [16]. However, the modern concept of the EAHE, as we know it today, began to take shape in the mid-20th century. Early research focused on understanding the fundamental principles of heat transfer between the ground and air. Pioneer studies have investigated the thermal properties of different types of soil, the impact of pipe configuration, and the optimal design parameters for EAHE systems. These early investigations laid the foundation for the development of practical applications of EAHE technology.

The number of articles devoted to the assessment of EAHE systems for buildings has been increasing in recent years, and several extensive review papers on the topic can be found in the literature [17–25], allowing us to conclude that interest in this subject is rising.

By taking advantage of the relatively constant temperature of the subsurface, EAHE systems offer a sustainable and efficient means of heating and cooling buildings. Earth-to-air heat exchangers (EAHEs) have found wide applications in various sectors of construction and industry. In the building sector, EAHE systems have found widespread application in residential and commercial properties. From single-family homes to large-scale commercial complexes, EAHEs contribute to improved thermal comfort and reduced energy consumption. Their applications are diverse and include space heating and cooling [26–29], ventilation [20–22], process cooling [30], and agricultural climate control [31]. By pre-heating or pre-cooling the incoming air, the EAHE reduces the load on conventional HVAC systems. In addition, they provide fresh filtered air, improving the indoor air quality. According to [32–34], the implementation of EAHE systems demonstrates significant potential as an effective passive cooling and heating technology, particularly for lightweight buildings located in high-altitude cold regions. Their research findings indicate a substantial improvement in indoor air quality, attributed to increased ventilation rates and reduced pollutant concentrations. Numerous studies have shown the effectiveness of EAHE in reducing energy consumption, improving indoor comfort, and reducing greenhouse gas emissions [35–39].

Recent studies have demonstrated the feasibility of integrating EAHE with solar energy, geothermal energy, and air heat pumps [40–45]. For instance, [46–51] investigated the performance of a hybrid system combining an EAHE, achieving significant energy savings compared with conventional HVAC systems. These types of solutions have the potential to increase the efficiency of energy buildings and reduce CO_2 emissions.

To optimize the performance of EAHE systems, researchers have conducted numerous studies to investigate the impact of these variables [52–57]. Additionally, advancements in modeling and simulation tools have enabled more accurate predictions of the performance of EAHE systems, thereby facilitating their design and optimization.

As technology continues to advance, EAHE systems have become increasingly sophisticated and efficient. Innovative design approaches, such as the use of advanced materials and intelligent control systems, further enhance the performance of these systems. Additionally, ongoing research is focused on optimizing the integration of EAHE with other building systems, such as HVAC and ventilation systems, to achieve optimal energy performance [51].

Despite the promising potential of earth-air heat exchangers (EAHEs), several significant challenges hinder their widespread adoption. These include substantial initial capital expenditure, sensitivity of system performance to variations in soil conditions, and the potential for gradual system degradation over time. To overcome these obstacles and maximize the long-term viability of EAHE technology, future research efforts should focus on developing innovative strategies to mitigate these challenges and enhance the long-term performance and reliability of EAHE systems.

Although numerous studies have investigated the performance and applications of EAHE systems, a comprehensive review that integrates the latest advances in the field is needed. This review provides a comprehensive overview of EAHE technologies, focusing on their applications in improving the energy efficiency and sustainability of buildings. By synthesizing the latest research findings, the significant potential of EAHE as a sustainable and energy-efficient solution for building air conditioning is demonstrated. This review will analyze different EAHE system configurations, including their integration with other renewable energy technologies, such as air-source heat pumps, photovoltaic systems, and solar chimneys. The analysis will assess the performance, environmental impact, and economic viability of these integrated systems. Finally, the review will identify key challenges and opportunities for future research and development, contributing to the development of sustainable building practices and informing the design and implementation of more energy-efficient and environmentally friendly building solutions. This review will cover the following key areas:

- the basic principles of heat transfer in EAHE;
- different configurations of EAHE systems, including open- and closed-loop systems;
- integration of EAHE systems with other building systems and renewable energy sources;
- challenges and future research directions.

By addressing these key areas, this review will contribute to the development of sustainable building technologies and support the move toward more energy-efficient buildings.

2. Materials and Methods

This study employed a comprehensive literature review approach to explore state-of-the-art EAHE technology. The primary sources of information included peer-reviewed articles indexed in reputable databases such as Scopus, Google Scholar, and Web of Science. Relevant keywords such as "earth-air heat exchanger", "ground-source heat pump", "renewable energy", and "energy efficiency" were used to identify a large pool of potential studies.

A rigorous filtering process was applied to select the most relevant and up-to-date research articles [57]. Criteria such as publication date, research methodology, and alignment with the study objectives were considered. This process resulted in the selection of 16,324 scientific papers that formed the foundation of this review.

To gain deeper insight into the trends and patterns within the field of EAHE research, a bibliographic analysis was conducted. This analysis involved mapping the citation network, identifying influential authors and institutions, and analyzing the most frequently cited keywords [58].

The selected articles were then subjected to a thorough qualitative analysis to extract key findings, identify research gaps, and discern emerging trends in EAHE technology. This analysis enabled a comprehensive understanding of the current state-of-the-art and future directions in this field.

3. EAHE Fundamentals

3.1. Classification of EAHE Systems

Earth-air heat exchangers represent an innovative solution in the field of energy efficiency. By leveraging the relatively stable temperature of the subsurface, EAHE systems pre-condition outdoor air before it enters a building.

Earth-air heat exchangers (EAHEs) have evolved significantly over the years, with advances in materials, design, and control strategies. This classification scheme provides a framework for understanding the various types of EAHE systems that have emerged (Figure 1).

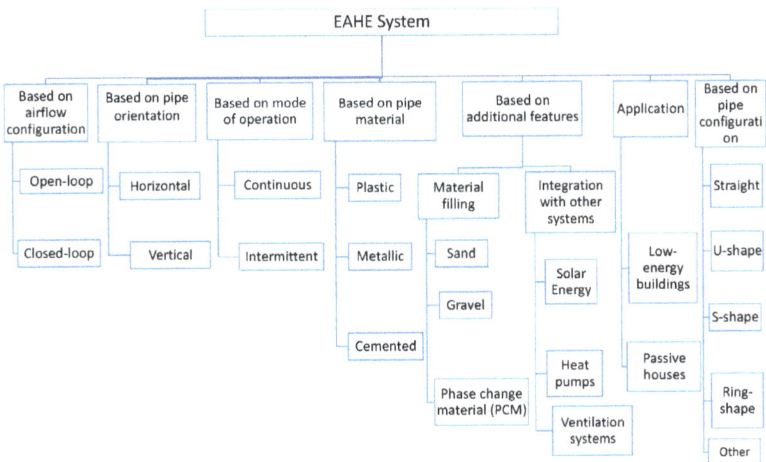

Figure 1. Classification of EAHE systems [24,35,59–82].

For different classifications, it is possible to appreciate the diversity of approaches to harnessing the thermal energy of the ground and to identify emerging trends in EAHE technology. Many researchers choose a hybrid configuration by coupling it with different passive techniques to improve the performance of the system in their respective research work using experimental, analytical, numerical, or some other studies [59].

The fundamental principle of the operation of EAHE is based on the conductive heat transfer between the air flowing through underground pipes and the surrounding soil. Below a certain depth, the ground temperature exhibits relative stability year-round. Consequently, the energy requirement for conditioning ventilation air can be substantially reduced by leveraging low-grade natural energy reserves. As illustrated in Figure 2, ambient air is forced into the pipe, where it undergoes heat exchange with the adjacent soil and is subsequently delivered to the indoor environment for ventilation purposes.

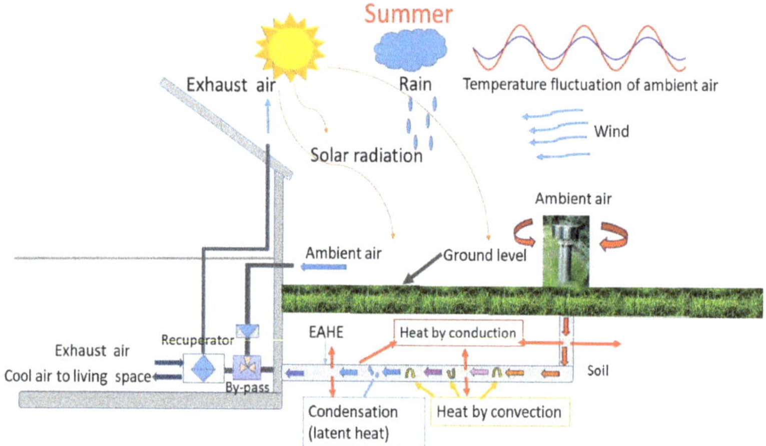

(a)

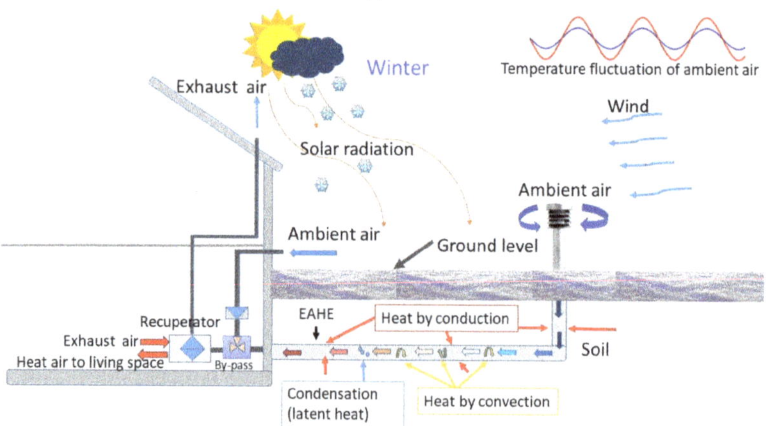

(b)

Figure 2. Working Principle of earth-air heat exchangers (EAHE) coupled with heat recovery units: (**a**) in summer; (**b**) in winter [83,84].

During summer (Figure 2a), warm ventilation air dissipates heat to the surrounding soil and undergoes a pre-cooling process before entering the indoor space. Conversely, in winter (Figure 2b), cool ventilation air can absorb heat from the surrounding soil, resulting in pre-heating. Although the ground temperature remains relatively stable throughout the year, seasonal variations occur, particularly in transitional seasons such as spring and autumn. These fluctuations influence the effectiveness of heat exchange and require consideration during system design and operation. During these transitional periods, the EAHE system provides moderate cooling or heating depending on the prevailing outdoor temperature. Horizontal EAHE systems, characterized by pipes laid horizontally in trenches, offer certain advantages. They are generally easier to install in areas with shallow groundwater levels and may require less excavation. However, their thermal performance can be influenced by variations in soil temperature and moisture content, which can fluctuate more significantly near the surface [85].

Vertical systems are often preferred for urban or built-up areas where land is limited and for applications requiring higher heat transfer rates [86]. They involve drilling boreholes into the ground and circulating air through a buried pipe network [87]. Vertical Ground-to-Air Heat Exchangers (VEAHEs) offer several key advantages over their horizontal counterparts by accessing deeper ground temperatures [88]. With increasing depth, ground temperatures exhibit reduced sensitivity to external environmental factors, thereby minimizing the impact of surface temperature variations caused by diurnal and seasonal changes. This results in more stable and reliable heat transfer to the VEAHE throughout the year, improving the overall system performance [87,89,90]. However, vertical EAHE systems present some challenges (Table 1).

Table 1. The key differences between vertical and horizontal earth-air heat exchangers.

Feature	Vertical EAHEs	Horizontal EAHEs
Ground access	deeper ground temperatures	near-surface ground temperatures
Temperature fluctuations	less susceptible to near-surface temperature variations (diurnal and seasonal)	more susceptible to near-surface temperature variations (diurnal and seasonal)
Impact of weather	less affected by weather conditions (e.g., solar radiation, precipitation, soil moisture)	more significantly affected by weather conditions
Installation	typically requires more complex and potentially more expensive installation	generally easier and less expensive to install
Land use	requires less land area	requires more land area
Maintenance	may be more challenging to access and maintain	generally easier to access and maintain

As shown in Table 1, drilling boreholes typically incur higher installation costs compared to trenching for horizontal systems, which can significantly increase the initial investment. Furthermore, borehole stability can be a concern in certain geological conditions, requiring careful site investigation and appropriate installation techniques to mitigate the risks of borehole collapse. Additionally, vertical systems may not be feasible in areas with high groundwater levels, limited space for borehole installation, or regions with high seismic activity. In conclusion, both horizontal and vertical EAHE systems offer viable solutions for improving energy efficiency. The optimal choice depends on a variety of factors, including site-specific conditions, project budget, desired thermal performance, and environmental considerations. A thorough site assessment and comprehensive cost-benefit analysis are crucial for selecting the most suitable EAHGE system for a given application.

3.2. Heat and Mass Transfer in Earth-Air Heat Exchangers

A thorough understanding of the heat and mass transfer processes within earth-air heat exchangers (EAHEs) is paramount for the design and optimization of these systems.

Heat and mass transfer within earth-air heat exchangers (EAHEs) involves complex interactions between buried pipes, surrounding soil, and supply air (Figure 3).

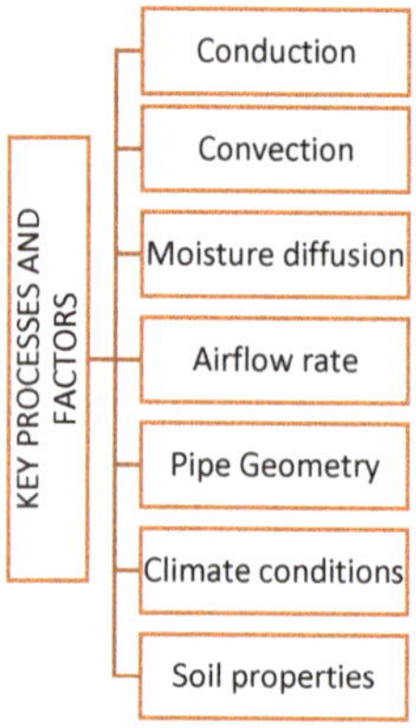

Figure 3. Key mechanisms and parameters affecting heat and mass transfer in EAHEs.

The interactions between the heat exchanger, soil, and supply air involve two areas of heat and mass transfer: the outer and inner surfaces of the heat exchanger pipe (Figure 4).

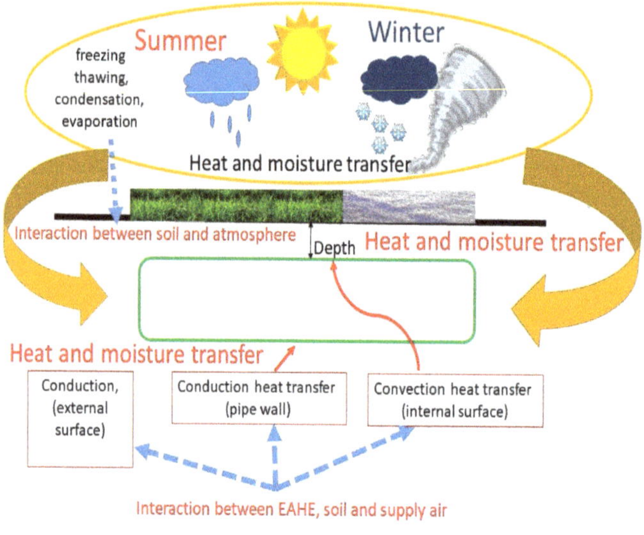

Figure 4. Heat and mass transfer in the EAHE.

At the outer surface of the buried pipe, heat is conducted from the wall of the pipe to the surrounding soil and governed by Fourier's law of heat conduction [91,92]. The rate of heat transfer is influenced by the thermal conductivity of the soil, its moisture content [93,94], and the temperature gradient between the pipe and the undisturbed soil. Additionally, moisture can diffuse between the soil and pipe wall depending on the relative humidity and temperature gradients. At the inner surface, heat is transferred convectively between the wall of the flowing air and the pipe [95]. The rate of convective heat transfer is influenced by factors such as air velocity, temperature difference between air and the pipe wall, and fluid properties. Moisture transfer is driven by concentration gradients and is influenced by relative humidity and temperature. Mass transfer, primarily in the form of moisture diffusion, can also occur between the air and the wall of the pipe, affecting the humidity of the supply air. The performance of earth-air heat exchangers (EAHEs) is influenced by a complex interplay of climatic factors, including temperature, precipitation, and surface conditions. Seasonal variations in temperature and precipitation directly affect the thermal properties of the subsurface and, consequently, the heating and cooling capacities of EAHEs. The interaction between the atmosphere and the subsurface, as well as the specific characteristics of the ground surface, can further modify these effects. For example, vegetation cover can reduce surface temperatures through evapotranspiration, while frozen ground can reduce thermal conductivity. Understanding these interactions is crucial for optimizing the design and operation of EAHE systems in various climatic regions.

Accurate modeling of heat and mass transfer in earth-air heat exchangers is essential for their effective design and optimization [96]. While significant progress has been made in this area, challenges remain due to the complex interplay of factors such as soil heterogeneity, moisture migration, and variable boundary conditions.

3.3. *Factors Influencing for Performance of Earth-Air Heat Exchangers*

The performance of earth-air heat exchangers (EAHE) is contingent upon the capacity of the surrounding soil to store and release thermal energy. A multitude of factors influence the ability to serve as a thermal energy reservoir. These factors can be broadly categorized into soil properties, environmental factors, ground temperature profiles, and system design parameters (Figure 5).

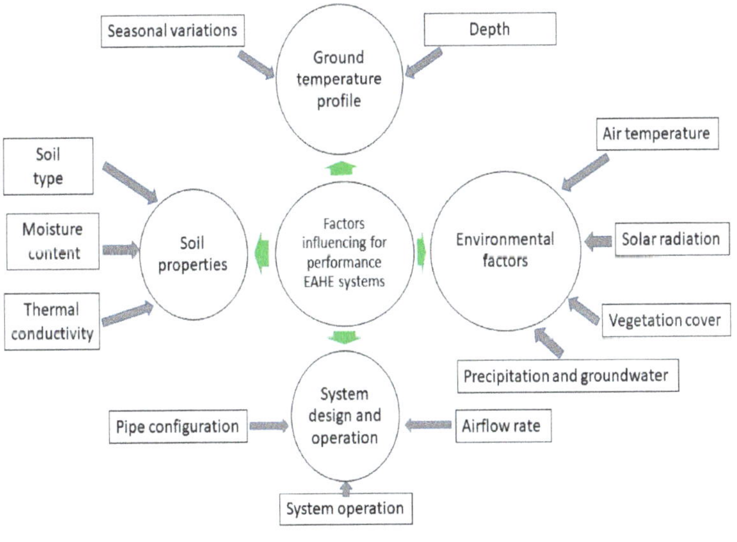

Figure 5. Factors influencing ground energy storage for EAHE system operations [97].

3.3.1. Effect of Soil Properties on EAHE Performance

Soil properties significantly influence heat transfer within the ground. The type of soil, such as clay, sand, or loam, plays a crucial role in determining its thermal properties. Clay soils, with their high clay content, typically exhibit higher thermal conductivity compared to sandy soils. Recently, articles have drawn attention to such issues, showing how soils with higher thermal conductivity, such as clay-rich soils, generally exhibit better heat transfer compared to sandy soils [71,72,98]. However, the specific properties of each soil type can vary significantly depending on factors such as particle size distribution, porosity, and mineral composition. Moisture content significantly influences the thermal conductivity. The presence of water within the soil matrix generally enhances heat transfer due to water's high thermal conductivity [99,100]. However, excessive moisture can have detrimental effects. A high moisture content can lead to increased soil density, which can hinder heat transfer. Moreover, it can contribute to the corrosion of buried pipes, compromising longevity and efficiency [101].

3.3.2. Effect of Burial Depth on EAHE Performance

Earth-air heat exchangers take advantage of the relatively stable ground temperature to provide heating and cooling. However, the temperature of the ground exhibits significant variations depending on the depth [102,103]. Determining the optimal burial depth for EAHE systems is crucial for maximizing system performance while minimizing installation costs. Numerous studies [20,104,105] have observed a distinct temperature gradient, with temperatures decreasing as the depth increases. This phenomenon creates a "zone of constant temperature" at a specific depth, usually ranging from 5 to 15 m [106]. This zone offers a stable thermal reservoir for EAHE systems, as temperatures remain relatively constant throughout the year. While deeper burial generally leads to more stable ground temperatures, studies have shown that the marginal increase in performance diminishes beyond a certain depth. Kaushal [107] observed that depths greater than 3.5 m offer minimal incremental performance gains compared to 3.5 m. However, the optimal burial depth varies depending on the location and site-specific factors. For example, ref. [108] found an optimal depth of 1 m in Malaysia, while Babar et al. [109] reported 3.96 m as suitable for Sahiwal, Pakistan, and Khan et al. [110] suggested 4.5 m for Lahore, Pakistan. These variations can be attributed to differences in soil properties, climatic conditions, and local ground temperature profiles. Ideally, the burial depth should be sufficient to reach a zone of relatively stable temperature while minimizing excavation costs. Therefore, the optimal burial depth should be determined based on a cost-benefit analysis, considering factors such as soil properties, ground temperature variations, and construction costs.

3.3.3. Effect of Climatic Conditions on EAHE Performance

The performance of earth-air heat exchangers is strongly influenced by climatic conditions. The ground temperature exhibits pronounced seasonal variations, with higher temperatures during summer and lower temperatures during winter. These fluctuations are primarily driven by changes in solar radiation and air temperature. During the heating season, when the ground temperature is higher than the desired indoor temperature, heat is extracted from the ground, while during the cooling season, excess heat is rejected to the ground. The magnitude of these heat fluxes is influenced by the temperature difference between the heat carrier fluid and the surrounding ground. Studies have shown that seasonal variations in temperature and precipitation significantly impact the heating and cooling potential of these systems. In [51,94,111], the performance of EAHE systems beneath grass and concrete surfaces was compared under different climates. Their findings demonstrated that vegetation cover, through evapotranspiration, can lead to lower subsurface temperatures and improved cooling performance compared to bare soil or concrete surfaces. Warmer climates typically offer greater heating potential, while colder climates provide better cooling performance. Additionally, the specific characteristics of the local climate, such as the annual temperature range and the frequency of extreme weather events, can influence the effectiveness of EAHEs. In addition to climatic factors, the nature of the ground surface also plays a crucial role. Studies have shown that different surface conditions, such as vegetation cover and soil type, can affect the thermal properties of the subsurface and, consequently, the performance of EAHEs [112]. The interaction between the ground surface and atmospheric temperatures, as emphasized by [113,114], exhibits significant seasonal variability due to changes in thermal properties and energy partitioning. For example, studies in various climatic zones have shown that different characteristics of the ground surface can significantly influence the performance of EAHE.

However, these studies have focused on warmer climates. In colder climates, seasonal variations in ground surface conditions, such as snow cover, soil freezing and thawing, and changes in soil moisture due to snowmelt, can significantly impact subsurface temperatures and, consequently, the performance of EAHE systems.

In cold climates, seasonal changes in ground surface conditions, such as freezing and thawing, can significantly impact EAHE performance. The formation of a frozen layer can reduce the thermal conductivity and alter the heat transfer mechanisms in the subsurface [115]. Moreover, snow cover can act as an insulating layer, affecting the heat exchange between the ground and the atmosphere. A study by Zajch et al. [116] demonstrated that in regions with distinct seasons, factors such as snow cover, soil moisture, and freeze-thaw cycles can significantly impact subsurface temperatures, directly affecting the performance of earth-air heat exchangers (EAHEs). Moreover, recent numerical models developed by Zajch et al. [117] provide more accurate predictions of the impacts of climate change scenarios on ground temperature and EAHE efficiency. Despite these advances, more research is needed to fully understand the effects of extreme weather events and changes in land use on EAHE performance.

3.3.4. The Influence of Design Parameters on the Performance of Earth-Air Heat Exchangers

The performance of earth-air heat exchangers (EAHEs) is significantly influenced by various design parameters, including pipe material, dimensions, spacing, soil type, burial depth, and airflow rate. These parameters must be carefully selected to ensure optimal heat exchange and energy efficiency. Recent years have witnessed a surge in research dedicated to optimizing the energy efficiency and design of earth-air heat exchangers (EAHEs). Modeling has proven to be an invaluable tool for predicting the influence of operational parameters, such as pipe length, radius, burial depth, and airflow rate, on the thermal performance of earth-air heat exchangers (EAHEs). While various computational tools are available for modeling EAHE systems, such as EnergyPlus (Version 8.7.0) and TRNSYS (version 17.0), these analytical tools are often cumbersome for rapid design iterations. Computational Fluid Dynamics (CFD) has emerged as a popular choice for researchers, offering a versatile platform for simulating EAHE performance. CFD's ability to discretize the system into a fine mesh and solve governing equations numerically enables rapid and cost-effective analysis. Numerous studies have highlighted the central role of geometric parameters, such as pipe diameter, depth, and length, in influencing system performance [118–125]. For instance, Haitham Sghiouri et al. [126] proposed a multi-objective optimization framework to design efficient and cost-effective earth-air heat exchanger (EAHE) systems. Their framework was validated through an experimental exchanger model, key design parameters were identified, and a genetic algorithm was employed for optimization. The optimization aimed to minimize the life-cycle cost (LCC) while maximizing the cooling potential across three diverse Moroccan climates. Multi-criteria decision-making methods were then applied to determine the optimal configurations for each location. The results emphasized the importance of site-specific design, as optimal configurations varied depending on factors such as the ground temperature gradient. For example, the optimal design for Marrakesh and Oujda was a 160 mm diameter, 49 m long pipe buried at 3 m depth, while Errachidia required a similar pipe buried at 4 m due to its higher ground temperature gradient. The EAHE systems delivered cooling potentials of 1447 kWh/year, 1172 kWh/year, and 1739 kWh/year with corresponding LCCs of $4122, $4091, and $4073 over 50 years for Marrakesh, Oujda, and Errachidia, respectively. The normalized life-cycle cost (NLCC) was lowest for Errachidia ($0.234/kWh), followed by Marrakesh ($0.285/kWh) and Oujda ($0.349/kWh).

Despite advancements in earth-air heat exchanger (EAHE) technology, the impact of channel cross-sectional geometry on the thermal performance and indoor air quality of earth-air heat exchangers (EAHE) has been a relatively understudied area [127]. Recent studies have indicated that the thermal performance of EAHEs is directly correlated with the heat transfer surface area within the channels. Wei et al. [128] investigated rectangular channels in straight EAHEs, finding that this configuration enhanced heating and cooling performance, providing more stable channel wall temperatures and reduced soil thermal disturbances compared to circular channel configurations. Jahanbin [129] examined elliptical channels in vertical U-shaped EAHE and observed a 17% improvement in reducing the system's thermal resistance compared to circular U-shaped channels. Benhammou et al. [130] conducted a case study on EAHEs for cooling applications, demonstrating that rectangular and elliptical channels outperformed conventional circular channels. Specifically, rectangular and elliptical channels provided approximately 88% and 93% more heat transfer surface area between the air and soil, respectively.

Numerous studies have investigated the influence of the airflow rate on the thermal performance of earth-air heat exchangers (EAHEs). Although higher airflow rates can improve heat transfer, they can also reduce system efficiency due to shorter contact times between the air and the ground [131]. The optimal airflow rate is a complex interplay of factors such as soil properties, climate conditions, and system design. While these parameters interact in complex ways, the specific relationship can vary depending on the soil conditions, climate, and system design. Benrachi et al. [132] provided insights into this complexity by showing that increasing wind speed from 2 m/s to 2.5 m/s resulted in a substantial decrease in cooling effectivity, from 60% to 33%. Moreover, ref. [112] explored the impact of airflow velocity on heat transfer rates, demonstrating that a decrease in velocity and an increase in diameter result in a reduced pressure drop along the length of the pipe during the airflow. However, Mihalakou et al. [133] observed a decrease in the temperature difference between the inlet and outlet air with increasing pipe diameter at a constant airflow rate. In conclusion, a review of the literature revealed that while airflow velocity and pipe diameter play an important role in determining EAHE performance, the performance of EAHE optimal design parameters depends on a complex interaction of factors, including soil properties, climate conditions, and system configuration.

Multi-pipe EAHEs have emerged as a promising technology for enhancing building energy efficiency. While offering advantages such as reduced pressure losses compared to single-pipe systems [134], multi-pipe EAHEs present design challenges, particularly regarding uneven airflow distribution among individual pipes.

Numerous studies have investigated the impact of different configurations on EAHE performance [135–138] and have demonstrated that U-shaped configurations generally exhibit superior thermal performance compared to Z-shaped designs. These findings were corroborated by Qi et al. [139], who also observed improved thermal performance and airflow uniformity in L- and U-shaped configurations.

Despite these advantages, achieving optimal performance in multi-pipe EAHEs remains a challenge. Recent studies [140] have revealed significant disparities in airflow distribution among individual pipes, leading to suboptimal performance and potential inefficiencies. This uneven flow distribution can result in substantial variations in airflow rates between different branches [141], and the static pressure differential between the inlet and outlet of the final pipes can significantly affect the airflow.

Although U-shaped configurations have shown promise in mitigating these airflow disparities, neither U-shaped nor Z-shaped designs have limitations. Experimental and CFD studies [142] have demonstrated that U-shaped structures generally exhibit higher thermal efficiency due to a more uniform airflow distribution and reduced pressure losses.

However, CFD simulations [143] have indicated that U-shaped configurations can exhibit lower outlet temperatures, potentially impacting the overall system performance. This finding contradicts the results [144], which, using a simplified CFD simulation of a 5-branch EAHE, observed lower outlet temperatures for Z-shaped configurations. Furthermore, Z-shaped configurations exhibit greater variability in airflow distribution among parallel branches compared to U-shaped configurations of similar complexity. The total pressure drop is higher for 90-degree connections compared to 45-degree connections, leading to an increased overall pressure drop in the Z-shaped configurations. An uneven airflow distribution can have significant consequences, with [141] reporting potential thermal efficiency losses of up to 20% in multitube EAHEs. In conclusion, while multi-pipe EAHEs offer significant potential for improving building energy efficiency, their performance is highly dependent on the design and operating conditions. Future research should focus on developing more accurate models for predicting the performance of these systems, exploring innovative design solutions to address the challenges associated with uneven airflow distribution, and optimizing system operation to maximize energy efficiency.

The influence of installation depth and length has also been extensively investigated. Muehleisen et al. [145–147] proposed an optimal burial depth of less than 5 m for EAHE, considering the trade-off between construction costs and energy efficiency.

Table 2 outlines the key factors that significantly impact the performance of earth-air heat exchangers (EAHEs), encompassing climatic conditions, ground surface characteristics, and various design parameters.

Table 2. Effect of key parameters on the performance of EAHE.

Parameter	Effect on Performance	References
Climatic conditions:		
temperature		
	In regions with significant temperature variations between summer and winter, EAHEs can exhibit high performance in both cooling and heating modes. In hot and arid regions, EAHE systems can effectively cool incoming air. In cold climates, EAHE systems can preheat incoming air, reducing the heating load. However, extreme temperatures can lead to ground overheating or freezing in the vicinity of the collectors, potentially compromising long-term system efficiency.	[11,25,113–116],
precipitation and extreme weather events		
	Rainfall and snowmelt can influence soil moisture content, affecting thermal conductivity and heat transfer.	[115–117]
Ground surface conditions:		
vegetation cover		
	Vegetation cover can influence ground temperature by shading the soil and reducing the impact of solar radiation.	[51,94,111,113,114]
soil type		
	Impacts the thermal conductivity and heat capacity of the soil.	[71,72,98–100]
Design Parameters:		
pipe geometry (diameter, length, shape)		
	Increased diameter generally leads to higher heat transfer rates but can also increase pressure drop; longer pipes generally provide greater heat exchange area; rectangular and elliptical shapes can offer increased heat transfer surface area compared to circular shapes.	[126,128–130,133]
burial depth		
	Optimal depth varies depending on ground temperature gradient and site-specific conditions.	[145–147]
airflow rate		
	Higher airflow rates can improve heat transfer but also reduce efficiency due to shorter contact times. The optimal airflow rate depends on various factors.	[112,132]
pipe configuration (multi-pipe systems)		
	U-shaped configurations generally exhibit superior thermal performance compared to Z-shaped designs; uneven airflow distribution can significantly impact performance.	[135–144]

In recent decades, there has been a surge in interest in EAHE systems, driven by increasing concerns about energy efficiency and environmental sustainability. Researchers have explored various configurations and applications of EAHE, including hybrid systems that combine EAHE with other renewable energy technologies, such as solar, geothermal, and wind energy. The growing interest in hybrid EAHE systems is driven by several factors, including stringent energy efficiency standards, advancements in renewable energy technologies, and climate change mitigation.

4. Integration of EAHE with Renewable Energy Sources

4.1. EAHEs Integrating with Air-Source Heat Pumps (ASHPs)

The integration of Earth-to-Air Heat Exchangers (EAHE) with Air-Source Heat Pumps (ASHPs) has emerged as a promising strategy to enhance the energy efficiency and sustainability of building heating and cooling systems. This synergistic approach leverages the stable ground temperature provided by EAHEs to pre-condition the inlet air to the ASHP, leading to improvements in the system's coefficient of performance (COP) and an extended operating range. Recent research has explored various aspects of this integration, including system performance optimization, economic feasibility, and environmental impact. However, the performance of EAHE-ASHP systems is influenced by complex interactions between soil properties, climatic conditions, and system design parameters. Further research is needed to fully understand these interactions and optimize the design and operation of these hybrid systems.

Recent research has demonstrated the potential of integrating EAHEs with ASHPs to improve the energy efficiency of building heating and cooling systems. Studies such as those conducted by Congedo et al. [43], Guo et al. [148], Do et al. [149], Mahmoud et al. [150], and Cavazzini et al. [77] have shown promising results under various climatic conditions. Two primary integration methods have been explored: open-loop and closed-loop integration.

Open-loop systems involve a direct interface between the ambient air and the ground (Figure 6). When ambient air passes through buried pipes, heat exchange occurs with a relatively stable ground temperature, preconditioning the air before entering the ASHP [151]. This preconditioning significantly enhances the heat pump's coefficient of performance (COP), particularly during extreme weather conditions. Cavazzini et al. [77] conducted a comprehensive study to evaluate the performance of an air-source heat pump integrated with an earth-air heat exchanger (EAHE) in a single-family residence located in northern Italy. This study aimed to assess the system's efficiency under real-world operating conditions and quantify the benefits of coupling an EAHE with an ASHP. Although ASHPs traditionally transfer heat between air and water for space heating and cooling, this novel approach involves preconditioning the air by passing it through an EAHE before it enters the ASHP.

The EAHE, in turn, leverages the thermal inertia of the ground to mitigate fluctuations in air temperature, enabling the ASHP to operate more efficiently. The researchers employed a detailed simulation model developed in MATLAB (r2019a) to replicate the system's behavior. The model was calibrated and validated using the year's worth of operational data from the building. The analysis focused on two primary objectives: understanding system performance under real-world conditions and quantifying the advantages of integrating an EAHE. The results demonstrated that the integration of an EAHE significantly improved system performance. EAHE contributed approximately 50% of the total heating demand during winter, reducing the reliance on auxiliary heating sources. Furthermore, the EAHE ensured a relatively stable supply of air throughout the year, leading to a more efficient operation of the heat pump. A detailed analysis of the energy consumption revealed that the system with the EAHE achieved a 25% reduction in energy consumption compared to a system without an EAHE. This was primarily due to the reduced operating hours of the heat pump and auxiliary heater.

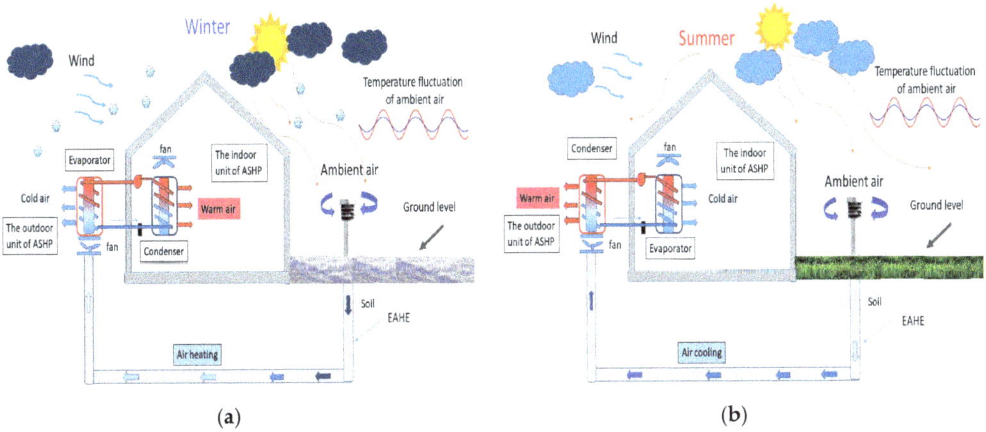

Figure 6. Open-loop EAHE integrated with an air-source heat pump: (**a**) in winter and (**b**) in summer.

Closed-loop systems harness Earth's low-grade thermal energy by circulating air through an underground network of pipes [80,152]. Schematic diagrams of the closed-loop EAHE system coupled with an air-source heat pump are shown in Figures 7a and 7b, respectively. Do et al. [149] integrated a closed-loop earth-air heat exchanger (EAHE) with an Air-source heat pump (ASHP) to cool residential buildings in the hot and humid climate zones of Texas, USA. The potential energy savings of this hybrid EAHE/ASHP system were evaluated by comparing simulations of ASHPs with and without buried pipes. The results indicated that the incorporation of an EAHE system yielded annual cooling energy savings of approximately 9.6% and 13.8% for Houston and Dallas, respectively. Despite these promising findings, the authors noted that their analysis was limited to energy consumption and savings and recommended further research to quantify the increased installation costs and reduced operating expenses of the system, thereby determining the extent to which these hybrid systems can be implemented in a cost-effective manner.

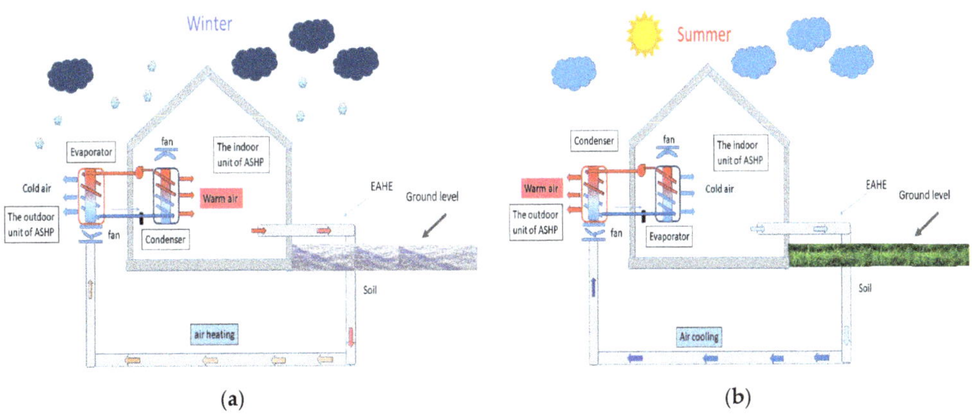

Figure 7. Closed-loop EAHE integrated with an air-source heat pump: (**a**) in winter and (**b**) in summer.

Although both systems offer distinct advantages, more research is needed to comprehensively evaluate their performance under various climatic conditions and develop optimized design guidelines. Factors such as pipe material, burial depth, and flow rate significantly influence the overall system efficiency and economic viability. Moreover, long-term monitoring and data analysis are essential for assessing the durability and reliability of these integrated systems.

4.2. EAHEs Integrating with Photovoltaic Systems

The integration of EAHEs with photovoltaic (PV) systems is a promising strategy for achieving significant improvements in building energy efficiency. Recent research has extensively explored the potential benefits of this synergistic approach, as evidenced by the studies conducted by Baglivo et al. [78], Kundu et al. [153], Dokmak et al. [154], Yadav et al. [155], Hraibet et al. [156], and Zhao et al. [157]. These hybrid systems offer several potential applications, including pre-cooling/pre-heating of supply air, temperature control for PV panels, the creation of hybrid energy systems, and the development of Combined Cooling, Heating, and Power (CCHP) systems (Figure 8).

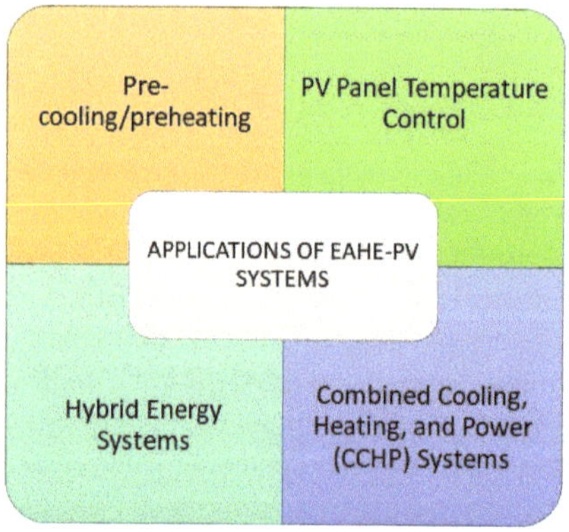

Figure 8. Applications of EAHE-PV systems.

A key advantage of integrating EAHEs with PV systems lies in their ability to pre-cool or preheat the supply air [158]. During the hot summer months, EAHEs can significantly reduce the load on building HVAC systems by providing pre-cooled air. On the contrary, during colder periods, preheated air from the EAHE can be used, minimizing the energy required for space heating. This preconditioning process demonstrably improves the overall efficiency of the HVAC system, leading to a substantial decrease in energy consumption [78]. The hybrid system described by Soares et al. [48] is aligned with the principles of sustainable building. The standalone photovoltaic system powers fans that circulate air through underground EAHE pipes, leveraging a stable ground temperature for preconditioning the outdoor air. This approach not only yields substantial energy savings but also enhances a building's independence from external energy sources.

Building-Integrated Photovoltaics (BIPVs) refer to photovoltaic modules that are integrated into the building envelope, such as the roof or facade. Unlike traditional photovoltaic systems, BIPV not only generates electricity, but also serves as a building component (Figure 9).

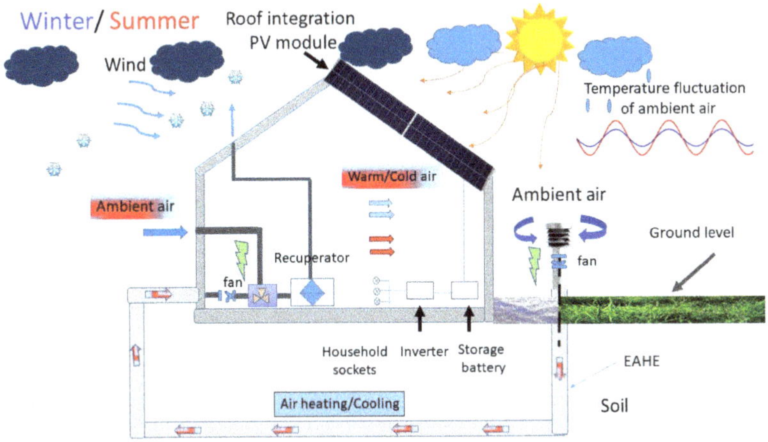

Figure 9. Hybrid system combining earth-air heat exchanger (EAHE) and roof-integrated photovoltaic systems (RIPV).

This dual functionality offers several advantages, including improved aesthetics, reduced installation costs, and enhanced energy efficiency [159,160]. However, BIPV systems are susceptible to overheating, which can significantly reduce their efficiency. The proposed hybrid system combines the benefits of both technologies. The EAHE pre-cools the supply air, which is then passed over the rear of the BIPV modules. This reduces the operating temperature of the photovoltaic cells, leading to increased electrical efficiency. Additionally, cooled air can be circulated within the building to provide thermal comfort. This approach not only improves the energy performance of the building but also reduces the overall environmental impact (Figure 10).

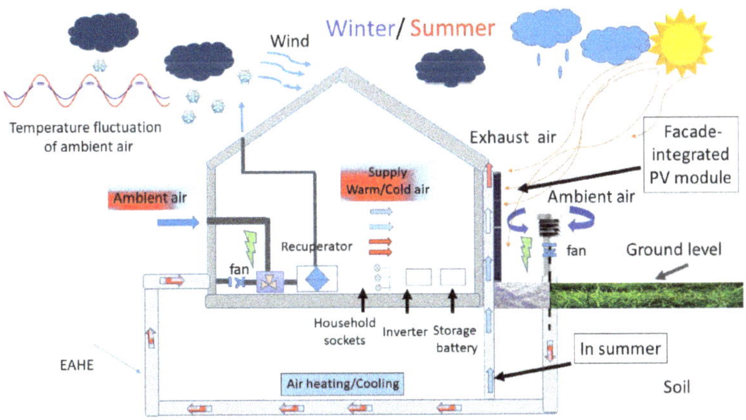

Figure 10. Hybrid system combining earth-air heat exchanger (EAHE) and fasade-integrated photovoltaics (FIPVs).

Figure 11 illustrates a hybrid system integrating an earth-air heat exchanger (EAHE) with phase change material (PCM) ceiling boards for passive cooling. This system effectively regulates indoor temperature, improves energy efficiency, and enhances occupant comfort. Intended for regions where natural ventilation is not effective due to high outdoor temperatures, this system is suitable for hot and humid climates.

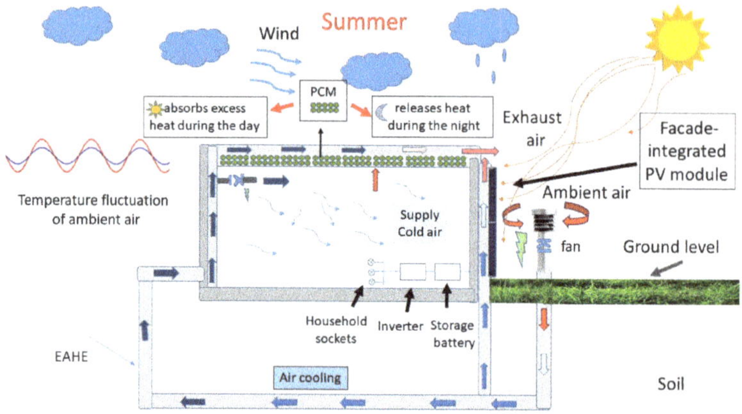

Figure 11. A hybrid system combining an earth-air heat exchanger (EAHE), fasade-integrated photovoltaics (FIPV), and PCM.

As shown in Figure 11, the EAHE pre-cools the incoming air by drawing it through underground pipes, where it exchanges heat with the cooler ground. This pre-cooled air is then circulated through the building. PCM embedded in the ceiling boards absorbs excess heat during the day, preventing overheating, and releases this stored heat during the night, reducing temperature fluctuations. A fan facilitates air circulation, enabling heat exchange between the EAHE, PCM, and indoor environment. The system operates in various modes, including solely utilizing PCM for moderate heat gain, relying solely on the EAHE during cooler periods, and a combined mode for peak loads. This hybrid approach offers enhanced cooling capacity, improved indoor comfort, and reduced energy consumption by leveraging natural cooling sources and minimizing reliance on mechanical refrigeration.

Shahsavar et al. [75] evaluated the performance of a hybrid system that combines an earth-air heat exchanger (EAHE) and a building-integrated photovoltaic/thermal system incorporated with a phase change material (PCM). The system was designed to satisfy the heating, cooling, and electrical demands of a building located in Kermanshah, Iran. A multi-objective genetic algorithm (MOGA) was used to optimize system parameters such as the length (L), width (W), and depth (D) of the PCM, mass flow rate (a), and depth and length of the EAHE (Di_{EAHE}, L_{EAHE}), with the aim of maximizing thermal and electrical outputs. The optimal photovoltaic/thermal (HPT) system exhibited the highest electrical output in July and the lowest in December. The optimized system could supply 26,536.1 kWh of cooling load, 132,693.3 kWh of heating load, and 9242.2 kWh of electrical load to the building annually.

Numerous studies have investigated the performance of EAHE-PV hybrid systems. Photovoltaic panels exhibit reduced efficiency at higher temperatures. Using the cooling capacity of the EAHE, the ambient temperature surrounding the photovoltaic panels can be reduced, leading to increased power output and improved overall system efficiency [161]. For example, Dhaidan et al. [162] conducted a comprehensive review of the application of geothermal energy in agricultural solar greenhouses, highlighting the potential of EAHE to improve the overall system performance.

Jakhar et al. [163] conducted a comprehensive numerical investigation into the thermal performance of EAHEs integrated with photovoltaic thermal (PV/T) systems across three distinct locations: Pilani, Ajmer (India), and Las Vegas (USA). Their validated thermodynamic model demonstrated significant performance enhancements when incorporating cooling measures. Notably, the integrated system exhibited substantial increases in thermal power across all locations compared to the PV/T system alone. This study highlighted the significant potential of these hybrid systems for improving building energy efficiency under diverse climatic conditions.

Finally, Lopez-Pascual et al. [164] proposed a compact cooling system for commercial photovoltaic panels based on low-enthalpy geothermal cooling. A single-phase closed-loop cooling system was used to dissipate heat from the solar panel using the constant and low temperature of the natural underground heat sink. A prototype incorporating a single-axis solar tracking mechanism was assembled and tested outdoors in Alcala de Henares, Madrid, Spain, in June 2022. Experimental results demonstrated that with a coolant flow rate of 1.8 L/min per square meter of panel surface, the cooling system reduced the temperature of the cooled panel by up to 20 °C, leading to a real increase in panel efficiency of up to 13.8%.

4.3. Energy Systems EAHEs Integrated with Solar Energy

The increasing demand for sustainable and energy-efficient building solutions has spurred the development of innovative technologies. Hybrid systems that integrate renewable energy sources with traditional building systems have attracted considerable attention. Among these, solar chimney (SC) systems coupled with photovoltaic panels and earth-air heat exchangers (SC-PV-EAHE) offer a promising avenue for achieving thermal comfort while reducing reliance on conventional cooling systems (Figure 12).

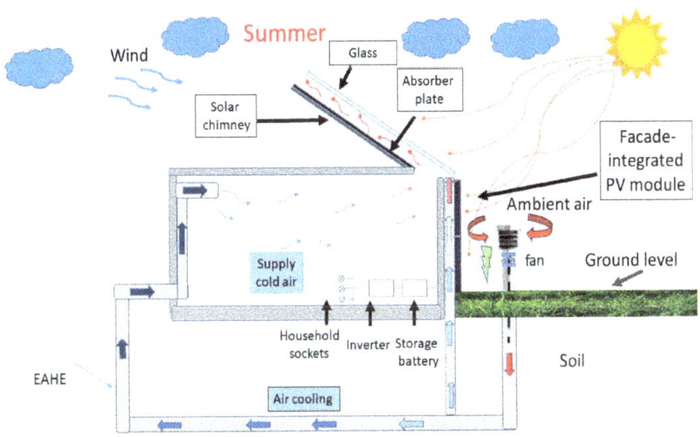

Figure 12. A Hybrid system combining an earth-air heat exchanger (EAHE), fasade-integrated photovoltaics (FIPV), and SC.

In [165,166], experimental results from a solar greenhouse (SG) integrated with an EAHE and photovoltaic system were presented. In these studies, cooled air from the EAHE was circulated through the SG to create a more comfortable environment for plant growth, while simultaneously cooling the PV modules mounted on the greenhouse structure. Their findings demonstrated that the circulation of air through the EAHE before directing it to the SG significantly reduced the ambient temperature within the greenhouse, leading to a notable improvement in the PV panel efficiency. For example, Yildiz et al. [167] reported a temperature reduction of approximately 8 °C between the inlet and outlet of the EAHE, resulting in 31% energy savings. The integrated system not only improved the efficiency of the PV panels but also provided a sustainable cooling solution for the greenhouse.

Alkaragoly et al. [168] made a notable contribution to this field by demonstrating the feasibility and effectiveness of such a system in a residential setting. Their study revealed that the SC-PV-EAHE system could effectively satisfy the cooling loads while simultaneously generating a substantial amount of electricity. This research investigated the performance of a novel hybrid system designed to provide thermal comfort and electricity generation for residential buildings. A comprehensive numerical analysis was conducted to evaluate the performance of the SC-PV-EAHE system under various climatic conditions. The study focused on two Middle Eastern cities, Tehran and Baghdad, which are known for their hot and arid climates. The EAHE system was modeled considering three primary heat transfer processes: convective heat transfer between the air flowing through the buried pipe and its inner surface, conductive heat transfer through the pipe wall, and conductive heat transfer between the pipe's outer surface and the surrounding soil. The optimal dimensions for the solar chimney and photovoltaic array were calculated based on the climatic conditions of each city. For instance, in Tehran and Baghdad, a chimney width of 2.4 m and a length of 4.216 m was recommended, accommodating 16 photovoltaic panels. Their findings underscored the system's effectiveness in achieving thermal comfort, with cooling demands ranging from 116 to 1500 W. Furthermore, the integrated photovoltaic panels generated a significant amount of electricity, with peak outputs reaching 1060 Watts in Tehran and 781 Watts in Baghdad. The system exhibited high electrical efficiency, with values of 9.7% for Tehran and 8.9% for Baghdad. The generated electricity can be used to power the building's electrical systems and improve the overall energy efficiency. These results not only validate the feasibility of such hybrid systems but also highlight their potential to reduce reliance on conventional cooling systems and contribute to sustainable building design.

The research presented in [52] offers a comprehensive evaluation of a hybrid solar chimney and EAHE system for passive building cooling in arid climates. This study investigated the performance of a novel hybrid system that combines solar chimneys and earth-air heat exchangers (EAHEs) for passive cooling in buildings located in hot, arid regions. Through a combination of experimental and numerical analyses, researchers have explored the intricate relationship between ventilation rates, solar chimney design parameters, EAHE geometry, pressure differentials, and climatic conditions. The study focused on two case scenarios: a baseline configuration with a solar chimney and EAHE, and a hybrid system that incorporates passive and active ventilation components. The results indicated that the proposed hybrid system significantly improved indoor thermal comfort, reducing zonal temperatures by up to 9 °C compared to the baseline case. The integration of electric fans into the hybrid system further improved the cooling performance and energy savings. Moreover, the system demonstrated substantial reductions in annual energy consumption and CO_2 emissions. Economic analysis revealed a relatively short payback period, highlighting the system's long-term economic viability.

Recent research by Li et al. [169] explored the potential of integrating solar chimneys, earth-air heat exchangers (EAHEs), and photovoltaic panels into a hybrid system for building applications. MATLAB (2017b (64-bit)) was used for the numerical modeling of the buried tubes, solar chimney, and photovoltaic panels, while the building model was developed using the TRNSYS 18 software, incorporating weather data and building modules. By systematically varying the number of EAHE tubes and solar chimney-photovoltaic modules, the researchers quantified the impact on key performance indicators, including ventilation rate, outlet air temperature, and electrical efficiency. A salient finding of this study was the substantial reduction in the peak daily ventilation rate achieved by the hybrid system compared to a standalone solar chimney system. Specifically, the integration of EAHEs led to a 22.05% decrease in the peak ventilation rate. Furthermore, the hybrid system demonstrated improved thermal performance, achieving an average reduction of 0.12 °C in the outlet air temperature. While indoor temperatures exhibited only minor variations between the hybrid and conventional ventilation systems, the hybrid system proved to be more energy efficient, showing a 1.34% enhancement in minimum electrical efficiency compared to an independent photovoltaic system. Parametric analyses revealed distinct influences of the system components. Increasing the number of EAHE tubes resulted in a significant increase in forced ventilation (over 36.43%) and a concomitant decrease in the temperature of the outlet air. In contrast, increasing the number of solar chimney-photovoltaic modules primarily affected the supply air and outlet air temperature, with a negligible impact on the efficiency of the photovoltaic panels.

4.4. Advanced Applications: Cooling, Heating, and Power Systems

With increasing concerns about climate change and the environmental impact of conventional air conditioning systems, there is a growing demand for sustainable and energy-efficient cooling solutions. Several published studies in the literature have focused on improving cooling systems without efficiency reductions, including evaluations of their economic feasibility. Jang et al. [170] investigated a building-integrated EAHE retrofitted with a supplementary groundwater system in a cooling-dominated climate (Figure 13).

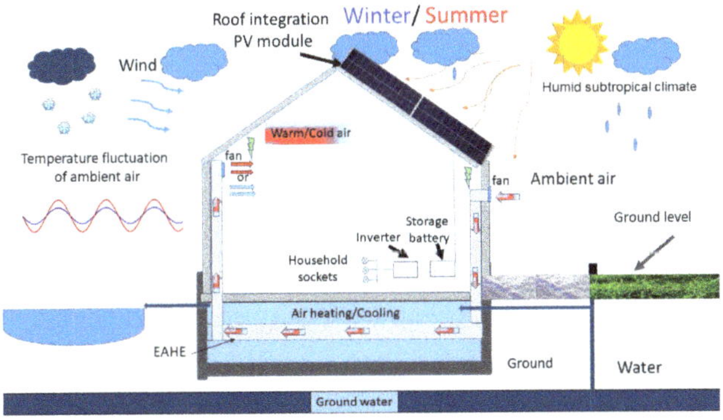

Figure 13. A hybrid system combining foundation-integrated water based EAHE and Facade-integrated photovoltaics (FIPV).

This study investigated a foundation integrated water-based EAHE (FIWEAHE) system enhanced by circulating supplementary groundwater from a residential well. Experimental and numerical results demonstrated significant performance improvements. After groundwater integration, the cooling capacity increased from 3.21 kW to 4.84 kW, and moisture removal rose from 1.97 kg/h to 4.24 kg/h during summer. The system effectively maintained indoor comfort, with summer outlet air temperatures between 24 and 26 °C and winter temperatures of 20–21 °C. The mean summer and winter COPs were 10.5 and 1.74, respectively. This study developed design relation curves to optimize pipe geometry and airflow velocity, providing valuable guidelines for the future design of the FIWEAHE system.

Research conducted by Ahmadi et al. [171] has shown that a hybrid system integrating EAHE and a water spray channel offers a promising solution for sustainable building cooling in the Tehran climate. In this study, the authors critically analyzed the system's effectiveness in providing thermal comfort and its environmental benefits within the context of the Tehran climate. According to the results, the cooling effectiveness of the proposed hybrid system exceeds 100%, indicating the system's capability to decrease the air dry-bulb temperature below the inlet ambient wet-bulb temperature. By effectively using the ground as a reliable source of alternative energy, this cooling system can be considered both ecofriendly and energy efficient. Consequently, the introduced system presents a viable alternative to conventional evaporative coolers or mechanical vapor compression systems.

The industrial applications of EAHEs are diverse and encompass a wide range of sectors. These include pre-cooling for air-cooled condensers and air compressors, dry cooling for thermal power plants, and cooling applications for advanced power generation systems, such as those employing supercritical Rankine cycles and gas turbines. Figure 14 shows the integration of an EAHE with a gas turbine power plant. The performance is significantly influenced by the ambient temperature, particularly during the summer months, which can lead to imbalances in the power supply and demand [172,173]. Inlet air cooling is crucial for maintaining high-efficiency power generation. Common inlet air cooling methods include direct evaporative cooling and electric refrigeration. Although electric refrigeration offers a significant reduction in temperature, it requires substantial energy consumption. In contrast, direct evaporative cooling cools air through water evaporation, but its effectiveness is highly dependent on the ambient air temperature and relative humidity. In hot and arid regions, direct evaporative cooling offers economic advantages over electric refrigeration.

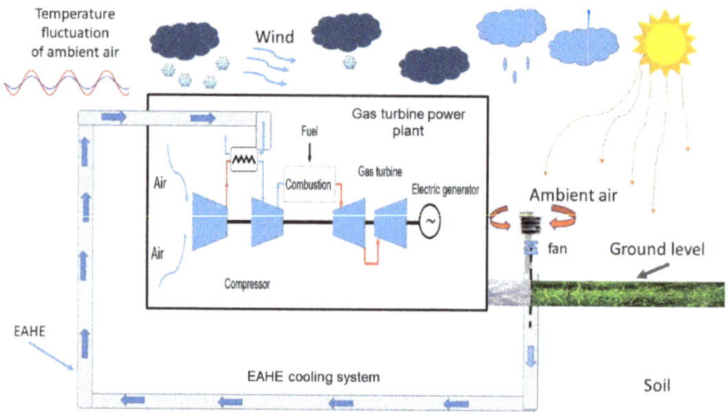

Figure 14. Gas turbine power plant with integrated EAHE for interstage air cooling.

Barakat et al. [174] investigated a novel hybrid inlet air cooling system for gas turbines, integrating an EAHE with a fogging system. Their comparative study in a 125 MW gas turbine unit demonstrated significant performance improvements. The annual average power output of the gas turbine equipped with the hybrid cooling system exhibited a 9.8% increase compared to the uncooled unit, while the fogging system alone resulted in a 6.6% increase. In particular, the hybrid system achieved a 50% reduction in water consumption compared to the fogging system during periods of high ambient temperatures.

A previous study [175] introduced a novel approach to cooling solar thermal power plants by utilizing earth-cooling air tunnels (Earth CATs) as a means to reduce water consumption. This research leverages the principle of existing earth-air heat exchangers (EAHEs) by exploiting the stable and relatively low underground temperature to pre-cool ambient air before it enters the plant's air condenser. A comprehensive computational fluid dynamics (CFD) analysis was conducted to assess the feasibility of this innovative cooling system and investigate the sensitivity of various geometric and flow parameters to optimize performance. The results demonstrated that a configuration employing pipes with a diameter of 0.5 m and an approximate length of 300 m is technically viable and offers a promising zero-water alternative to conventional water-cooling technologies. This study contributes to a growing body of research on sustainable and water-efficient cooling solutions for thermal power plants.

Benhammou et al. introduced a novel design of a passive cooling system consisting of an EAHE supported by a wind tower (Figure 15). The results demonstrated that the ambient air passing through the wind tower connected to the EAHE exhibited lower temperatures compared to the air exiting a conventional cooling tower.

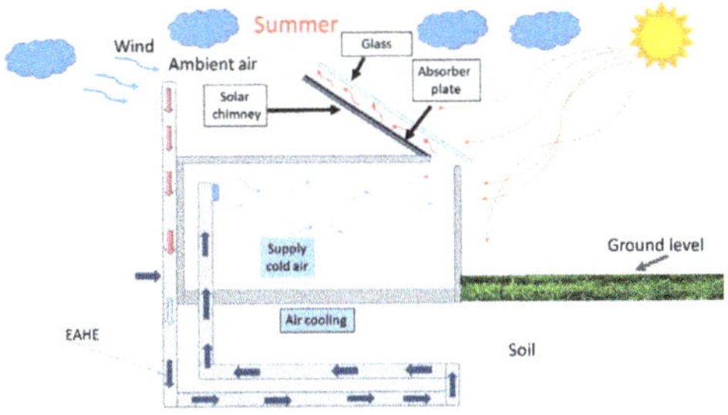

Figure 15. Integrated EAHE and wind tower for building cooling.

Shahsavar and Arici [176] investigated a novel hybrid system that integrated photovoltaic thermal (PVT) panels and an earth-air heat exchanger (EAHE) for building heating and cooling. This system harnesses both solar and geothermal energy to condition the outdoor air before it enters the building's HVAC system, while simultaneously generating electricity. The study used a genetic algorithm to optimize the annual energy and exergy output of the system. The proposed system consists of a PVT panel and an EAHE.

During the cooling season, outdoor air is passed through the EAHE to reduce its temperature, while return air from the building is used to cool the PVT panels. In contrast, during the heating season, the outdoor air is preheated by passing through both the PVT panels and the EAHE. The research findings revealed that while the combined PVT-EAHE system produced slightly less annual energy (93,925.6 kWh) compared to a standalone PVT-EAHE system (96,448.6 kWh), it exhibited a higher exergy output (10,904.5 kWh) compared to the standalone system (10,015.5 kWh). This indicates that the hybrid system offers improved thermodynamic efficiency despite a slight reduction in overall energy production.

5. Conclusions

This comprehensive review delves into the latest advancements in earth-air heat exchanger (EAHE) technology, with a particular focus on their application in buildings and industrial settings. By synthesizing the findings of recent studies, this review highlights the significant potential of EAHEs as a sustainable and energy-efficient solution for building climate control. Based on the analysis of the different EAHE integration studies summarized in Table 3, it is clear that the integration of EAHE with other renewable energy technologies is a promising way to achieve high energy efficiency and significantly reduce the environmental impact of building operations. EAHE systems offer a viable and sustainable alternative to traditional HVAC systems by leveraging the relatively constant temperature of the subsurface. By pre-cooling or pre-heating ventilation air, EAHEs can significantly reduce energy consumption and lower operational costs. Moreover, the integration of EAHEs with other renewable energy sources, such as air-source heat pumps, photovoltaic thermal (PVT) panels, wind towers, fogging systems, water spray channels, solar chimneys, and photovoltaic systems, has demonstrated enhanced performance and reduced environmental impact. The performance of EAHE systems is influenced by a multitude of factors, including soil properties, climatic conditions, system design, and operational parameters. A thorough understanding of these factors is essential to optimize the design of the system and predict performance accurately.

Table 3. Summary of EAHE systems and integrating technologies.

Type of System	Reference	Location	Configuration Parameters	Primary Benefit	Main Results	Improvement	Energy Efficiency
standalone EAHE; open loop, without external devices; experimental techniques	[177]	Bechar (arid regions located in the Southwest of Algeria)	PVC pipe, 66 m long and 110 mm in diameter; buried at a depth of 1.5 m in an agricultural area; annual undisturbed subsoil temperature at a 1.5 m depth of 28 °C	space heating/cooling	raise the air temperature by 10 °C in the heating regime and reduce it by 11.9 °C in the cooling regime	dependent on ground temperature, depth, and system design	significant performance capabilities, achieving a maximum temperature rise of 10 °C in heating mode and a maximum temperature drop of 11.9 °C in cooling mode; a maximum relative humidity rise of 19% in humidification mode and a maximum drop of 27% in dehumidification mode; different daily operating regimes: heating and dehumidification predominant in colder periods (00:00–08:00 and 18:00–23:00) and cooling and humidification predominant in warmer daytime hours (09:00–17:00); effective climate control in different environmental conditions
standalone EAHE, open loop; experimental and numerical modeling techniques	[17]	Bucaramanga, Colombia (tropical climate)	simulations used a PVC pipe (k = 0.185 W/m·K) buried at 1.5 m depth; reference case with a 60 m length, 0.2 m diameter, and 1240 kg/h flow rate was established; each parameter (mass flow rate: 620, 1240, 1860 kg/h; diameter: 0.1, 0.2, 0.3 m; length: 90, 150, 200 m) was varied independently while others remained constant	reduction of building cooling load	optimal EAHE configuration determined: 252 m length (7 pipes in parallel), 0.2 m diameter; achieved air temperatures between 20.9 °C and 24.1 °C; meeting thermal comfort requirements	dependent on local climatic conditions and soil properties; potential for wider adoption of EAHE technology in similar tropical environments	the maximum cooling capacity is 11.9 °C and the maximum heating capacity is 10 °C; with a maximum COP of 160 and an energy-efficient fan system (0.12 kW per motor), the system significantly outperforms conventional HVAC systems; the effectiveness of the EAHE system in reducing cooling loads will depend on the specific climatic conditions and soil characteristics of the location where it is installed; variations in these factors can influence the system's performance and may require adjustments to the design or operating parameters to achieve optimal results

Table 3. *Cont.*

Type of System	Reference	Location	Configuration Parameters	Primary Benefit	Main Results	Improvement	Energy Efficiency
Vertical Earth-Air Heat Exchanger (VEAHE) with integrated tubular phase change material (PCM); experimental and numerical modeling techniques	[62]	China (subtropical monsoon climate)	vertical borehole configuration with integrated PCM tubes; tube depth was varied within a range of 12 m to 24 m; PCM container diameter: 50 mm and 150 mm	enhanced thermal performance of the VEAHE system; reduced air temperature fluctuations at the outlet; increased average cooling capacity; smaller occupied area compared to traditional VEAHE systems	significant reduction in air temperature peak and fluctuation at the outlet with increasing tube depth; PCM thermal conductivity had a noticeable impact on performance at larger container diameters (150 mm); optimal container length for the system was found to be 12 m	- improved thermal stability of the system; - enhanced system performance through reduced air temperature fluctuations and increased cooling capacity	demonstrated significant performance enhancements, reducing peak air temperature at the outlet by 4.73 °C (from 25.74 °C to 21.01 °C) and significantly diminishing temperature fluctuations from 3.59 °C to 0.62 °C as the tube depth increased from 12 m to 24 m, improving the system's efficiency and reducing temperature fluctuations, the integrated PCM system has the potential to reduce energy consumption associated with cooling operations;
EAHE and ASHP	[77]	single-family building in North Italy (continental climate)	passive house construction with high insulation levels; ASHP unit (model: Nilan Compact P); EAHE for pre-heating/pre-cooling; passive heat recovery exchanger; electric heater for peak demand. monitoring system	- reduced energy consumption compared to operation without the EAHE; - enhanced indoor comfort by maintaining stable indoor temperatures	the significant impact of air humidity on heat pump performance	- enhanced system performance and efficiency through the integration of the EAHE; - indoor comfort by providing more stable and consistent indoor temperatures	significantly enhanced system performance, reducing energy consumption by 30% in winter and ensuring adequate cooling capacity during summer compared to a scenario without the heat exchanger while maintaining unit efficiency close to design values
EAHE and ASHP; (closed-loop EAHE integrated with the return-air duct system); simulation software DOE-2.1e version 119	[149]	Texas, USA (residential buildings in hot and humid climates)	- integration of the return-air duct system with the ground; simulation model varied parameters such as EAHE pipe length, buried pipe depth, - airflow velocity; EAHE system: pipe length from 15.2 to 91.4 m; buried depth: 1.8 m, 2.7 m, 3.7 m, 4.6 m; airflow velocity from 5.8 m/s to 2.9 m/s; validated the simulation model by comparison with experimental data from previous studies	reduced cooling energy consumption in residential buildings	longer pipe lengths and deeper burial depths generally increased cooling savings; reduced airflow velocity led to increased cooling energy savings; deeper burial depths beyond 3.7–4.6 m offered diminishing returns; locations with lower ground temperatures (like Dallas) showed greater potential for energy savings	enhanced energy efficiency of residential cooling systems; reduced reliance on conventional cooling methods, potentially leading to lower energy costs and reduced environmental impact	demonstrated significant annual cooling energy savings of 9.6% in Houston and 13.8% in Dallas when integrating a closed-loop EAHE system with a residential ASHP, with the greatest savings observed during the early cooling season (May–July); the EAHE system's cooling performance was improved by utilizing longer pipe lengths, deeper ground levels, and slower airflow velocities;

Table 3. *Cont.*

Type of System	Reference	Location	Configuration Parameters	Primary Benefit	Main Results	Improvement	Energy Efficiency
EAHE and PV	[163]	Pilani and Ajmer (India), Las Vegas (U.S.)	integration of an EAHE with a rooftop PV/T system	combined power and heat generation from the PV/T system	improved EAHE heating capacity with PV/T integration; significant reduction in PV cell temperature with cooling; feasibility of hybrid system in different climatic conditions	enhanced thermal performance of the EAHE system, increased system efficiency, and reduced reliance on conventional heating sources	the integration of PV/T collectors with the EAHE system resulted in a substantial reduction in cell temperature, with a maximum decrease of 10.9 °C observed in Las Vegas; the heating power increase varied significantly between locations, ranging from 23.47 to 298.74 Wh in Pilani and 41.43 to 270.75 Wh in Las Vegas; improved energy efficiency by utilizing solar energy to enhance the performance of the EAHE; potential for reduced reliance on fossil fuels for space heating;
PV-assisted EAHE for greenhouse cooling	[167]	solar Energy Institute, Ege University, Izmir, Turkey	- PV cell capacity 0.9 kW; air blower capacity 736 W; air flow rate 5300 m^3/h	increased PV efficiency, greenhouse cooling	average EAHE exergetic efficiency 56.3%; average PV cell exergetic efficiency 4.94%; average air blower exergetic efficiency 63.1%; overall system exergetic efficiency 23.6%; maximum EAHE exergetic efficiency: 59.6%; maximum air blower exergetic efficiency 70.7%; maximum overall system exergy efficiency 32.2%	improved system efficiency through new technologies; 12-month performance monitoring recommended	The experimental results demonstrated significant energy savings, achieving 2.84 kWh and a 31% reduction in energy consumption using a 0.9 kW solar photovoltaic system. The overall system exergy efficiency reached a maximum of 32.2% at 02:20 PM, with component efficiencies of 59.6% for the EAHE, 70.7% for the air blower, and 4.94% for the solar photovoltaic cells; further research is needed to optimize system performance and explore avenues for improving exergetic efficiency through advancements in component technologies
SC-PV-EAHE hybrid system	[168]	Baghdad and Tehran climates	- 16 PV panels integrated into the solar chimney; EAHE system (Tehran: 1 system with 25 m length and 0.5 m radius pipe; Baghdad: 2 systems with 25 m length and 0.5 m radius pipe); solar chimney dimensions: width 2.4 m length of 2 PV panels; length 4.216 m (width of eight panels); air channel width: 0.3 m.	achieve thermal comfort in buildings, produce electricity, and utilize renewable energy sources	the SC-PV-EAHE system can achieve thermal comfort for cooling demands ranging from 116 W to 1500 W; electricity generation at 12:00: Tehran: 1042 W; Baghdad: 768 W; the feasibility of the system in hot climates	the study provides a foundation for further research and optimization of the SC-PV-EAHE system for different climates and building types	this configuration effectively met the cooling demands of the residential room, ranging from 116 to 1500 W, while generating peak electrical power of 1060 W in Tehran and 781 W in Baghdad, demonstrating the system's potential for both thermal comfort and electricity generation; the system effectively utilizes solar energy for both cooling and electricity generation; further research is needed to assess the system's performance in other climates and building types

Table 3. *Cont.*

Type of System	Reference	Location	Configuration Parameters	Primary Benefit	Main Results	Improvement	Energy Efficiency
EAHE and solar chimney	[169]	numerical study, applicable to various locations	- number of EAHE pipes; quantity of solar chimney-photovoltaic modules	simultaneous ventilation, cooling, and electricity generation for buildings	reduced peak daytime ventilation compared to a system without PV integration; slight reduction in outlet air temperature; minimal difference in indoor temperatures between the hybrid and ventilation-only systems; improved electrical efficiency compared to standalone PV systems; increasing EAHE pipes significantly increases ventilation volume and slightly improves electrical efficiency; increasing solar chimney-PV modules increases ventilation and outlet air temperature, but has minimal impact on temperature and electrical efficiency of PV panels	optimizing the design and operation of the hybrid system by considering factors like the number of EAHE pipes and solar chimney-photovoltaic modules	the hybrid system demonstrated a significant reduction in peak daytime ventilation rate by 22.05% compared to the standalone solar chimney system while simultaneously improving the minimum electrical efficiency by 1.34% compared to an independent photovoltaic system; potential for efficient energy utilization by combining ventilation, cooling, and electricity generation; improved electrical efficiency compared to standalone PV systems
EAHE and groundwater (FIWEAHE)	[170]	Taiwan (Cooling-dominated climate)	- the system utilizes a blower with a volumetric flow rate of 738.4 m^3/h and a power consumption of 0.46 W; four EAHE are employed, each with an average length of 40 m and a diameter of 0.2 m; the pipes are spaced 0.1 m apart and buried at a depth of 1.3 m; a gradient of 0.3–0.05 m is incorporated for proper drainage; the pipes are configured in a U-shape for efficient water flow	improved building cooling performance; reduced energy consumption for air conditioning; enhanced indoor thermal comfort	summer COP from 10.5 to average, maximum 17.4; winter COP 1.74; outlet air temperature: 24–26 °C (summer) and 20–21 °C (winter); non-uniform air distribution observed at low airflow velocities	reduced excavation costs by integrating EAHE with the building foundation	high summer COP (10.5); significant improvement in cooling capacity and moisture removal with water supplementation

Table 3. *Cont.*

Type of System	Reference	Location	Configuration Parameters	Primary Benefit	Main Results	Improvement	Energy Efficiency
EAHE and fogging system	[174]	New Damietta power plant, coastal region of Egypt	EAHE: bank of parallel pipes with length and inner diameter buried underground; fogging system: atomizes water into micro-scaled droplets; hybrid system: combines EAHE with a subsequent fogging stage.	enhance gas turbine performance by cooling inlet air; increase power output and efficiency, and reduce water consumption compared to fogging-only systems	hybrid system boosts average annual energy output by 9.8% compared to 8% for fogging and 6.6% for EAHE alone; reduces water consumption by 50% compared to the fogging system; most efficient in regions with temperatures exceeding 30 °C and relative humidity below 60%	the potential of hybrid cooling systems for improved gas turbine performance and reduced water consumption	the hybrid cooling system exhibited the highest performance, increasing average annual energy output by 9.8% compared to 8% for fogging and 6.6% for the EAHE system alone while simultaneously reducing water consumption by 50% compared to fogging. This system is particularly well-suited for regions with ambient temperatures exceeding 30 °C and relative humidity below 60%; the impact of various factors (e.g., soil properties, groundwater levels, ambient conditions) on system performance may require further investigation
- PVT-HP-EAHE system (hybrid building-integrated PVT-heat pipe and EAHE)	[176]	General applicability	PE system (PVT-Heat Pipe only); PHE system (PVT-Heat Pipe + EAHE)	simultaneous heating/cooling of building air and electricity generation; improved energy and exergy efficiency compared to individual systems	PE system (PVT-Heat Pipe only): higher thermal energy and thermal exergy outputs, lower electrical energy output compared to PHE, annual useful energy: 96,448.6 kWh, annual useful exergy: 10,015.5 kWh; PHE System (PVT-Heat Pipe + EAHE): higher electrical energy output, lower thermal energy and thermal exergy outputs compared to PE, annual useful energy: 93,925.6 kWh, annual useful exergy: 10,904.5 kWh	higher exergy than standalone PVT; hybrid approach combining PVT-HP and EAHE for enhanced system performance	the PHE system demonstrated higher electrical energy output (10,126.0 kWh) compared to the PE system (9223.3 kWh); the PE system exhibited superior thermal and exergy performance, achieving 70,828.2 kWh of thermal energy and 10,015.5 kWh of exergy compared to 65,797.9 kWh and 10,904.5 kWh for the PHE system, respectively

The complex interactions between the heat and mass transfer within EAHEs have been explored in detail, providing valuable insights for future research and development. The key findings of this review underscore the potential of EAHEs to contribute to a more sustainable built environment. However, more research is needed to address challenges such as optimizing the system design for various climatic conditions, developing accurate predictive models, and evaluating the long-term performance and durability of EAHE systems. Building on the foundation of existing research, the field of EAHE technology can continue to advance and offer innovative solutions for sustainable building design.

Funding: This research received no external funding.

Conflicts of Interest: The author declares no conflicts of interest. The funders had no role in the design of the study; in the collection, analyses, or interpretation of data; in the writing of the manuscript; or in the decision to publish the results.

References

1. Wingrove, K.; Heffernan, E.; Daly, D. Increased home energy use: Unintended outcomes of energy efficiency focused policy. *Build. Res. Inf.* **2024**, *52*, 577–595. [CrossRef]
2. Deroubaix, A.; Labuhn, I.; Camredon, M.; Gaubert, B.; Monerie, P.A.; Popp, M.; Siour, G. Large uncertainties in trends of energy demand for heating and cooling under climate change. *Nat. Commun.* **2021**, *12*, 5197. [CrossRef]
3. United Nations Environment Programme. Global Status Report for Buildings and Construction: Beyond Foundations: MAINSTREAMING Sustainable Solutions to Cut Emissions from the Buildings Sector. Nairobi 2024. Available online: https://wedocs.unep.org/20.500.11822/45095 (accessed on 4 December 2024).
4. Arowoiya, V.A.; Onososen, A.O.; Moehler, R.C.; Fang, Y. Influence of Thermal Comfort on Energy Consumption for Building Occupants: The Current State of the Art. *Buildings* **2024**, *14*, 1310. [CrossRef]
5. IEA. *Tracking Clean Energy Progress 2023*; IEA: Paris, France, 2023; Available online: https://www.iea.org/reports/tracking-clean-energy-progress-2023 (accessed on 4 December 2024).
6. IEA. *Southeast Asia Energy Outlook 2024*; IEA: Paris, France, 2024; Available online: https://www.iea.org/reports/southeast-asia-energy-outlook-2024 (accessed on 4 December 2024).
7. Santamouris, M.; Vasilakopoulou, K. Present and future energy consumption of buildings: Challenges and opportunities towards decarbonization. *E-Prime—Adv. Electr. Eng. Electron. Energy* **2021**, *1*, 100002. [CrossRef]
8. Damm, A.; Köberl, J.; Prettenthaler, F.; Rogler, N.; Töglhofer, C. Impacts of + 2 °C global warming on electricity demand in Europe. *Clim. Serv.* **2017**, *7*, 12–30. [CrossRef]
9. IEA. *World Energy Outlook*; IEA: Paris, France, 2022; Available online: https://www.iea.org/reports/world-energy-outlook-2022 (accessed on 15 May 2023).
10. Gonçalves, A.C.R.; Costoya, X.; Nieto, R.; Liberato, M.L.R. Extreme weather events on energy systems: A comprehensive review on impacts, mitigation, and adaptation measures. *Sustain. Energy Res.* **2024**, *11*, 4. [CrossRef]
11. Li, D.H.W.; Yang, L.; Lam, J.C. Impact of climate change on energy use in the built environment in different climate zones—A review. *Energy* **2012**, *42*, 103–112. [CrossRef]
12. Salata, F.; Falasca, S.; Ciancio, V.; Curci, G.; de Wilde, P. Climate-change related evolution of future building cooling energy demand in a Mediterranean Country. *Energy Build.* **2023**, *290*, 113112. [CrossRef]
13. Gao, K.; Fong, K.F.; Lee, C.K.; Lau, K.K.-L.; Ng, E. Balancing thermal comfort and energy efficiency in high-rise public housing in Hong Kong: Insights and recommendations. *J. Clean. Prod.* **2024**, *437*, 140741. [CrossRef]
14. Santamouris, M. Chapter 2—Energy Consumption and Environmental Quality of the Building Sector. In *Minimizing Energy Consumption, Energy Poverty and Global and Local Climate Change in the Built Environment: Innovating to Zero*; Santamouris, M., Ed.; Elsevier: Amsterdam, The Netherlands, 2019.
15. Xu, X.; Yu, H.; Sun, Q.; Tam, V.W.Y. A critical review of occupant energy consumption behavior in buildings: How we got here, where we are, and where we are headed. *Renew. Sustain. Energy Rev.* **2023**, *182*, 113396. [CrossRef]
16. Bisoniya, T.S.; Kumar, A.; Baredar, P.V. Experimental and analytical studies of earth–air heat exchanger (EAHE) systems in India: A review. *Renew. Sustain. Energy Rev.* **2013**, *19*, 238–246. [CrossRef]
17. Basok, B.; Davydenko, B.; Pavlenko, A.M. Numerical network modeling of heat and moisture transfer through capillary-porous building materials. *Materials* **2021**, *14*, 1819. [CrossRef] [PubMed]
18. Bordoloi, N.; Sharma, A.; Nautiyal, H.; Goel, V. An intense review on the latest advancements of Earth Air Heat Exchangers. *Renew. Sustain. Energy Rev.* **2018**, *89*, 261–280. [CrossRef]

19. Bughio, M.; Ba-hale, S.; Mahar, W.A.; Schuetze, T. Parametric Performance Analysis of the Cooling Potential of Earth-to-Air Heat Exchangers in Hot and Humid Climates. *Energies* **2022**, *15*, 7054. [CrossRef]
20. Zhang, C.; Wang, J.; Li, L.; Wang, F.; Gang, W. Utilization of Earth-to-Air Heat Exchanger to Pre-Cool/Heat Ventilation Air and Its Annual Energy Performance Evaluation: A Case Study. *Sustainability* **2020**, *12*, 8330. [CrossRef]
21. Mohammadi, S.; Jahang-ir, M.H. Numerical investigation of the saturating soil layers' effect on air temperature drops along the pipe of Earth-Air Heat Exchanger systems in heating applications. *Geothermics* **2024**, *123*, 103109. [CrossRef]
22. Yue, Y.; Yan, Z.; NI, P.; Lei, F.; Qin, G. Enhancing the performance of earth-air heat exchanger: A flexible multi-objective optimization framework. *Appl. Therm. Eng.* **2024**, *244*, 122718. [CrossRef]
23. Yue, Y.; Yan, Z.; Ni, P.; Lei, F.; Yao, S. Machine learning-based multi-performance prediction and analysis of Earth-Air Heat Exchanger. *Renew. Energy* **2024**, *227*, 120550. [CrossRef]
24. Lapertot, A.; Cuny, M.; Kadoch, B.; Le Métayer, O. Optimization of an earth-air heat exchanger combined with a heat recovery ventilation for residential building needs. *Energy Build.* **2021**, *235*, 110702. [CrossRef]
25. Xiao, J.; Wang, Q.; Wang, X.; Hu, Y.; Cao, Y.; Li, J. An earth-air heat exchanger integrated with a greenhouse in cold-winter and hot-summer regions of northern China: Modeling and ex-perimental analysis. *Appl. Therm. Eng.* **2023**, *232*, 120939. [CrossRef]
26. Dewangan, C.; Shukla, A.K.; Salhotra, R.; Dewan, A. Analysis of an earth-to-air heat exchanger for enhanced residential thermal comfort. *Environ. Prog. Sustain. Energy* **2024**, *43*, e14346. [CrossRef]
27. Ahsan, T.A.; Rahman, M.S.; Ahamed, M.S. Geothermal energy application for greenhouse microclimate management: A review. *Geothermics* **2025**, *127*, 103209. [CrossRef]
28. Kazem, H.A.; Chaichan, M.T.; Al-Waeli, A.H.; Sopian, K.; Alnaser, N.W.; Alnaser, W.E. Energy Enhancement of Building-Integrated Photovoltaic/Thermal Systems: A Systematic Review. *Sol. Compass* **2024**, *12*, 100093. [CrossRef]
29. Ma, Q.; Qian, G.; Yu, M.; Li, L.; Wei, X. Performance of windcatchers in improving indoor air quality, thermal comfort, and energy efficiency: A review. *Sustainability* **2024**, *16*, 9039. [CrossRef]
30. Hu, Z.; Yang, Q.; Tao, Y.; Shi, L.; Tu, J.; Wang, Y. A review of ventilation and cooling systems for large-scale pig farms. *Sustain. Cities Soc.* **2023**, *89*, 104372. [CrossRef]
31. Castro, R.P.; Dinho da Silva, P.; Pires, L.C.C. Advances in Solutions to Improve the Energy Performance of Agricultural Greenhouses: A Comprehensive Review. *Appl. Sci.* **2024**, *14*, 6158. [CrossRef]
32. Vakiloroaya, V.; Samali, B.; Fakhar, A.; Pishghadam, K. A review of dif-ferent strategies for HVAC energy saving. *Energy Convers. Manag.* **2014**, *77*, 738–754. [CrossRef]
33. Zhong, K.; Meng, Q.; Liu, X. A Ventilation Experimental Study of Thermal Performance of an Urban Underground Pipe Rack. *Energy Build.* **2021**, *241*, 110852. [CrossRef]
34. Li, J.; Jimenez-Bescos, C.; Calautit, J.K.; Yao, J. Evaluating the Energy-Saving Potential of Earth-Air Heat Exchanger (EAHX) for Passivhaus Standard Buildings in Different Cli-mates in China. *Energy Build.* **2023**, *288*, 113005. [CrossRef]
35. Pikra, G.; Darmanto, P.S.; Astina, I.M. A review of solar chimney-earth air heat exchanger (SCEAHE) system integration for thermal comfort building. *J. Build. Eng.* **2024**, *98*, 111484. [CrossRef]
36. Azzi, A.; Tabaa, M.; Chegari, B.; Hachimi, H. Balancing Sustainability and Comfort: A Holistic Study of Building Control Strategies That Meet the Global Standards for Efficiency and Thermal Comfort. *Sustainability* **2024**, *16*, 2154. [CrossRef]
37. Acharya, P.; Sharma, A.; Singh, Y.; Sirohi, R. Passive Cooling Strategies Towards the Sustainability of Livestock Building-An Overview. *Indian Vet. J.* **2024**, *101*, 17–26. [CrossRef]
38. Shukla, B.K.; Pandey, A.K.; Singh, A.; Singh, D.; Anas, M. A holistic review on energy-efficient green buildings considering environmental, economical, and technical factors. In *Clean Energy*; CRC Press: Boca Raton, FL, USA, 2024; pp. 68–85.
39. Bosu, I.; Mahmoud, H.; Ookawara, S.; Hassan, H. Applied single and hybrid solar energy techniques for building energy consumption and thermal comfort: A comprehensive review. *Sol. Energy* **2023**, *259*, 188–228. [CrossRef]
40. Chel, A.; Tiwari, G.N. Stand-alone photovoltaic (PV) integrated with earth to air heat exchanger (EAHE) for space heating/cooling of adobe house in New Delhi (India). *Energy Convers. Manag.* **2010**, *51*, 393–409. [CrossRef]
41. Ríos-Arriola, J.; Velázquez-Limón, N.; Aguilar-Jiménez, J.A.; Corona, J.A.; Islas, S.; Reyes-López, J.A.; Luna, A. Air Conditioning of an Off-Grid Remote School with an Earth to air Heat Exchanger Coupled Indirectly to a Solar Cooling System. *Int. J. Environ. Res.* **2024**, *18*, 112. [CrossRef]
42. Qin, D.; Liu, J.; Zhang, G. A novel solar-geothermal system integrated with earth–to–air heat exchanger and solar air heater with phase change material—Numerical modelling, experimental calibration and parametrical analysis. *J. Build. Eng.* **2021**, *35*, 101971. [CrossRef]
43. Congedo, P.M.; Baglivo, C.; Bonuso, S.; D'Agostino, D. Numerical and experimental analysis of the energy performance of an air-source heat pump (ASHP) coupled with a horizontal earth-to-air heat exchanger (EAHX) in different climates. *Geothermics* **2020**, *87*, 101845. [CrossRef]
44. Baglivo, C.; Congedo, P.M.; Laforgia, D. Air cooled heat pump coupled with horizontal air-ground heat exchanger (haghe) for zero energy buildings in the mediterranean climate. *Energy Procedia* **2017**, *140*, 2–12. [CrossRef]

45. Akhtari, M.R.; Shayegh, I.; Karimi, N. Techno-economic assessment and optimization of a hybrid renewable earth-air heat exchanger coupled with electric boiler, hydrogen, wind and PV configurations. *Renew. Energy* **2020**, *148*, 839–851. [CrossRef]
46. Soni, S.K.; Pandey, M.; Bartaria, V.N. Hybrid ground coupled heat exchanger systems for space heating/cooling applications: A review. *Renew. Sustain. Energy Rev.* **2016**, *60*, 724–738. [CrossRef]
47. Yang, L.H.; Huang, B.H.; Hsu, C.Y.; Chen, S.L. Performance analysis of an earth–air heat exchanger integrated into an agricultural irrigation system for a greenhouse environmental temperature-control system. *Energy Build.* **2019**, *202*, 109381. [CrossRef]
48. Soares, N.; Rosa, N.; Monteiro, H.; Costa, J.J. Advances in standalone and hybrid earth-air heat exchanger (EAHE) systems for buildings: A review. *Energy Build.* **2021**, *253*, 111532. [CrossRef]
49. Anshu, K.; Kumar, P.; Pradhan, B. Numerical simulation of stand-alone photovoltaic integrated with earth to air heat exchanger for space heating/cooling of a residential building. *Renew. Energy* **2023**, *203*, 763–778. [CrossRef]
50. Ali, B.M.; Akkaş, M. The Green Cooling Factor: Eco-Innovative Heating, Ventilation, and Air Conditioning Solutions in Building Design. *Appl. Sci.* **2023**, *14*, 195. [CrossRef]
51. D'agostino, D.; Greco, A.; Masselli, C.; Minichiello, F. The employment of an earth-to-air heat exchanger as pre-treating unit of an air conditioning system for energy saving: A comparison among different worldwide climatic zones. *Energy Build.* **2020**, *229*, 110517. [CrossRef]
52. Serageldin, A.A.; Abdeen, A.; Ahmed, M.M.; Radwan, A.; Shmroukh, A.N.; Ookawara, S. Solar chimney combined with earth to-air heat exchanger for passive cooling of residential buildings in hot areas. *Sol. Energy* **2020**, *206*, 145–162. [CrossRef]
53. Agarwal, S.K.; Batista, R.C. Enhancement of building thermal performance: A comparative analysis of integrated solar chimney and geothermal systems. *J. Sustain. Energy* **2023**, *2*, 91–108. [CrossRef]
54. El-Said, E.M.; Sharaf, M.A.; Aljabr, A.; Marzouk, S.A. Enhancing the performance of an earth air heat exchanger with novel pipe configurations. *Int. J. Heat Fluid Flow* **2024**, *110*, 109630. [CrossRef]
55. Rosa, N.; Soares, N.; Costa, J.J.; Santos, P.; Gervásio, H. Assessment of an earth-air heat exchanger (EAHE) system for residential buildings in warm-summer Mediterranean climate. *Sustain. Energy Technol. Assess.* **2020**, *38*, 100649. [CrossRef]
56. Hasan, N.; Arif, M.; Khader, M.A. Earth Air Tunnel Heat Exchanger for Building Cooling and Heating. In *Heat Transfer-Design, Experimentation and Applications*; IntechOpen: London, UK, 2021.
57. Mitha, S.B.; Omarsaib, M. Emerging technologies and higher education libraries: A bibliometric analysis of the global literature. *Libr. Hi Tech* **2024**. *ahead of print*. [CrossRef]
58. Van Eck, N.; Waltman, L. Software survey: VOSviewer, a computer program for bibliometric mapping. *Scientometrics* **2010**, *84*, 523–538. [CrossRef]
59. Ahmad, S.N.; Prakash, O. A review on modelling, experimental analysis and parametric effects of earth–air heat exchanger. *Model. Earth Syst. Environ.* **2022**, *8*, 1535–1551. [CrossRef]
60. Agrawal, K.K.; Misra, R.; Agrawal, G.D.; Bhardwaj, M.; Jamuwa, D.K. Effect of different design aspects of pipe for earth air tunnel heat exchanger system: A state of art. *Int. J. Green Energy* **2019**, *16*, 598–614. [CrossRef]
61. Noman, S.; Tirumalachetty, H.; Athikesavan, M.M. A comprehensive review on experimental, numerical and optimization analysis of EAHE and GSHP systems. *Environ. Sci. Pollut. Res.* **2022**, *29*, 67559–67603. [CrossRef] [PubMed]
62. Liu, Z.; Sun, P.; Li, S.; Yu, Z.J.; El Mankibi, M.; Roccamena, L.; Zhang, G. Enhancing a vertical earth-to-air heat exchanger system using tubular phase change material. *J. Clean. Prod.* **2019**, *237*, 117763. [CrossRef]
63. Ahmed, S.F.; Khan, M.M.K.; Amanullah, M.T.O.; Rasul, M.G.; Hassan, N.M.S. Performance assessment of earth pipe cooling system for low energy buildings in a subtropical climate. *Energy Convers. Manag.* **2015**, *106*, 815–825. [CrossRef]
64. Maytorena, V.M.; Moreno, S.; Hinojosa, J.F. Effect of operation modes on the thermal performance of EAHE systems with and without PCM in summer weather conditions. *Energy Build.* **2021**, *250*, 111278. [CrossRef]
65. Niu, F.; Yu, Y.; Yu, D.; Li, H. Investigation on soil thermal saturation and recovery of an earth to air heat exchanger under different operation strategies. *Appl. Therm. Eng.* **2015**, *77*, 90–100. [CrossRef]
66. Sakhri, N.; Menni, Y.; Ameur, H. Effect of the pipe material and burying depth on the thermal efficiency of earth-to-air heat exchangers. *Case Stud. Chem. Environ. Eng.* **2020**, *2*, 100013. [CrossRef]
67. Lekhal, M.C.; Benzaama, M.H.; Kindinis, A.; Mokhtari, A.M.; Belarbi, R. Effect of geo-climatic conditions and pipe material on heating performance of earth-air heat exchangers. *Renew. Energy* **2021**, *163*, 22–40. [CrossRef]
68. Xiao, J.; Li, J. Influence of different types of pipes on the heat exchange performance of an earth-air heat exchanger. *Case Stud. Therm. Eng.* **2024**, *55*, 104116. [CrossRef]
69. Peretti, C.; Zarrella, A.; De Carli, M.; Zecchin, R. The design and environmental evaluation of earth-to-air heat exchangers (EAHE). *A Lit. Rev. Renew. Sustain. Energy Rev.* **2013**, *28*, 107–116. [CrossRef]
70. Pavlenko, A.M.; Usenko, B.O.; Koshlak, H.V. Analysis of thermal peculiarities of alloying with special properties. *Metall. Min. Ind.* **2014**, *2*, 15–20.
71. Agrawal, K.K.; Misra, R.; Agrawal, G.D. Improving the thermal performance of ground air heat exchanger system using sand-bentonite (in dry and wet condition) as backfilling material. *Renew. Energy* **2020**, *146*, 2008–2023. [CrossRef]

72. Besler, M.; Cepiński, W.; Kęskiewicz, P. Direct-contact air, gravel, ground heat exchanger in air treatment systems for cowshed air conditioning. *Energies* **2021**, *15*, 234. [CrossRef]
73. Liu, Z.; Sun, P.; Xie, M.; Zhou, Y.; He, Y.; Zhang, G.; Qin, D. Multivariant optimization and sensitivity analysis of an experimental vertical earth-to-air heat exchanger system integrating phase change material with Taguchi method. *Renew. Energy* **2021**, *173*, 401–414. [CrossRef]
74. Zhou, T.; Xiao, Y.; Huang, H.; Lin, J. Numerical study on the cooling performance of a novel passive system: Cylindrical phase change material-assisted earth-air heat exchanger. *J. Clean. Prod.* **2020**, *245*, 118907. [CrossRef]
75. Shahsavar, A.; Azimi, N. Performance evaluation and multi-objective optimization of a hybrid earth-air heat exchanger and building-integrated photovoltaic/thermal system with phase change material and exhaust air heat recovery. *J. Build. Eng.* **2024**, *90*, 109531. [CrossRef]
76. Myroniuk, K.; Furdas, Y.; Zhelykh, V.; Adamski, M.; Gumen, O.; Savin, V.; Mitoulis, S.A. Passive Ventilation of Residential Buildings Using the Trombe Wall. *Buildings* **2024**, *14*, 3154. [CrossRef]
77. Cavazzini, G.; Zanetti, G.; Benato, A. Analysis of a domestic air heat pump integrated with an air-geothermal heat exchanger in real operating conditions: The case study of a single-family building. *Energy Build.* **2024**, *315*, 114302. [CrossRef]
78. Baglivo, C.; Bonuso, S.; Congedo, P.M. Performance Analysis of Air Cooled Heat Pump Coupled with Horizontal Air Ground Heat Exchanger in the Mediterranean Climate. *Energies* **2018**, *11*, 2704. [CrossRef]
79. Li, H.; Ni, L.; Yao, Y.; Sun, C. Annual performance experiments of an earth-air heat exchanger fresh air-handling unit in severe cold regions: Operation, economic and greenhouse gas emission analyses. *Renew. Energy* **2020**, *146*, 25–37. [CrossRef]
80. Ascione, F.; D'Agostino, D.; Marino, C.; Minichiello, F. Earth-to-air heat exchanger for NZEB in Mediterranean climate. *Renew. Energy* **2016**, *99*, 553–563. [CrossRef]
81. Thiers, S.; Peuportier, B. Thermal and environmental assessment of a passive building equipped with an earth-to-air heat exchanger in France. *Sol. Energy* **2008**, *82*, 820–831. [CrossRef]
82. Wojtkowiak, J. Experimental flow characteristics of multi-pipe earth-to-air heat exchangers. *Found. Civ. Environ. Eng.* **2012**, *15*, 5–18.
83. Bojic, M.; Trifunovic, N.; Papadakis, G.; Kyritsis, S. Numerical simulation, technical and economic evaluation of air-to-earth heat exchanger coupled to a building. *Energy* **1997**, *22*, 1151–1158. [CrossRef]
84. Li, H.; Ni, L.; Liu, G.; Zhao, Z.; Yao, Y. Feasibility study on applications of an Earth-air Heat Exchanger (EAHE) for preheating fresh air in severe cold regions. *Renew. Energy* **2019**, *133*, 1268–1284. [CrossRef]
85. Lattieff, F.A.; Atiya, M.A.; Lateef, R.A.; Dulaimi, A.; Jweeg, M.J.; Abed, A.M.; Talebizadehsardari, P. Thermal analysis of horizontal earth-air heat exchangers in a subtropical climate: An experimental study. *Front. Built Environ.* **2022**, *8*, 981946. [CrossRef]
86. Molina-Rodea, R.; Wong-Loya, J.A.; Pocasangre-Chávez, H.; Reyna-Guillén, J. Experimental evaluation of a "U" type earth-to-air heat exchanger planned for narrow installation space in warm climatic conditions. *Energy Built Environ.* **2024**, *5*, 772–786. [CrossRef]
87. Ahmed, S.F.; Liu, G.; Mofijur, M.; Azad, A.K.; Hazrat, M.A.; Chu, Y.M. Physical and hybrid modelling techniques for earth-air heat exchangers in reducing building energy consumption: Performance, applications, progress, and challenges. *Sol. Energy* **2021**, *216*, 274–294. [CrossRef]
88. Tasdelen, F.; Dagtekin, I. A numerical study on heating performance of horizontal and vertical earth-air heat exchangers with equal pipe lengths. *Therm. Sci.* **2022**, *26 Pt A*, 2929–2939. [CrossRef]
89. Salhein, K.; Kobus, C.J.; Zohdy, M.; Annekaa, A.M.; Alhawsawi, E.Y.; Salheen, S.A. Heat Transfer Performance Factors in a Vertical Ground Heat Exchanger for a Geothermal Heat Pump System. *Energies* **2024**, *17*, 5003. [CrossRef]
90. Liu, Z.; Yu, Z.J.; Yang, T.; Li, S.; El Mankibi, M.; Roccamena, L.; Zhang, G. Experimental investigation of a vertical earth-to-air heat exchanger system. *Energy Convers. Manag.* **2019**, *183*, 241–251. [CrossRef]
91. Koshlak, H.; Pavlenko, A. Heat and Mass Transfer During Phase Transitions in Liquid Mixtures. *Rocz. Ochr. Srodowiska* **2019**, *21*, 234–249.
92. Basok, B.; Davydenko, B.; Koshlak, H.; Novikov, V. Free Convection and Heat Transfer in Porous Ground Massif during Ground Heat Exchanger Operation. *Materials* **2022**, *15*, 4843. [CrossRef]
93. Go, G.H.; Lee, S.R.; Nikhil, N.V.; Yoon, S. A new performance evaluation algorithm for horizontal GCHPs (ground coupled heat pump systems) that considers rainfall infiltration. *Energy* **2015**, *83*, 766–777. [CrossRef]
94. Rodríguez-Vázquez, M.; Xamán, J.; Chávez, Y.; Hernández-Pérez, I.; Simá, E. Thermal potential of a geothermal earth-to-air heat exchanger in six climatic conditions of México. *Mech. Ind.* **2020**, *21*, 308. [CrossRef]
95. Bisoniya, T.S. Design of earth–air heat exchanger system. *Geotherm. Energy* **2015**, *3*, 18. [CrossRef]
96. Basok, B.; Pavlenko, A.; Nedbailo, A.; Bozhko, I.; Novitska, M.; Koshlak, H.; Tkachenko, M. Analysis of the Energy Efficiency of the Earth-To-Air Heat Exchanger. *Rocz. Ochr. Sr.* **2022**, *24*, 202–213. [CrossRef]
97. Greco, A.; Gundabattini, E.; Solomon, D.G.; Singh Rassiah, R.; Masselli, C. A Review on Geothermal Renewable Energy Systems for Eco-Friendly Air-Conditioning. *Energies* **2022**, *15*, 5519. [CrossRef]

98. Peñaloza Peña, S.A.; Jaramillo Ibarra, J.E. Potential applicability of earth to air heat exchanger for cooling in a colombian tropical weather. *Buildings* **2021**, *11*, 219. [CrossRef]
99. Malek, K.; Malek, K.; Khanmohammadi, F. Response of soil thermal conductivity to various soil properties. *Int. Commun. Heat Mass Transf.* **2021**, *127*, 105516. [CrossRef]
100. Cuny, M.; Lin, J.; Siroux, M.; Magnenet, V.; Fond, C. Influence of coating soil types on the energy of earth-air heat exchanger. *Energy Build.* **2018**, *158*, 1000–1012. [CrossRef]
101. Di Sipio, E.; Bertermann, D. Factors influencing the thermal efficiency of horizontal ground heat exchangers. *Energies* **2017**, *10*, 1897. [CrossRef]
102. Kumar Singh, R.; Sharma, R.V. Mathematical Investigation of Soil Temperature Variation for Geothermal Applications. *Int. J. Eng.* **2017**, *30*, 1609–1614.
103. Faridi, H.; Arabhosseini, A.; Zarei, G.; Okos, M. Degree-Day Index for Estimating the Thermal Requirements of a Greenhouse Equipped with an Air-Earth Heat Exchanger System. *J. Agric. Mach.* **2021**, *11*, 83–95.
104. Ascione, F.; Bellia, L.; Minichiello, F. Earth-to-air heat exchangers for Italian climates. *Renew. Energy* **2011**, *36*, 2177–2188. [CrossRef]
105. Vidhi, R. A review of underground soil and night sky as passive heat sink: Design configurations and models. *Energies* **2018**, *11*, 2941. [CrossRef]
106. Singh, R.K.; Sharma, R.V. Numerical analysis for ground temperature variation. *Geotherm. Energy* **2017**, *5*, 22. [CrossRef]
107. Kaushal, M. Geothermal cooling/heating using ground heat exchanger for various experimental and analytical studies: Comprehensive review. *Energy Build.* **2017**, *139*, 634–652. [CrossRef]
108. Sanusi, A.N.; Shao, L.; Ibrahim, N. Passive ground cooling system for low energy buildings in Malaysia (hot and humid climates). *Renew. Energy* **2013**, *49*, 193–196. [CrossRef]
109. Babar, H.; Ali, H.M.; Haseeb, M.; Abubaker, M.; Muhammad, A.; Irfan, M. Experimental investigation of the ground coupled heat exchanger system under the climatic conditions of Sahiwal, Pakistan. In Proceedings of the World Congress on Engineering, London, UK, 4–6 July 2018; pp. 4–6.
110. Khan, T.A.; Shabbir, K.; Khan, M.M.; Siddiqui, F.A.; Taseer, M.Y.R.; Imtiaz, S. Earth-tube system to control indoor thermal environment in residential buildings. *Tech. J.* **2020**, *25*, 33–40.
111. Alves, A.B.M.; Schmid, A.L. Cooling and heating potential of underground soil according to depth and soil surface treatment in the Brazilian climatic regions. *Energy Build.* **2015**, *90*, 41–50. [CrossRef]
112. Bhandari, R. Sustainable cooling solutions for building environments: A comprehensive study of earth-air cooling systems. *Adv. Mech. Eng.* **2024**, *16*, 16878132241272209. [CrossRef]
113. Pollack, H.N.; Smerdon, J.E.; van Keken, P.E. Variable seasonal coupling between air and ground temperatures: A simple representation in terms of subsurface thermal diffusivity. *Geophys. Res. Lett.* **2005**, *32*. [CrossRef]
114. Smerdon, J.E.; Pollack, H.N.; Cermak, V.; Enz, J.W.; Kresl, M.; Safanda, J.; Wehmiller, J.F. Air-ground temperature coupling and subsurface propagation of annual temperature signals. *J. Geophys. Res. Atmos.* **2004**, *109*. [CrossRef]
115. Kane, D.L.; Hinkel, K.M.; Goering, D.J.; Hinzman, L.D.; Outcalt, S.I. Non-conductive heat transfer associated with frozen soils. *Glob. Planet. Change* **2001**, *29*, 275–292. [CrossRef]
116. Zajch, A.; Gough, W.A. Seasonal sensitivity to atmospheric and ground surface temperature changes of an open earth-air heat exchanger in Canadian climates. *Geothermics* **2021**, *89*, 101914. [CrossRef]
117. Zajch, A.; Gough, W.; Chiesa, G. Earth–Air Heat Exchanger Potential Under Future Climate Change Scenarios in Nine North American Cities. In *Sustainability in Energy and Buildings: Proceedings of SEB 2019*; Springer: Singapore, 2020; pp. 109–119.
118. Singh, B.; Kumar, R.; Asati, A.K. Influence of parameters on performance of earth air heat exchanger in hot-dry climate. *J. Mech. Sci. Technol.* **2018**, *32*, 5457–5463. [CrossRef]
119. Liu, Z.; Yu, Z.J.; Yang, T.; Roccamena, L.; Sun, P.; Li, S.; El Mankibi, M. Numerical modeling and parametric study of a vertical earth-to-air heat exchanger system. *Energy* **2019**, *172*, 220–231. [CrossRef]
120. Hasan, M.I.; Noori, S.W.; Shkarah, A.J. Parametric study on the performance of the earth-to-air heat exchanger for cooling and heating applications. *Heat Transf. —Asian Res.* **2019**, *48*, 1805–1829. [CrossRef]
121. Ali, S.; Muhammad, N.; Amin, A.; Sohaib, M.; Basit, A.; Ahmad, T. Parametric optimization of earth to air heat exchanger using response surface method. *Sustainability* **2019**, *11*, 3186. [CrossRef]
122. Agrawal, K.K.; Misra, R.; Agrawal, G.D. To study the effect of different parameters on the thermal performance of ground-air heat exchanger system: In situ measurement. *Renew. Energy* **2020**, *146*, 2070–2083. [CrossRef]
123. Mahach, H.; Benhamou, B. Extensive parametric study of cooling performance of an earth-to-air heat exchanger in hot semi-arid climate. *J. Therm. Sci. Eng. Appl.* **2021**, *13*, 031006. [CrossRef]
124. Zhang, Z.; Sun, J.; Zhang, Z.; Jia, X.; Liu, Y. Numerical Research and Parametric Study on the Thermal Performance of a Vertical Earth-to-Air Heat Exchanger System. *Math. Probl. Eng.* **2021**, *2021*, 5557280. [CrossRef]

125. Yu, W.; Chen, X.; Ma, Q.; Gao, W.; Wei, X. Modeling and assessing earth-air heat exchanger using the parametric performance design method. *Energy Sources Part A Recovery Util. Environ. Eff.* **2022**, *44*, 7873–7894. [CrossRef]
126. Sghiouri, H.; Wakil, M.; Charai, M.; Mezrhab, A. Development of a Multi-Objective optimization framework for Earth-to-Air heat Exchanger Systems: Enhancing thermal performance and economic viability in Moroccan climates. *Energy Convers. Manag.* **2024**, *321*, 119024. [CrossRef]
127. de Andrade, I.R.; dos Santos, E.D.; Zhang, H.; Rocha, L.A.O.; Razera, A.L.; Isoldi, L.A. Multi-Objective Numerical Analysis of Horizontal Rectilinear Earth–Air Heat Exchangers with Elliptical Cross Section Using Constructal Design and TOPSIS. *Fluids* **2024**, *9*, 257. [CrossRef]
128. Cuny, M.; Lin, J.; Siroux, M.; Fond, C. Influence of an improved surrounding soil on the energy performance and the design length of earth-air heat exchanger. *Appl. Therm. Eng.* **2019**, *162*, 114320. [CrossRef]
129. Jahanbin, A. Thermal performance of the vertical ground heat exchanger with a novel elliptical single U-tube. *Geothermics* **2020**, *86*, 101804. [CrossRef]
130. Benhammou, M.; Sahli, Y.; Moungar, H. Investigation of the impact of pipe geometric form on earth-to-air heat exchanger performance using Complex Finite Fourier Transform analysis. Part I: Operation in cooling mode. *Int. J. Therm. Sci.* **2022**, *177*, 107484. [CrossRef]
131. Darius, D.; Misaran, M.S.; Rahman, M.M.; Ismail, M.A.; Amaludin, A. Working parameters affecting earth-air heat exchanger (EAHE) system performance for passive cooling: A review. *IOP Conf. Ser. Mater. Sci. Eng.* **2017**, *217*, 012021. [CrossRef]
132. Benrachi, N.; Ouzzane, M.; Smaili, A.; Lamarche, L.; Badache, M.; Maref, W. Numerical parametric study of a new earth-air heat exchanger configuration designed for hot and arid climates. *Int. J. Green Energy* **2020**, *17*, 115–126. [CrossRef]
133. Mihalakakou, G.; Lewis, J.O.; Santamouris, M. On the heating potential of buried pipes techniques—Application in Ireland. *Energy Build.* **1996**, *24*, 19–25. [CrossRef]
134. Amanowicz, Ł.; Wojtkowiak, J. Comparison of single-and multipipe earth-to-air heat exchangers in terms of energy gains and electricity consumption: A case study for the temperate climate of central europe. *Energies* **2021**, *14*, 8217. [CrossRef]
135. Mihalakakou, G.; Souliotis, M.; Papadaki, M.; Halkos, G.; Paravantis, J.; Makridis, S.; Papaefthimiou, S. Applications of earth-to-air heat exchangers: A holistic review. *Renew. Sustain. Energy Rev.* **2022**, *155*, 111921. [CrossRef]
136. Qi, D.; Li, S.; Zhao, C.; Xie, W.; Li, A. Structural optimization of multi-pipe earth to air heat exchanger in greenhouse. *Geothermics* **2022**, *98*, 102288. [CrossRef]
137. Amanowicz, Ł.; Wojtkowiak, J. Thermal performance of multi-pipe earth-to-air heat exchangers considering the non-uniform distribution of air between parallel pipes. *Geothermics* **2020**, *88*, 101896. [CrossRef]
138. Minaei, A.; Rabani, R. Development of a transient analytical method for multi-pipe earth-to-air heat exchangers with parallel configuration. *J. Build. Eng.* **2023**, *73*, 106781. [CrossRef]
139. Qi, D.; Li, A.; Li, S.; Zhao, C. Comparative analysis of earth to air heat exchanger configurations based on uniformity and thermal performance. *Appl. Therm. Eng.* **2021**, *183*, 116152. [CrossRef]
140. Brum, R.S.; Ramalho, J.V.; Rodrigues, M.K.; Rocha, L.A.; Isoldi, L.A.; Dos Santos, E.D. Design evaluation of Earth-Air Heat Exchangers with multiple ducts. *Renew. Energy* **2019**, *135*, 1371–1385. [CrossRef]
141. Amanowicz, Ł.; Wojtkowiak, J. Approximated flow characteristics of multi-pipe earth-to-air heat exchangers for thermal analysis under variable airflow conditions. *Renew. Energy* **2020**, *158*, 585–597. [CrossRef]
142. Amanowicz, Ł. Influence of geometrical parameters on the flow characteristics of multi-pipe earth-to-air heat exchangers–experimental and CFD investigations. *Appl. Energy* **2018**, *226*, 849–861. [CrossRef]
143. Badescu, V.; Isvoranu, D. Pneumatic and thermal design procedure and analysis of earth-to-air heat exchangers of registry type. *Appl. Energy* **2011**, *88*, 1266–1280. [CrossRef]
144. Sofyan, S.E.; Riayatsyah, T.M.I.; Hu, E.; Tamlicha, A.; Pahlefi, T.M.R.; Aditiya, H.B. Computational fluid dynamic simulation of earth air heat exchanger: A thermal performance comparison between series and parallel arrangements. *Results Eng.* **2024**, *24*, 102932. [CrossRef]
145. Muehleisen, R.T. Simple design tools for earth-air heat exchangers. *Proc. SimBuild* **2012**, *5*, 723–730.
146. Mehdid, C.E.; Benchabane, A.; Rouag, A.; Moummi, N.; Melhegueg, M.A.; Moummi, A.; Brima, A. Thermal design of Earth-to-air heat exchanger. Part II a new transient semi-analytical model and experimental validation for estimating air temperature. *J. Clean. Prod.* **2018**, *198*, 1536–1544. [CrossRef]
147. Ali, M.H.; Kurjak, Z.; Beke, J. Investigation of earth air heat exchangers functioning in arid locations using Matlab/Simulink. *Renew. Energy* **2023**, *209*, 632–643. [CrossRef]
148. Guo, X.; Wei, H.; He, X.; Du, J.; Yang, D. Experimental evaluation of an earth–to–air heat exchanger and air source heat pump hybrid indoor air conditioning system. *Energy Build.* **2022**, *256*, 111752. [CrossRef]
149. Do, S.L.; Baltazar, J.C.; Haberl, J. Potential cooling savings from a ground-coupled return-air duct system for residential buildings in hot and humid climates. *Energy Build.* **2015**, *103*, 206–215. [CrossRef]

150. Mahmoud, M.; Abdelkareem, M.A.; Olabi, A.G. Earth air heat exchangers. In *Renewable Energy-Volume 2: Wave, Geothermal, and Bioenergy*; Academic Press: Cambridge, MA, USA, 2024; pp. 163–179.
151. Yan, T.; Xu, X. Utilization of ground heat exchangers: A review. *Curr. Sustain./Renew. Energy Rep.* **2018**, *5*, 189–198. [CrossRef]
152. Qi, X.; Yang, D.; Guo, X.; Chen, F.; An, F.; Wei, H. Theoretical modelling and experimental evaluation of thermal performance of a combined earth-to-air heat exchanger and return air hybrid system. *Renew. Energy* **2024**, *236*, 121418. [CrossRef]
153. Kundu, A. Application of Geothermal Energy-Based Earth-Air Heat Exchanger in Sustainable Buildings. In *Heat Transfer Enhancement Techniques: Thermal Performance, Optimization and Applications*; John Wiley & Sons: Hoboken, NJ, USA, 2025; pp. 221–232.
154. Dokmak, H.; Faraj, K.; Faraj, J.; Castelain, C.; Khaled, M. Geothermal systems classification, coupling, and hybridization: A recent comprehensive review. *Energy Built Environ.* 2024, in press. [CrossRef]
155. Yadav, S.; Panda, S.K.; Tiwari, G.N.; Al-Helal, I.M.; Hachem-Vermette, C. Periodic theory of greenhouse integrated semi-transparent photovoltaic thermal (GiSPVT) system integrated with earth air heat exchanger (EAHE). *Renew. Energy* **2022**, *184*, 45–55. [CrossRef]
156. Hraibet, M.K.; Ismaeel, A.A.; Mohammed, M.J. Performances evaluation of different photovoltaic-thermal solar collector designs for electrical and hot air generation technology. *AIP Conf. Proc.* **2024**, *3002*, 060005.
157. Zhao, J.; Huang, B.; Li, Y.; Zhao, Y. Comprehensive review on climatic feasibility and economic benefits of Earth-to-Air Heat Exchanger (EAHE) systems. *Sustain. Energy Technol. Assess.* **2024**, *68*, 103862. [CrossRef]
158. Gorjian, S.; Singh, R.; Shukla, A.; Mazhar, A.R. On-farm applications of solar PV systems. In *Photovoltaic Solar Energy Conversion*; Academic Press: Cambridge, MA, USA, 2020; pp. 147–190.
159. Aljashaami, B.A.; Ali, B.M.; Salih, S.A.; Alwan, N.T.; Majeed, M.H.; Ali, O.M.; Shcheklein, S.E. Recent improvements to heating, ventilation, and cooling technologies for buildings based on renewable energy to achieve zero-energy buildings: A systematic review. *Results Eng.* **2024**, *23*, 102769. [CrossRef]
160. Hepbasli, A. Low exergy modelling and performance analysis of greenhouses coupled to closed earth-to-air heat exchangers (EAHEs). *Energy Build.* **2013**, *64*, 224–230. [CrossRef]
161. Shaaban, A.; Mosa, M.; El Samahy, A.; Abed, K. Enhancing the Performance of Photovoltaic Panels by Cooling: A Review. *Int. Rev. Autom. Control* **2023**, *16*, 26–43. [CrossRef]
162. Dhaidan, N.S.; Al-Shohani, W.A.; Abbas, H.H.; Rashid, F.L.; Ameen, A.; Al-Mousawi, F.N.; Homod, R.Z. Enhancing the thermal performance of an agricultural solar greenhouse by geothermal energy using an earth-air heat exchanger system: A review. *Geothermics* **2024**, *123*, 103115. [CrossRef]
163. Jakhar, S.; Soni, M.S.; Boehm, R.F. Thermal modeling of a rooftop photovoltaic/thermal system with earth air heat exchanger for combined power and space heating. *J. Sol. Energy Eng.* **2018**, *140*, 031011. [CrossRef]
164. Lopez-Pascual, D.; Valiente-Blanco, I.; Manzano-Narro, O.; Fernandez-Munoz, M.; Diez-Jimenez, E. Experimental characterization of a geothermal cooling system for enhancement of the efficiency of solar photovoltaic panels. *Energy Rep.* **2022**, *8*, 756–763. [CrossRef]
165. Tiwari, G.N.; Singh, S.; Singh, Y.; Tiwari, A.; Panda, S.K. Enhancement of daily and monthly electrical power of off-grid greenhouse integrated semi-transparent photo-voltaic thermal (GiSPVT) system by integrating earth air heat exchanger (EAHE). *E-Prime-Adv. Electr. Eng. Electron. Energy* **2022**, *2*, 100074. [CrossRef]
166. Salem, H.H.; Hashem, A.L. Integration of Earth-air heat exchanger in buildings review for theoretical researches. *AIP Conf. Proc.* **2023**, *2787*, 030009.
167. Yildiz, A.; Ozgener, O.; Ozgener, L. Exergetic performance assessment of solar photovoltaic cell (PV) assisted earth to air heat exchanger (EAHE) system for solar greenhouse cooling. *Energy Build.* **2011**, *43*, 3154–3160. [CrossRef]
168. Alkaragoly, M.; Maerefat, M.; Targhi, M.Z.; Abdljalel, A. An innovative hybrid system consists of a photovoltaic solar chimney and an earth-air heat exchanger for thermal comfort in buildings. *Case Stud. Therm. Eng.* **2022**, *40*, 102546. [CrossRef]
169. Li, Y.; Chu, S.; Zhao, J.; Li, W.; Tan, B.; Lu, J. A numerical study on the performance of a hybrid ventilation and power generation system. *Appl. Therm. Eng.* **2024**, *238*, 122228. [CrossRef]
170. Yang, L.H.; Hu, J.W.; Chiang, Y.C.; Chen, S.L. Performance analysis of building-integrated earth-air heat exchanger retrofitted with a supplementary water system for cooling-dominated climate in Taiwan. *Energy Build.* **2021**, *242*, 110949. [CrossRef]
171. Ahmadi, S.; Irandoost Shahrestani, M.; Sayadian, S.; Maerefat, M.; Haghighi Poshtiri, A. Performance analysis of an integrated cooling system consisted of earth-to-air heat exchanger (EAHE) and water spray channel. *J. Therm. Anal. Calorim.* **2021**, *143*, 473–483. [CrossRef]
172. Radchenko, M.; Yang, Z.; Pavlenko, A.; Radchenko, A.; Radchenko, R.; Koshlak, H.; Bao, G. Increasing the efficiency of turbine inlet air cooling in climatic conditions of China through rational designing—Part 1: A case study for subtropical climate: General approaches and criteria. *Energies* **2023**, *16*, 6105. [CrossRef]
173. Liu, T.; Pang, H.; He, S.; Zhao, B.; Zhang, Z.; Wang, J.; Gao, M. Evaporative Cooling Applied in Thermal Power Plants: A Review of the State-of-the-Art and Typical Case Studies. *Fluid Dyn. Mater. Process.* **2023**, *19*, 2229–2266. [CrossRef]

174. Barakat, S.; Ramzy, A.; Hamed, A.M.; El-Emam, S.H. Augmentation of gas turbine performance using integrated EAHE and Fogging Inlet Air Cooling System. *Energy* **2019**, *189*, 116133. [CrossRef]
175. de la Rocha Camba, E.; Petrakopoulou, F. Earth-cooling air tunnels for thermal power plants: Initial design by CFD modelling. *Energies* **2020**, *13*, 797. [CrossRef]
176. Shahsavar, A.; Arıcı, M. Multi-objective energy and exergy optimization of hybrid building-integrated heat pipe photovoltaic/thermal and earth air heat exchanger system using soft computing technique. *Eng. Anal. Bound. Elem.* **2023**, *148*, 293–304. [CrossRef]
177. Sakhrı, N.; Mennı, Y.; Ameur, H.; Chamkha, A.J. Experimental study of a stand-alone earth to air heat exchanger for heating and cooling in arid regions. *J. Therm. Eng.* **2021**, *7*, 1206–1215. [CrossRef]

energies

MDPI

Review

A Review of CO_2 Clathrate Hydrate Technology: From Lab-Scale Preparation to Cold Thermal Energy Storage Solutions

Sai Bhargav Annavajjala [1], Noah Van Dam [1], Devinder Mahajan [2,3] and Jan Kosny [1,*]

[1] Department of Mechanical and Industrial Engineering, University of Massachusetts Lowell, Lowell, MA 01854, USA; saibhargav_annavajjala@student.uml.edu (S.B.A.); noah_vandam@uml.edu (N.V.D.)
[2] Department of Materials Science and Chemical Engineering, Stony Brook University, Stony Brook, NY 11794, USA; devinder.mahajan@stonybrook.edu
[3] Institute of Gas Innovation and Technology, Stony Brook, NY 11794, USA
* Correspondence: jan_kosny@uml.edu

Abstract: Carbon dioxide (CO_2) clathrate hydrate is gaining attention as a promising material for cold thermal energy storage (CTES) due to its high energy storage capacity and low environmental footprint. It shows strong potential in building applications, where space cooling accounts for nearly 40% of total energy use and over 85% of electricity demand in developed countries. CO_2 hydrates are also being explored for use in refrigeration, cold chain logistics, supercomputing, biomedical cooling, and defense systems. With the growing number of applications in mind, this review focuses on the thermal behavior of CO_2 hydrates and their environmental impact. It highlights recent efforts to reduce formation pressure and temperature using chemical promoters and surfactants. This paper also reviews key experimental techniques used to study hydrate properties, including Fourier transform infrared spectroscopy (FTIR), X-ray diffraction (XRD), high-pressure differential scanning calorimetry (HP-DSC), and the T-history method. In lifecycle comparisons, CO_2 hydrate systems show better energy efficiency and lower carbon emissions than traditional ice or other phase-change materials (PCMs). This review also discusses current commercialization challenges such as high energy input during formation and promoter toxicity. Finally, practical strategies to move CO_2 hydrate-based CTES from lab-scale studies to real-world cooling and temperature control applications are discussed.

Keywords: energy demands; cold thermal energy storage; hydrate formation; promoters; characterization techniques; lifecycle analysis

1. Introduction

The growth in global energy demand and the need for environmental protection have heightened the urgency of reducing carbon emissions (both CO_2 and CH_4). The United Nations has prioritized the reduction of carbon emissions as a key strategy to combat climate change. Buildings account for approximately 40% of the world's energy consumption, with heating, ventilation, and air conditioning (HVAC) systems being responsible for approximately half of that usage. Energy usage is rapidly growing as the number of residential, commercial, and industrial buildings continues to rise, making the building sector the largest consumer of energy globally [1].

HVAC systems are critical for maintaining indoor comfort, but their energy demands fluctuate throughout the day and across seasons, creating peak loads that can strain the power grid and increase energy costs. During extreme temperature conditions, cooling or heating demands spike, requiring additional capacity to meet these temporary surges in space conditioning demand [2]. Thermal energy storage (TES) technologies offer a promising solution by storing excess thermal energy during off-peak periods and releasing it when demand rises.

Beyond HVAC systems, refrigeration represents another significant portion of energy consumption. Commercial refrigeration in industries such as medicine, food storage, supermarkets, plant nurseries, transportation, as well as data centers, supercomputing, and microchip manufacturing processes can be very energy intensive. For example, HVAC and refrigeration systems account for 50 to 60 percent of the total energy consumption in supermarkets [3], including both space conditioning for the comfort of the occupants and the safety of the food products stored [4]. The integration of TES technologies with HVAC systems can enhance efficiency, reduce operational costs, and mitigate peak energy demand. TES systems can operate on either sensible heat storage (SHS) or latent heat storage (LHS). SHS relies on the temperature change of the storage material, such as water, stone, or concrete, to store and release energy [5,6]. It is simple and cost-effective but requires large volumes of material due to the relatively low volumetric energy storage density. LHS, on the other hand, utilizes phase-change materials (PCMs) to store energy during phase transitions (e.g., melting and solidification), offering higher volumetric energy storage density and maintaining nearly constant operating temperatures. Individual PCMs are limited in their temperature range, but the wide variety of PCMs can be used to address the desired operating conditions.

Figure 1 (adapted from Awan et al. [7]) illustrates the thermal behavior of PCM during heating and cooling cycles. During heating, the solid phase initially undergoes sensible heating, where its temperature rises proportionally to the energy absorbed. Upon reaching the melting temperature, the material transitions from a solid to a liquid. During this process, energy is absorbed as latent heat while maintaining a constant temperature, as shown by the horizontal section of the graph. Once the phase-change process is complete, the liquid PCM undergoes further sensible heating, with the temperature increasing proportionally to the energy input. The additional latent heat storage mechanism provides significant energy density benefits but limits the operating temperature range of an individual PCM. Thus, PCMs must be carefully selected to match the expected operating temperatures. In addition, some PCMs may degrade over time, and heat transfer rates can be lower during phase transitions. Furthermore, some PCMs undergo changes in volume during phase-change processes, adding complexity to the PCM system design.

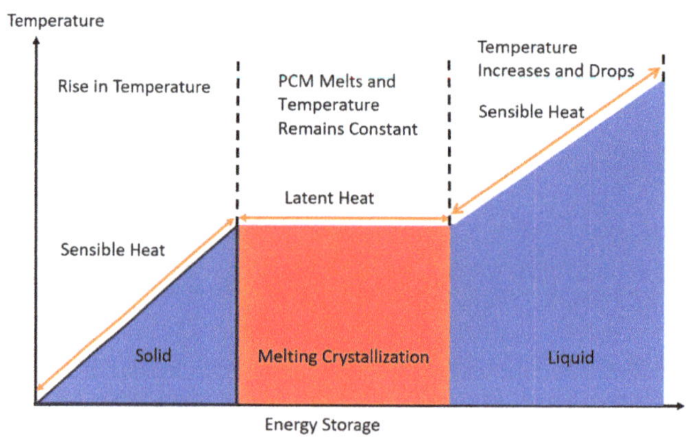

Figure 1. Thermal behavior during the phase-change cycle of a PCM. Reproduced with permission from a publication by Energies, 2023 [7].

There are several different types of PCMs used today, including organic, inorganic, and eutectic substances. Organic PCMs boast a wide range of phase-change temperatures, a high latent heat of melting, and compatibility with construction materials. They also exhibit long-term stability and eliminate supercooling issues. However, their thermal conductivity is relatively low, they have high thermal expansion coefficients, and they pose flammability risks. Inorganic PCMs offer high heat storage capacity, excellent thermal conductivity, and low volumetric expansion [8]. However, some of them are prone to supercooling, they may cause corrosion, and present incongruent melting behavior. Eutectic PCMs provide a compromise in performance between organic and inorganic substances, but they often suffer from lower phase-change enthalpy and tend to be more costly [9].

Pure water ice is one of the oldest and most widely used PCMs. However, producing ice requires specialized low-temperature refrigeration equipment and consumes significant amounts of energy, which in turn directly impacts the costs. High temperature differences between the processing temperature and the ice also negatively impact the system's efficiency, increasing the overall costs [10].

Gas hydrates, particularly CO_2 clathrate hydrates, have emerged as promising TES materials due to their higher phase transition temperature compared to that of water (improving efficiency in space conditioning applications) and superior dissociation energy. CO_2 hydrates and their slurries exhibit phase-change temperatures between 278 K and 285 K and latent enthalpies between 350 and 520 kJ/kg [11], which are ideal for air-conditioning needs. Unlike conventional PCMs, CO_2 hydrates offer environmental benefits, as they are biodegradable and non-corrosive. Beyond thermal energy storage, CO_2 hydrates also show potential for a usage in wastewater treatment and desalination, showcasing their versatility in energy and environmental applications [12].

A gas clathrate hydrate generally represents a structure where one component, known as the host (water), forms cages that enclose the molecules of another component, referred to as the guest (gas). The crystalline lattice of the host is a thermodynamically metastable phase, and the presence of guest molecules in its cavities stabilizes it. Under specific temperature and pressure conditions, the clathrate hydrate achieves thermodynamic stability when a certain proportion of its cavities is occupied by guest molecules [13].

The structure of the gas hydrate is determined by the type of guest gas molecules it contains, and it is capable of storing large amounts of gas. For example, one volume of methane hydrate can theoretically accommodate up to 172 volumes of methane gas [14]. Gas hydrates are typically categorized into three distinct structural forms: sI, sII, and sH. The sI and sII structures are both cubic crystal arrangements with lattice parameters of approximately 11.6 Å and 12.0 Å, respectively [15]. These structures are formed by varying arrangements of water molecules that create different types of cages, which encapsulate gas molecules. The size and shape of these cages significantly influence the properties of the hydrate, including its stability and gas storage capacity. Small guest molecules consisting of molecular diameters ranging between 0.4 and 0.55 nm form an sI structure, and larger guest molecules consisting of 0.6 to 0.7 nm molecular diameter help in the formation of the sII structure. The sH structure needs both a small guest molecule and a large guest molecule ranging between 0.75 and 0.9 nm [16]. As depicted in Figure 2, the clathrate hydrate structures are shaped by five distinct types of polyhedral: 5^{12}, $5^{12}6^2$, $5^{12}6^4$, $4^35^66^3$, and $5^{12}6^8$, where, for example, the designation 5^{12} indicates the presence of 12 pentagonal faces, while $5^{12}6^2$ represents a structure with 12 pentagonal faces and 2 hexagonal faces. The unit cell of the sI structure has the form $2(5^{12})\cdot 6(5^{12}6^2){:}46H_2O$, the sII structure has the form $16(5^{12})\cdot 8(5^{12}6^4){:}136H_2O$, and the sH has the form $3(5^{12})\cdot 2(4^35^66^3)\cdot 1(5^{12}6^8){:}34H_2O$ [17,18].

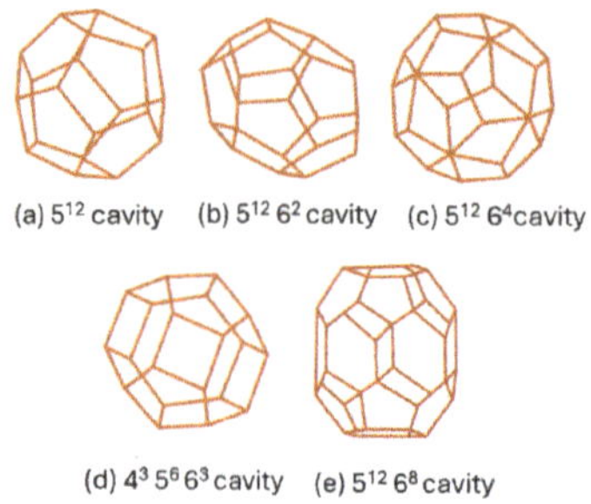

Figure 2. Polyhedral cage structures in gas hydrate formations—(**a**) 5^{12} Cavity–sI structure; (**b**) $5^{12}6^2$ Cavity–sII structure; (**c**) $5^{12}6^4$ Cavity–sII structure; (**d**) $4^35^66^3$ Cavity–sH structure; (**e**) $5^{12}6^8$ Cavity–sH structure. Reproduced with permission from a study published by Elsevier, 2006 [18].

In crystal structures of CO_2 hydrates, water and CO_2 molecules are bonded by weak van der Waals forces. The structure becomes stable only when a guest substance CO_2 is present, and only under specific pressures and temperatures. Otherwise, it turns into a normal ice structure. The water molecule content in each structure is different, with 46, 136, and 34 water molecules for sI, sII, and sH structure hydrates, respectively [19].

Following the current developments in the fields of building energy efficiency, refrigeration, and thermal energy storage, this review provides a unique discussion of CO_2 hydrate technologies, as future cold energy storage materials and thermofluids. While earlier reviews focus on limited aspects, such as formation kinetics or a narrow class of promoters, this paper connects multiple areas of CO_2 hydrate research into a coherent narrative. It begins with a discussion of phase-change material applications, gas hydrate formation, phase behavior, and structural thermodynamics, then involves formation enhancers, including some of the most recent developments, such as metal organic frameworks (MOFs) and aziridine. The authors also address safety and environmental regulations surrounding additive use. This paper looks beyond the lab-scale experiments by considering scalability challenges, energy performance, and the full system lifecycle. By pulling together these many strands, this review supports diverse aspects to move from bench-scale hydrate studies to field-ready cold thermal energy storage that aligns with current energy performance and decarbonization goals.

2. CO_2 Hydrate Formation

CO_2 clathrate hydrate, which resembles shaved ice or *Granita Siciliana* desserts, is a crystalline substance composed of water ice and carbon dioxide. The first evidence for the existence of CO_2 hydrates dates back to 1882. It was associated with carbonic acid research performed by Polish physicist and chemist Zygmunt Florenty Wróblewski. Wroblewski was also the first to provide the approximate chemical formula for CO_2 hydrates-$CO_2{\cdot}8H_2O$ [20].

2.1. When and Where Does Gas Hydrate Formation Occur?

The gas hydrate formation process begins with nucleation and is followed by hydrate growth. Nucleation is the fundamental step in the formation of gas hydrates and is formed at the gas–water interface. This process begins with tiny clusters of water molecules forming around gas molecules, creating incomplete crystal structures. These early formations continuously appear and disappear due to fluctuations in temperature, pressure, and mass. Only the most stable clusters, those that reach a critical size, can grow into a full hydrate structure. Hydrates can form in two primary ways: spontaneously, without impurities (homogeneous nucleation), or with the help of impurities such as dust or pipe/reactor walls (heterogeneous nucleation). The formation of hydrates is influenced by several factors, including temperature, pressure, and the presence of impurities. The rate at which hydrates form is affected by these conditions, with lower temperatures and higher pressures usually promoting faster formation [21]. In the real world, hydrate formation occurs through heterogeneous nucleation, where the presence of interfaces or foreign surfaces significantly lowers the energy barrier for nucleation, compared to homogeneous nucleation. Surfaces with moderate hydrophobicity, such as stainless steel, silica, and carbon-based substrates, have been shown to promote hydrate nucleation by providing sites that support the alignment of water and gas molecules into pre-nucleation clusters [22]. Experimental visualization techniques such as high-pressure microscopy and cryogenic scanning electron microscopy (Cryo-SEM) have shown that hydrate formation often initiates at the solid–liquid–gas interface, especially at the microscopic surface defects such as scratches and crevices, which act as energy wells for nucleation. Nanoscale textures increase the available surface area and localize gas molecules by enhancing the probability of nucleation events Surface-treated substrates, such as hydrophobically modified silica or polydopamine-coated carbon fibers, have been studied to reduce induction time and increase nucleation density, which indicates that surface chemistry and morphology directly influence hydrate nucleation kinetics [23].

In addition to solid surfaces, as discussed further in the upcoming sections, the presence of particulate additives such as graphene oxide, activated carbon, and metal–organic frameworks (MOFs) has been used to enhance nucleation by providing additional surface area and gas-trapping sites, which improve local gas concentration and water structuring. These materials not only serve as nucleation sites but also influence the growth patterns of hydrates, helping to make formation processes more efficient.

2.2. Memory Effect in Gas Hydrate Formation

It is a challenging task to investigate the nucleation process, both through theoretical/numerical analysis, and in situ laboratory experiments. One of the most theoretically demanding, and still not fully explained, aspects associated with hydrate formation is called a memory effect. Gas–water systems that have previously undergone hydrate formation tend to form hydrates more rapidly upon re-exposure to similar conditions. Residual water after initial hydrate formation and dissociation is called memory water.

There are several physical phenomena which are considered responsible for this increase in nucleation rate. The short list of the best-known theoretical explanations of the memory effect origins includes the following nano- and micro-scale mechanisms:

- The residual structure mechanism;
- The gas supersaturation mechanism;
- The impurity imprinting theorem (for situations when hydrate formation takes place in porous media, i.e., fixed-bed continuous-flow reactors).

In thermal energy-oriented CO_2 hydrate applications such as thermofluids and cold storage, the first two mechanisms are the most important.

The residual structure mechanism theory assumes that the hydrate-like molecular structure remains in the decomposed solution and can provide nucleation sites for hydrate reformation [24].

The gas supersaturation mechanism theory considers that, after the initial hydrate formation and dissociation, the retained gas nanobubbles can affect the thermodynamic state of the water–gas system and the following nucleation process of the hydrate [25].

In general, the above behavior causing the memory effect is attributed to the existence of residual clathrate-like structures and/or trapped gas nanobubbles following the initial hydrate formation and dissociation process. They serve as nucleation sites for subsequent hydrate formation. Studies utilizing the classical nucleation theory have modeled this phenomenon, providing a framework for predicting and controlling hydrate formation in various systems [3]. It is important to highlight that in addition to the reduction in the hydrate formation time, the memory effect also influences hydrate morphology and transport properties. However, with prolonged thermal exposure, these residual structures or nanobubbles can dissolve effectively, erasing the memory effect and causing the system to behave similarly to fresh water. X-ray CT imaging revealed that hydrate reformation from memory water provides a more homogeneous distribution of hydrates, in contrast to the localized or clumped distributions typically observed during initial hydrate formation from fresh water [26].

2.3. Conditions Needed for CO_2 Hydrate Formation

The properties and formation of CO_2 clathrate hydrates have been explained by Sloan Jr. et al. [16] from the perspective of the physical reaction between CO_2 gas and water. In this process, gas molecules are trapped in the cavities of the network of water molecules. This typically takes place in the body-centered or diamond cubic lattice of structure II, which consists of 16 small cages with 5^{12} cages per unit cell and 8 large cages with $5^{12}6^4$ cages per unit cell. Each hydrate structure is occupied by at most one guest molecule. Both methane and CO_2 are gases with low molecular weight and form structure I, which consists of large cages ($5^{12}6^2$ cages per unit cell) and smaller cages ($5^{12}6^3$ cages per unit cell) in hydrate equilibria [16]. The cubic structure sI clathrate is the only stable structure that has been formed at 0.7 GPa, as reported by Bollengier et al. (2013) [27]. It has been observed that the transition phase of CO_2 sI hydrates to a high-pressure hydrate starts below 280 K and at pressures up to 1 GPa.

Structure sII is formed at higher pressures and has both a small cage ($5^{12}6^2$) and a large cage ($5^{12}6^3$). Due to the larger cage, sII can potentially store more guest molecules than sI. sII can be typically formed at a pressure of 300 MPa and a temperature of 249 K [28,29].

As mentioned before, CO_2 clathrate hydrates form under specific temperature and pressure conditions. This process is often influenced by the local CO_2 concentration and by the presence of a variety of additives, such as nucleation and crystal formation stimulants. The CO_2 hydrate formation process involves CO_2 interacting with water molecules to create a cage-like structure, especially near the melting point of ice, where a quasi-liquid layer (QLL) plays a crucial role. Raman spectroscopy-based observations by Zhou et al. [30] indicate that hydrate formation is spatially heterogeneous within ice particles, with the quasi-liquid layer near the melting point of ice facilitating localized conversion. The observed variability in gas concentration is attributed to incomplete conversion and mass transfer limitations between the ice core and the outer layers.

Experimental studies have shown that electrolytes such as NaCl, KCl, $CaCl_2$, and their binary mixtures, as well as synthetic seawater solution, impact the equilibrium conditions of CO_2 hydrate formation. The equilibrium data collected by Dholobhai et al. 1993, across a temperature range of 259 K to 281 K and a pressure range of 0.9 to 4.1 MPa, revealed that the composition of electrolytes significantly inhibits hydrate formation, altering the equilibrium temperature at a given pressure [31].

The use of methane and CO_2 mixtures in hydrate formation experiments demonstrated the importance of specific pressure and temperature ranges for equilibrium points. Methanol solutions are also used as promoters in some studies, where methanol influences hydrate formation temperatures. The results align with predictions based on the formula developed by Hammerschmidt [32].

$$\Delta T = K\frac{1}{100 - I} \tag{1}$$

where *K* is a specific parameter dependent on the nature of the additive, and *I* is the amount of antifreeze in percent with respect to water. The agreement between the experimental data and the Hammerschmidt equation was reasonably good in the aqueous liquid—hydrate—gas region [33].

In the CO_2–H_2O system, the coexistence of gas, ice, and hydrates has been investigated by multiple researchers, such as Miller and Smythe [34], Adamson and Jones [35], Wendland [36], Yasuda and Ohmura [37]. Below the water melting point temperature, ice is more thermodynamically stable than liquid water, making the main type of equilibrium under such conditions a monovariant V-I-H (vapor-ice-hydrate) equilibrium. Studies by Kimura et al. 2021 confirmed that the difference in equilibrium pressures for hydrates with isotopic compositions of 12 CO_2 and 13 CO_2 is minimal, with the 12 CO_2 hydrate having slightly lower equilibrium pressure [38].

Experimental data indicate that even at temperatures below the freezing point of water, a metastable equilibrium involving vapor, supercooled liquid water, and hydrate (V–Lw*–H) can occur. This represents an extension of the typical vapor–liquid water–hydrate (V–Lw–H) equilibrium into colder temperatures. When hydrates decompose below 273 K, an ice layer forms on the surface of the hydrate particles, which slows down further decomposition. This phenomenon is known as the self-preservation of gas hydrates. Metastable supercooled water and gas hydrates can coexist, as seen in the dissociation of hydrates at sub-freezing temperatures for gases such as CO_2, propane, methane, and ethane. The presence of supercooled water during hydrate dissociation has been verified using Raman spectroscopy [39] and nuclear magnetic resonance relaxation spectroscopy [40].

Several studies have provided data on the vapor–liquid equilibrium of pure CO_2 and the three-phase equilibrium involving vapor, liquid CO_2, and hydrate (V–L CO_2–H), as well as vapor, liquid CO_2, and liquid water (V–L CO_2–Lw) in the CO_2–H_2O system. The ice melting curve in a CO_2 atmosphere (V-Lw-I equilibrium) shows a much steeper slope compared to the equilibrium line of pure water due to the high solubility of CO_2 in water, which lowers the equilibrium temperature of ice melting. A metastable three-phase equilibrium V–Lw*–I (vapor–supercooled liquid water–ice) can also occur at pressures above 1.04 MPa. The unique aspect of CO_2 as a hydrate-forming gas is that it can exist in both gas and liquid forms at typical hydrate formation conditions, unlike non-condensing gases. This complicates the phase behavior of the CO_2–H_2O system [41,42].

By plotting the phase diagram for CO_2 hydrate, we can clearly see the conditions under which CO_2 exists in different phases, such as ice, water, and hydrate. A comparative analysis shows that the experimental data for the two-phase equilibrium of pure CO_2 match well with the data for the three-phase equilibrium in the CO_2–H_2O system. Above these equilibrium lines, the stability of the CO_2 hydrate is governed by the three-phase equilibrium among liquid CO_2, liquid water, and hydrate (LCO_2–Lw–H). Reference data on the vapor–liquid equilibrium of pure CO_2 and experimental data on V–L CO_2–H (vapor–liquid CO_2–hydrate) and V–L CO_2–Lw (vapor–liquid CO_2–liquid water) three-phase equilibria for the CO_2–H_2O system have been provided by multiple studies [43,44]. Plotting the phase diagram of CO_2 hydrate provides a clear idea on the conditions under which CO_2 exits in different phases, such as ice, water, and hydrate. A comparative analysis shows that the experimental points for the two-phase equilibrium of pure CO_2 align with data for the three-phase equilibrium in the CO_2–H_2O system.

Following the combination of the experimental data analyzed by Semenov et al. [45], Figure 3 plots the phase diagram for the CO_2–H_2O system, showing various equilibrium lines and regions for the existence of liquid and gaseous CO_2. This phase diagram is essential for understanding the phase behavior of CO_2 hydrates and developing accurate thermodynamic models. The lower quadruple point Q1, where the lines of monovariant equilibria V–Lw–H, V–I–H, and V–Lw–I intersect, defines a non-variant equilibrium because four phases (vapor, liquid water, ice, and hydrate) coexist at fixed pressure and temperature values. Determining Q1 coordinates involves experimentally measuring equilibrium types and approximating the experimental points to find their intersection.

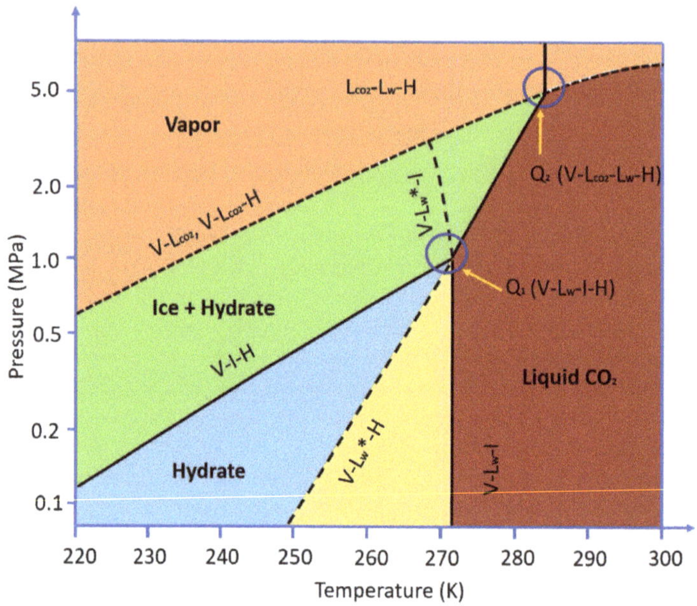

Figure 3. Phase diagram for CO_2–H_2O covering a temperature range from 220 to 300 K and a pressure range of 0.08–8 MPa [45].

In Figure 3, the solid line V–Lw–H indicates a three-phase equilibrium involving gas, aqueous solution, and gas hydrate. The dashed line V–Lw*–H represents a metastable three-phase equilibrium featuring gas, supercooled aqueous solution, and gas hydrate. Another solid line, V–I–H, corresponds to a three-phase equilibrium comprising gas, ice, and gas hydrate. Line V–Lw–I denotes a three-phase equilibrium involving gas, liquid aqueous solution, and ice. A dotted line, V–Lw*–I, represents a metastable three-phase equilibrium involving gas, supercooled aqueous solution, and ice. The solid line V-L CO_2 signifies a two-phase equilibrium of gaseous and liquid carbon dioxide. Overlapping with it is the solid line V–LCO_2–(H or Lw), indicating a three-phase equilibrium of gaseous and liquid carbon dioxide and either gas hydrate or aqueous solution. The solid line L CO_2–Lw–H signifies a three-phase equilibrium involving a liquid carbon dioxide-rich phase, aqueous solution, and gas hydrate. Another red star, the Q2 point (283 K, 4.494 MPa), denotes the intersection showing the coexistence of gas, liquid carbon dioxide-rich phase, aqueous solution, and gas hydrate [45].

The formation of the CO_2 clathrate hydrate mainly depends on temperature [46,47] and pressure [48] parameters, as pure CO_2 clathrate hydrates are formed under high pressures and low temperatures. The dissociation pressure of CO_2 clathrate hydrates can be lowered by using mixtures of gases or different additives. For example, adding methane (CH_4) or hydrogen (H_2) to CO_2 can reduce the pressure required for hydrate formation. The presence of certain additives/promoters, such as salts or surfactants, can also influence hydrate stability, reduce induction time, and lower dissociation pressure [49].

These promoters act as catalysts in accelerating the formation process by lowering the activation energy required for hydrate growth and enabling the hydrate to form at mild pressures and viable temperature conditions. Additional discussion of promoters is provided in Section 3.

2.4. Hydrate Induction Time

Hydrate induction is an important kinetic characteristic parameter in hydrate formation, and it can also be reduced by promoters, as they accelerate the formation process and then promote the growth rate. Induction time refers to the time that is needed for the formation of the first hydrate nuclei that reach a critical size, beyond which they can grow into macroscopic hydrate structures. Depending on the process duration and formation steps, induction time is categorized into two types, including microscopic and macroscopic times, which are typically used in hydrate formation analysis. The microscopic induction time refers to the duration required for the nucleation of hydrates, that is, the formation of hydrate critical nuclei before the system reaches a phase equilibrium state. The macroscopic induction time is more commonly used in experimental studies and is defined as the time required for the system to reach a phase equilibrium state, followed by a sudden change in the state parameters, such as temperature and pressure. This is observed as the interval between the moment the system achieves phase equilibrium temperature and the appearance of the first inflection point on the temperature–time curve. Figure 4 illustrates both induction time definitions, together with the changes in the crystal structures [50]. The formation process of gas hydrates begins with the interaction between water molecules and guest molecules. Initially, water molecules arrange themselves around the guest molecules, leading to the formation of labile clusters, which are unstable and dynamic. These clusters then undergo agglomeration, where they merge and stabilize, eventually forming nucleation sites. As nucleation progresses, the hydrate structure begins to grow, ultimately leading to the formation of solid hydrates. The last part of the figure shows a reactor developed by the SBU collaboration for controlled hydrate formation [51].

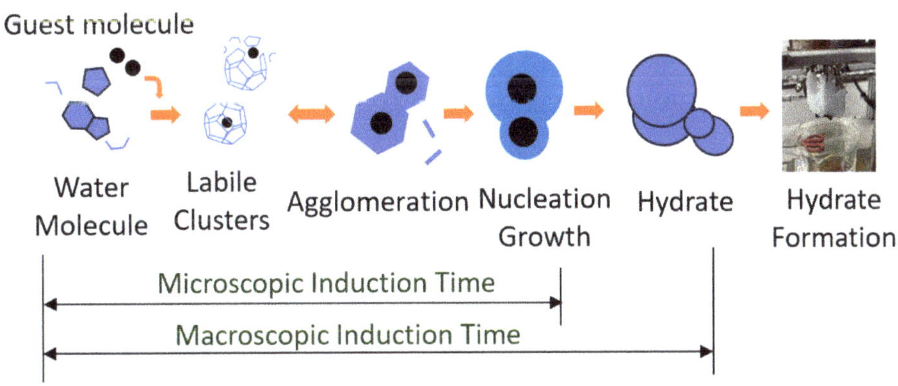

Figure 4. Induction time stages and hydrate crystal development during CO_2 hydrate formation. The arrows in the figure represent the progression of molecular and phase changes from water molecules to hydrate formation. Reproduced with permission from *RSC Advances*, 2019 [50]. Right-hand side photograph: original work by D. Mahajan.

3. Hydrate Formation Promoters

A variety of different types of promoters have already been evaluated for processing control in gas hydrate formation experiments, and almost every promoter behaves differently dueing gas hydrate formation. Some promoters can help with faster hydrate formation, while others might influence size and morphology and improve long-term stability during repeated cycles of formation and dissociation. In general, two basic groups of hydrate promotors are used to support CO_2 hydrate formation: (i) thermodynamic promotors and (ii) kinetic hydrate promoters.

In this chapter, the use of promoters and surfactants to enhance CO_2 hydrate formation is discussed. The following sections categorize promoter mechanisms and introduce emerging sustainable materials. They integrate thermodynamic and kinetic aspects of hydrate nucleation with quantitative performance metrics (e.g., induction time, formation rates) and environmental impact assessments.

3.1. Thermodynamic Promoters

Thermodynamic promoters enhance CO_2 hydrate formation through several mechanisms. They interact with water molecules at the water–CO_2 interface, modify the interfacial tension, and reduce the energy barrier for hydrate nucleation. These promoters selectively occupy the guest cages within the hydrate structure and create more favorable conditions for CO_2 molecules to fill the remaining cages. Promoters also influence the water molecule structure, making it more conducive to hydrate formation. They alter the thermodynamic properties of the system, such as the chemical potential of water or CO_2, helping make the formation of hydrates more energetically favorable [52]. Promoters also help lower the pressure and/or increase the temperature at which CO_2 hydrates are stable by effectively shifting the equilibrium curve to more favorable conditions for hydrate formation. For instance, when cyclopentane is added to a CO_2 hydrate system, there is a shift in the equilibrium curve, which indicates that hydrates can form at higher temperatures or lower pressures than would be possible without the promoter. This shift makes the hydrate formation process more energy-efficient and feasible under less extreme conditions, enabling practical applications. The most commonly used thermodynamic promoters are tetrabutylammonium bromide (TBAB), tetrahydrofuran (THF), cyclopentane (CP), cyclohexane, butane, propane, and argon. Figure 5 shows the structures of selected promoters.

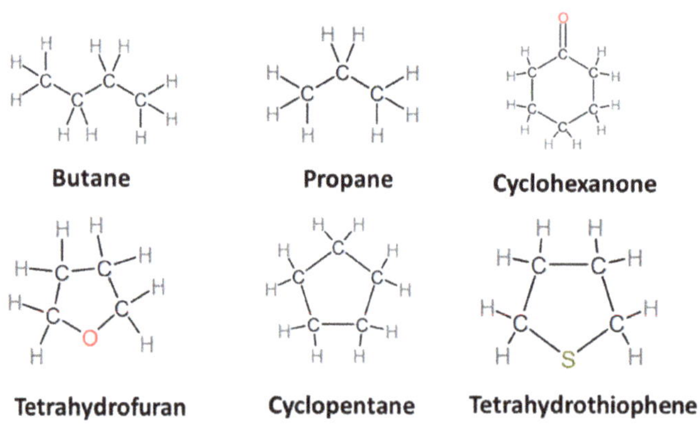

Figure 5. Structures of some commonly used hydrate promoters [53].

Table 1 presents various thermodynamic promoters that are used by different researchers and compares hydrate formation temperatures, pressure, solid fractions, and concentration ranges. Tetrabutylammonium bromide (TBAB) is effective in hydrate formation across temperatures from 279 to 291 K and pressures between 1.4 and 4.5 MPa, with concentrations varying from 0.1 to 4.0 mol%. This promoter is known for its ability to stabilize hydrates under moderate pressure conditions. Cyclopentane (CP) operates within a temperature range of 287 to 293 K and a pressure range of 0.5 to 2.6 MPa, with a water-to-CP ratio of 19:1. This shows that it can form hydrates at ambient temperatures. Cyclohexane functions at lower temperatures, from 275 to 278 K, and pressures between

0.95 and 1.81 MPa, typically at a concentration of 0.5 mol%. This promoter works under conditions similar to those of CP but at slightly lower temperatures. Tetrahydrofuran (THF) is effective over a wide temperature range of 279 to 291 K and pressures from 0.18 to 3.17 MPa, with concentrations between 4 and 10 mol%. THF is a commonly used promoter due to its efficiency in reducing the pressure required for hydrate formation. Propane operates within a temperature range of 274 to 282 K and pressures between 0.5 and 3.7 MPa, with concentrations ranging from 3 to 60 mol%. It is widely studied for its ability to facilitate hydrate formation under moderate conditions. Each promoter plays a crucial role in modifying the equilibrium conditions for CO_2 hydrate formation, enhancing the process's feasibility under specific environmental conditions.

Table 1. Thermodynamic promoters and their influence on CO_2 hydrate formation conditions.

Thermodynamic Promoter	Temperature (K)	Pressure (MPa)	Concentration (mol%)	References
Pure CO_2 hydrate	273.2 to 281	5.0 to 25	No promoters	[54]
Tetrabutylammonium bromide (TBAB)	279 to 291	1.4 to 4.5	0.1 to 4.0	[55,56]
Cyclopentane (CP)	287 to 293	0.5 to 2.6	Water: CP–19:1	[57,58]
Tetrahydrofuran (THF)	279 to 291	0.18 to 3.17	4 to 10	[57,59,60]
Propane	274 to 282	0.5 to 3.7	3 to 60	[61]

3.2. Kinetic Hydrate Promoters

Kinetic hydrate promoters are additives that accelerate the rate of hydrate formation without altering the equilibrium conditions. Unlike thermodynamic promoters, which interact with the crystal structure and shift the equilibrium curve, kinetic promoters primarily affect the nucleation and growth kinetics of hydrates. There are three common types of kinetic hydrate promoters: (i) surfactants, (ii) amino acids, and (iii) solid particle additives, each having its own specific mechanism and effects.

3.2.1. Surfactants

Surfactants such as fatty acids or alkyl sulfates play a crucial role in carbon dioxide (CO_2) hydrate systems. By lowering the surface tension between CO_2 gas and water, surfactants stabilize the interface, which is essential for the formation of hydrate crystals. This reduction in surface tension effectively decreases the energy barrier necessary for the nucleation and growth of hydrate crystals. They also influence the morphology and size distribution of these crystals. They achieve this by affecting the arrangement of water molecules at the gas–water interface. Through this molecular reorganization, surfactants facilitate the initial formation of hydrate crystals and impact their subsequent development and structural characteristics.

Typically, surfactants are classified according to the configuration of their polar head groups. Non-ionic surfactants have no charged groups in their head regions. Anionic surfactants are a type of surface-active agent that possess negatively charged ions, also known as anions, in their molecular structure. The head of an ionic surfactant carries either a net positive or net negative charge.

As shown in Figure 6, surfactants have hydrophobic and hydrophilic parts. The hydrophobic tail of the surfactant has a nonpolar long hydrocarbon chain and interacts with the gas phase or the nonpolar regions of the hydrate structure. In contrast, the hydrophilic head is polar and engages with the water phase through functional groups, such as -OH (hydroxyl), -COOH (carboxyl), or -SO_3H (sulfonic acid). These surfactants are further categorized into three classes, each with their own surfactant chemical, as shown in Table 2.

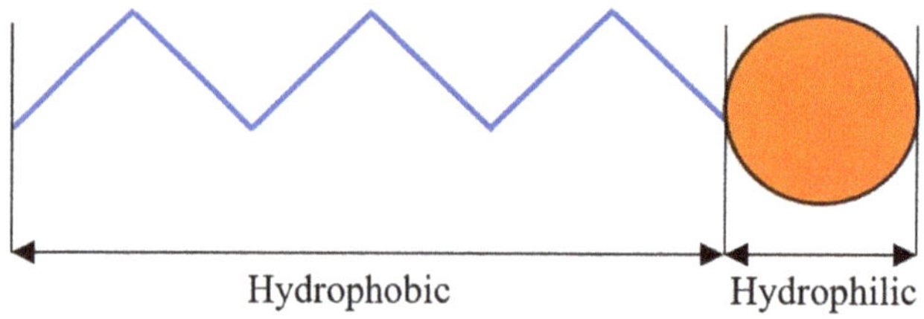

Figure 6. Surfactant molecular configuration with hydrophilic and hydrophobic domains. Reproduced with permission from *Energy and Fuels*, 2021 [53].

Table 2. Examples of some widely used surfactants.

Category	Surfactants	References
Anionic Surfactants	Sodium Dodecyl Sulfate (SDS), Sodium Dodecyl Benzene Sulfonate (SDBS)	[62]
Non-Ionic Surfactants	Tween 80, Oxyethylene nonanoic acid, Span 80	[63]
Cationic Surfactant	Dodecyl trimethyl ammonium bromide (DTAB), Hexadecyl trimethyl ammonium bromide (CTAB)	[64]

Table 3 shows the different surfactants used, and the formation conditions studied by different researchers.

Table 3. Surfactants and corresponding experimental hydrate formation conditions.

Surfactant	Temperature (K)	Pressure (MPa)	References
Sodium Dodecyl Sulfate (SDS)	270–277	3.0–15.0	[65,66]
Tween 80	274	3.55	[67]
Silica Gel	273	3.00	[65]
Silica Sand and zeolite 13X	274	3.00	[68]
Quiescent water	273	3.90	[69]

Sodium dodecyl sulfate (SDS) is the most widely used surfactant in the formation of CO_2 hydrates. It does not affect the equilibrium conditions but reduces the induction time. N. Molokitina et al. conducted experiments by varying the SDS concentration from 100 ppm to 5000 ppm and observed a decrease in the induction time [66,70].

3.2.2. Amino Acids

Amino acids serve in a similar way to kinetic promoters for CO_2 hydrate formation due to their unique ability to interact with both water and CO_2 molecules. Unlike traditional surfactants, amino acids are biodegradable and non-polluting biomass-based promoters. The effectiveness of amino acids in promoting hydrate formation is attributed to their molecular structure, which includes both hydrophobic and hydrophilic regions. As illustrated in Figure 7, the hydrophobic side chains preferentially associate with CO_2 molecules, while the polar functional groups interact with surrounding water molecules, contributing to the formation of pre-nucleation structures. This dual affinity promotes both nucleation and early-stage hydrate growth. Figure 8 presents common amino acids explored in recent studies, including L-leucine, L-methionine, and L-tryptophan, each exhibiting favorable kinetics under moderate pressure and temperature conditions. The corresponding hydrate formation conditions for these amino acids are summarized in Table 4.

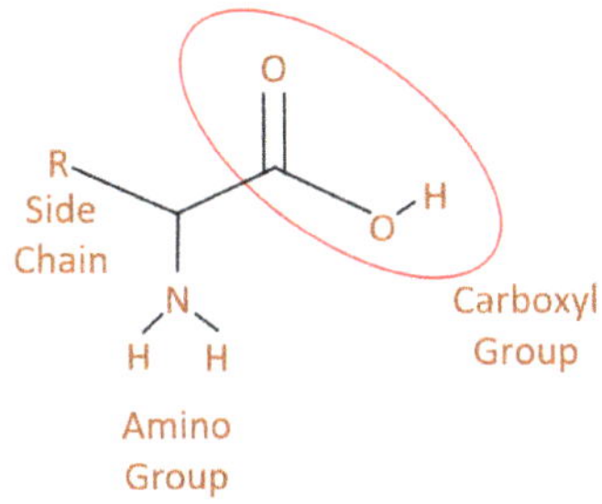

Figure 7. Molecular structure of the amino acid-based surfactant. Reproduced with permission from *Energy and Fuels*, 2021 [53].

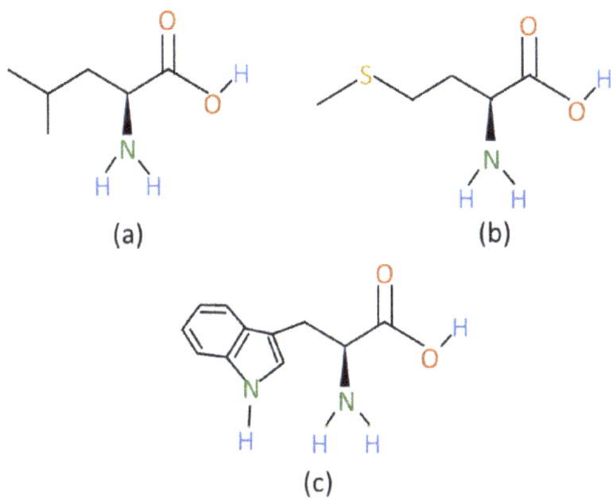

Figure 8. (**a**) L-leucine, (**b**) L-methionine, (**c**) L-tryptophan. Reproduced with permission from *Energy and Fuels*, 2021 [53].

Table 4. Amino acid-based promoters and hydrate formation conditions.

Amino Acid	Pressure (MPa)	Temperature (K)	References
L-leucine	3.3	275	[71]
L-methionine	3.3	275	[72]
L-norleucine	3.3	273	[72]
L-norvaline	3.3	273	[72]
L-tryptophan	3.0–3.5	273–278	[73]

3.2.3. Nanoparticles/Solid Particles

Nanoparticles have a size between 1 and 100 nm and have been proven to be highly effective kinetic promoters in CO_2 hydrate formation due to their unique properties, such as high surface area, relatively high thermal conductivity, and their ability to interact with hydrates, forming components at the molecular level. They accelerate the hydrate formation process by providing numerous nucleation sites, enhancing overall thermal conductivity, and preventing hydrate particle agglomeration [74,75].

Silica nanoparticles significantly reduce induction time by increasing nucleation sites, and carbon nanotubes (CNTs) promote rapid hydrate formation and improve stability through their high surface area and excellent thermal properties [76]. Metal oxide nanoparticles such as TiO_2 and Al_2O_3 are commonly used to enhance the thermal conductivity of hydrates, with efficient heat dissipation and sustained hydrate growth. Table 5 shows various nanomaterials used as hydrate-formation promoters.

Table 5. Nanoparticles and their impact on CO_2 hydrate formation temperature and pressure.

Nanomaterial	Pressure (MPa)	Temperature (K)	Reference
Nano graphite	2.5 to 6.5	272 to 277	[74,77]
Copper (Cu)	4.0	274	[78]
Zinc oxide (ZnO)	1.0	293 to 303	[79]
Silver (Ag)	2.08 to 3.24	273 to 275	[80]
Silica nanoparticles	0.6	260–276	[81]

While Tables 1–5 summarize various hydrate additives and their effects, this section provides a critical comparison of promoter performance metrics to support application-specific material selection. Table 6 shows the performance and environmental characteristics of performance-enhancing additives for CO_2 hydrates.

Table 6. Performance characteristics and environmental impact of hydrate-promoting additives.

Type	Induction Time	Formation Rate	Concentration Used	Effect	References
TBAB	Shortens induction time significantly (from 6 to 15 min to <5 min)	Moderate; accelerates hydrate nucleation and CO_2 uptake	0.29 mol% (approx. 3–5)	Environmentally friendly	[82,83]
THF	Strongly reduces induction time (less than 1 min at 5 mol%)	High; especially with co-additives such as SDS	1–5 mol%	Acutely toxic, volatile and corrosive to equipment	[84,85]
SDS	Increased with higher SDS concentration (as low as <1 wt%)	High; rapid hydrate growth, especially in the presence of stirring	0.01–1 wt%	Toxic, highly foamy, and difficult to recycle	[79,86]
Amino Acids	Moderate (some amino acids promote hydrate nucleation, others are neutral or inhibitory)	Moderate; up to 356 $mg \cdot g^{-1}$ uptake in 15 min with L-methionine	0.2 wt% (typical for amino acid studies)	Non-toxic and biocompatible nature	[72,87]

In addition to traditional nanoparticles such as silica and metal oxides, researchers have recently turned to metal–organic frameworks (MOFs) as promising solid-phase promoters for CO_2 hydrate formation. MOFs are crystalline porous materials made from metal ions coordinated with organic ligands, offering high surface areas, tunable pore structures, and exceptional gas adsorption capabilities. Their unique properties allow them to not only serve as nucleation platforms but also actively participate in the CO_2 capture and conversion processes at the molecular level [88]. Cao et al. [89] provided an overview of enzyme immobilization within MOFs for carbon dioxide (CO_2) conversion. MOFs such as the Zeolitic Imidazolate Framework (ZIF) series and the Materials of Institute Lavoisier

(MIL) series are discussed as porous carriers that provide a controlled microenvironment for enzyme stabilization and CO_2 diffusion. The functional groups on MOF surfaces, along with pore size and geometry, influence coenzyme regeneration (e.g., nicotinamide adenine dinucleotide, $NADH/NAD^+$) and substrate transport, which are important for catalytic efficiency in enzyme-driven CO_2 conversion pathways [89].

From a hydrate inhibition standpoint, Saikia et al. [90] examined the influence of MOFs on CO_2 hydrate formation kinetics in gas-dominated pipelines. They synthesized University of Oslo-66 (UiO-66) and its amine-functionalized counterpart, UiO-66-NH_2, via hydrothermal methods. The materials were characterized using Powder X-ray Diffraction (PXRD), Brunauer–Emmett–Teller (BET) surface area analysis, scanning electron microscopy (SEM), Fourier transform infrared spectroscopy (FTIR), and thermogravimetric analysis (TGA). The experiments, conducted in a sapphire rocking cell at subzero temperatures and pressures exceeding 30 bar, used MOF concentrations of 0.25 and 0.5% weight/volume (w/v). UiO-66-NH_2 showed higher inhibition efficiency than UiO-66, likely due to the presence of –NH_2 groups, which can engage in hydrogen bonding with water and enhance CO_2 adsorption near the hydrate-forming interface. A BET analysis showed surface areas around 1000–1500 m^2/g, indicating a high density of potential adsorption sites that could alter local phase behavior and nucleation timing. The observed hydrate inhibition was evaluated from a kinetic perspective, including shifts in induction time and suppression of crystal growth [90]. All these studies show that MOFs can serve dual roles: (1) enhancing the thermodynamic and kinetic aspects of CO_2 hydrate formation and (2) acting as smart delivery systems for other chemical promoters or catalytic agents. Compared to conventional solid nanoparticles, MOFs offer modularity in design, enabling the specific tailoring of surface chemistry, pore size, and thermal conductivity for targeted hydrate performance.

Kazemi et al. [91] synthesized a green and scalable method for fabricating Mixed-Metal MOF-74 using cobalt (Co^{2+}) and nickel (Ni^{2+}) by using environmentally friendly solvents (water and ethanol) and replacing hazardous organic solvents such as DMF (N,N-Dimethylformamide). DMF is a widely used polar aprotic solvent in conventional MOF synthesis because of its ability to dissolve both metal salts and organic linkers, supporting controlled crystal growth. However, it poses health and environmental risks due to its toxicity when inhaled or upon skin contact and is classified as a reproductive toxin. Its disposal requires careful handling, which makes it unsuitable for large-scale or environmentally conscious applications. To overcome these limitations, the researchers used ethanol–water mixtures along with metal acetates, eliminating the need for harmful solvents and nitrate salts. This process not only reduced the environmental burden but also enabled rapid synthesis at 90 °C for 1 h, forming MOF particles with well-defined, rod-like morphology and diameters below 100 nm. This structural feature enhances gas diffusion, facilitates faster hydrate nucleation, and increases accessibility to active sites critical for hydrate formation [91]. From an economic standpoint, the green synthesis route significantly reduces material and processing costs. Ethanol and water are inexpensive and readily available, unlike DMF and other toxic solvents that require specialized handling. The short synthesis time and low reaction temperature also improve scalability. The use of base metals such as cobalt and nickel, rather than rare or precious metals, improve economic feasibility. These advances collectively make MOF synthesis production more practical for integration into hydrate-based gas separation and storage technologies.

Stability under hydrate-forming conditions in water-rich, high-pressure environments is another important factor. Traditional MOFs suffer degradation due to hydrolysis of the metal–ligand coordination bonds in the presence of water. Water sensitivity remains a challenge for many MOFs, which can compromise their structural integrity and performance. However, the CoNi-MOF-74 synthesized by Kazemi et al. [91] exhibited excellent hydrothermal stability and sustained CO_2 adsorption capacity over ten adsorption–desorption cycles, including in flue gas conditions. The isosteric heat of adsorption was measured at 40.7 kJ/mol, indicating a strong yet reversible physisorption, which is ideal for hydrate promotion [92].

3.3. Promoters Efficiency and Process Scalability Challenges

Recent experimental work by Wang et al. [93] shows that, while kinetic promoters such as sodium p-styrene sulfonate accelerate CO_2 hydrate formation by attaining a two-stage CO_2 removal rate of 53.65%, the overall capture performance improved to 64.66% only when thermodynamic promoters (TBAB and cyclopentane) were present. This enhancement requires an additional formation stage, involving a trade-off between reducing pressure requirements (from 8.5 MPa to 1.5 MPa) and increasing process complexity. Their 4E (Energy, Exergy, Economy, Environment) analysis shows that the low-pressure formation with atmospheric dissociation (L-A) route offers the best overall performance, achieving the highest exergy efficiency (0.725) while minimizing indirect CO_2 emissions. These results suggest that although promoters can lower formation barriers and improve removal rates, scalable CO_2 hydrate technologies must carefully balance process complexity and overall cost with capture efficiency, number of stages, energy consumption, and environmental impacts through integrated system design [93]. Molecular simulations by Phan and Striolo et al. [94] show the promoter efficiency for CO_2 hydrate formation by using aziridine, a promoter that yields significant efficiency compared to traditional promoters such as THF and pyrrolidine in accelerating hydrate growth rates. At 269.1 K, the presence of aziridine more than doubled the CO_2 hydrate cage growth rate compared to a promoter-free system. This shows its strong kinetic enhancement effect. Promoter performance is governed by the sum of two key thermodynamic parameters: the free energy barrier for CO_2 desorption (ξ) and the binding free energy at the hydrate–aqueous interface (ΔG_bind). Aziridine had the most favorable combined values, correlating directly with faster and more stable hydrate formation, while THF reduced the hydrate formation pressure, and it was less effective kinetically. Ideal promoters for large-scale CO_2 hydrate applications must simultaneously promote nucleation kinetics and maintain thermodynamic stability, rather than focusing on just pressure reduction alone [94]. Experiments conducted by Pang et al. [95] show the advanced understanding of hydrate-based CO_2 sequestration scalability by constructing a 1695 L three-dimensional sediment model to mimic in situ reservoir conditions. Their study shows that while continuous hydrate formation can be achieved without mechanical wellbore blockage, large-scale operations pose challenges such as strong spatial heterogeneity in hydrate distribution, localized pore clogging, and progressive water depletion within the sediment matrix. These effects cumulatively lower the effective storage capacity and reduce CO_2 injectivity over time. Successful field-scale deployment will require dynamic control over injection rates, pressure profiles, and water management strategies to avoid early saturation and maximize reservoir utilization. While technical feasibility has been already experimentally validated at large volumes, the precise process engineering remains critical for converting laboratory-scale success into reliable field-scale CO_2 storage operations [95].

3.4. Regulatory Standards Related to CO_2 Hydrate Promoters and Surfactants

In hydrate-based carbon capture and storage (CCS) systems, the selection of suitable promoters is not only important for enhancing hydrate formation kinetics, but also for ensuring environmental safety and compliance with health regulations. While promoters such as tetrahydrofuran (THF), tetra-n-butylammonium bromide (TBAB), and cyclopentane (CP) are effective in facilitating hydrate formation, their environmental and health implications require careful consideration.

Tetrahydrofuran (THF) is a widely used thermodynamic promoter known for its ability to form structure II hydrates at relatively mild conditions. However, THF poses several health and environmental risks, as it is classified as a Group 2B carcinogen by the International Agency for Research on Cancer (IARC), showing possible carcinogenicity to humans [96]. THF is also highly flammable and volatile, capable of forming explosive peroxides upon prolonged exposure to air. Its high volatility and water solubility increase the risk of environmental contamination, particularly in aquatic systems. Regulatory agencies such as the U.S. Occupational Safety and Health Administration (OSHA) have established strict guidelines for THF handling and exposure limits to mitigate these risks [97]. Cyclopentane (CP) is another thermodynamic promoter that forms structure II hydrates and is valued for its ability to promote hydrate formation under mild conditions. However, CP is a volatile organic compound (VOC) with a relatively high vapor pressure, leading to rapid evaporation and potential atmospheric release during handling and hydrate dissociation processes. CP is also flammable and poses inhalation risks. Environmental concerns include its contribution to ground-level ozone formation and potential toxicity to aquatic life if released into water bodies. Mitigation strategies involve implementing closed-loop systems to capture and recycle CP, thereby minimizing environmental release and occupational exposure [98].

In contrast to these conventional promoters, amino acids have emerged as environmentally friendly alternatives for promoting gas hydrate formation. Amino acids such as L-methionine, L-threonine, and L-isoleucine have demonstrated efficacy as kinetic promoters, enhancing hydrate formation rates and gas storage capacities. L-methionine has been shown to reduce induction times and increase CO_2 uptake in hydrate formation processes, outperforming traditional promoters such as SDS in certain conditions [99]. Amino acids are biodegradable, non-toxic, and naturally occurring, which makes them attractive candidates for sustainable hydrate-based CCS applications.

The environmental and health concerns associated with conventional promoters can be mitigated by some strategies, including the development of closed-loop systems to prevent the release of volatile promoters, the implementation of recovery and recycling processes to minimize waste, and the adoption of alternative promoters with favorable environmental profiles. Regulatory agencies have to establish comprehensive guidelines for the use and disposal of hydrate promoters with regard to their potential environmental persistence and toxicity.

4. Experimental Setups Used in CO_2 Hydrate Formation Experiments

This section presents state-of-the-art developments in different types of experimental setups and methodologies performed by different researchers. A variety of potential gas hydrate applications, combined with different combinations of precursors and processing conditions, indicates a need for various experimental setups/configurations used in the preparation of CO_2 hydrates. These setups might vary based on reactor type, such as high-pressure stirred reactors or batch reactors, or they might vary based on driving forces such as pressure change, temperature change, and hydrate promoter addition. The following sub-sections discuss different types of reactor configurations, and some of the testing methodologies used by different researchers for each reactor type.

4.1. Stirred Reactors

Stirred reactors are the most widely used today to prepare gas hydrates at the lab scale and probably represent the easiest method to form the hydrates. However, mixing water and CO_2 while they turn into an icy solid structure can be challenging because, at some point, this mixture turns into a viscous slurry. At this stage, it is usually difficult to mix the slurry further, since it requires significantly more power than is applied at the beginning of the process. That is the reason why advancements in reactor designs are still needed before this technology can be scaled up for commercialization, which requires the continuous synthesis of hydrates. The same reactor setup usually works with different impellers and is designed to allow for variation in the volume of water added, stirring rate, impeller type, and the temperature–pressure conditions under which hydrates form. For example, Pivezhani et al. performed a numerical analysis and more than 50 experiments by changing the physical and experimental conditions, such as equipping four different types of impellers (three blades, six blades, and anchor types), adjusting stirring rates from 100 to 500 rpm, and changing the volume of water between 350 and 500 mL. The temperature and pressure varied between 274 K and 277 K and from 20.6 bar to 23 bar [100].

These types of reactors increase the interfacial area between the gas and liquid phases by promoting faster mass transfer in the formation of CO_2 hydrates. As described in numerous publications, they are best suited for lab-scale CO_2 hydrate production [101–103]. Because of the unique design of the impellers, stirred reactors provide better contact between CO_2 gas and water, which helps accelerate the formation of hydrate nuclei. The agitation process in this type of reactor ensures a uniform temperature profile within the reactor. By adjusting the agitation speed, control over the reaction kinetics can be attained by helping to set up a consistent and repeatable production [104]. Once the hydrates are formed inside the reactor, the agitator requires more power to overcome the viscosity of the resulting slurry. The basic components of a stirred reactor are a gas and water supply, agitator, cooling system, data logger, and an optional camera to enable visualization inside the reactor, as shown in Figure 9 [105].

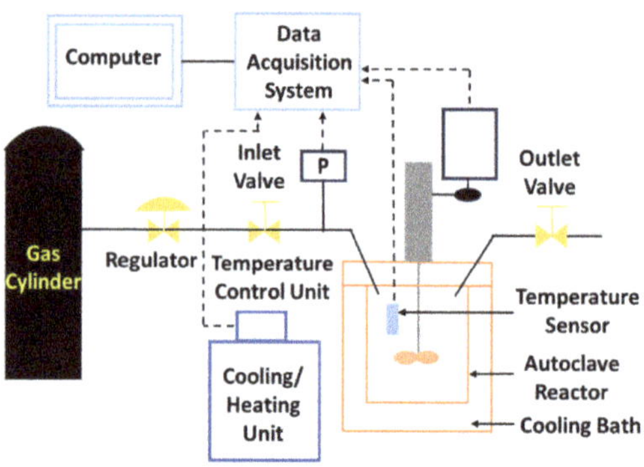

Figure 9. Stirred reactor schematic for laboratory-scale CO_2 hydrate formation. Reproduced with permission from *Journal of Natural Science and Engineering*, 2015 [105].

Figure 10 shows a 2000 mL reactor at University of Massachusetts Lowell (UML), which was introduced to address some of the challenges associated with traditional stirred reactors. Compared to commonly used benchtop reactors with a volume below 0.5 L, this reactor is designed to handle larger volumes of water and CO_2. The increased capacity allows for more efficient mixing. Aerogel blanket insulation helps reduce temperature stratification, allowing for better control of reaction conditions, which is essential for the formation of hydrates. The reactor has an impeller system to handle the viscous slurry formed during the hydrate formation process. This system provides good mixing efficiency and maintains performance even as the mixture becomes more viscous, reducing the risk of the impeller getting stuck.

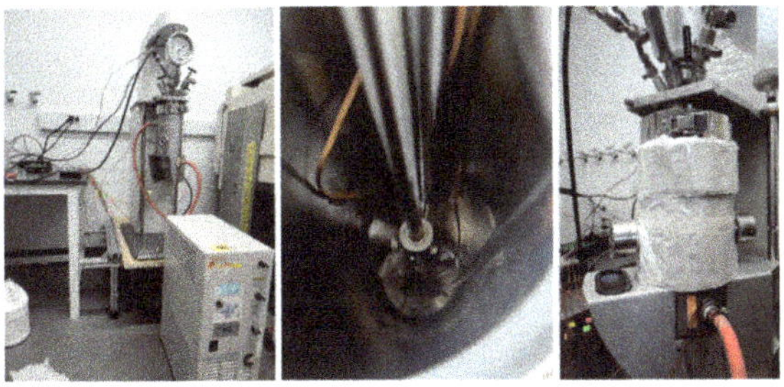

Figure 10. (**a**) Assembled reactor in UML lab environment; (**b**) internal view of the reactor showing the impeller and thermocouples; (**c**) insulated reactor setup.

Figure 11 shows a type of reactor that has a mixing blade (impeller) that is designed to draw gas into the liquid. This helps to mix the gas and liquid together better and prevents the hydrate crystals from sticking together. The gas is pushed around inside the liquid by the impeller. It flows down towards the center and then is captured by the blades. When the impeller spins, the gas is forced out into the liquid as tiny bubbles. While this method improves the reactor's performance, there is still a risk that the impeller could become stuck if the liquid becomes too thick with hydrate crystals [106].

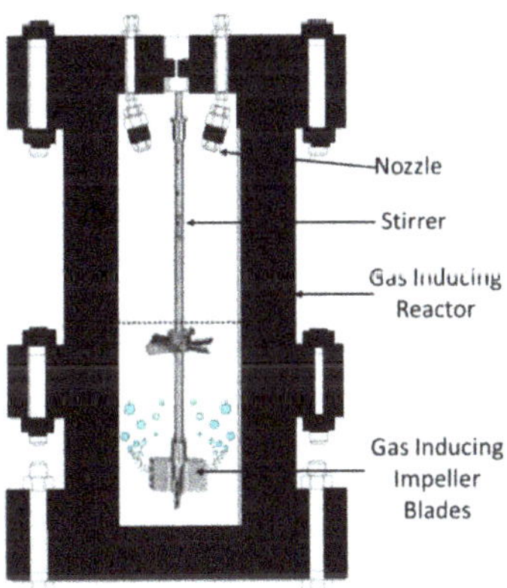

Figure 11. Gas-inducing reactor with impeller-based mixing design. Reproduced with permission from *Journal of Greenhouse Gas Control*, 2010 [106].

While stirred reactors are effective for CO_2 hydrate formation at the lab scale, their potential for scalability is challenging due to increasing slurry viscosity during hydrate growth. This increases the torque demand on the impeller, which leads to jamming or inefficient mixing. To overcome these issues, design innovations such as variable-speed motors with torque feedback, gas-inducing impeller geometries, and non-contact magnetic stirrers can be incorporated into the reactor. Pipe reactors with the screw mixing mechanism/conveyors can be used for continuous hybrid production. Self-lubricating coatings or Teflon-lined shafts may reduce frictional resistance during high-viscosity operations, providing better control under semi-solid flow conditions.

4.2. Bubble-Forming, Ejector, and Continuous-Flow Reactors

Bubble-forming reactors, also known as a bubble-tower reactors, are a type of reactor designed for gas–liquid reactions. They use advanced microbubble technology, proposed by Takahashi et al. [107] based on an ejector-type loop reactor (ELR). In this reactor, high-velocity liquid flow creates microbubbles within the fluid, leading to thorough mixing of gas and liquid, which is a fundamental criterion impacting the hydrate formation rate [107]. Figure 12 shows the ELR with the ejector schematics. One of the greatest advantages of this kind of reactor is its ability to speed up the hydrate formation process. It can handle the heat generated during this process more effectively than other reactor types. This means that less energy is needed to compress the gas, and this makes the operation more straightforward and more efficient [108]. However, removing the formed hydrates from the bubble-forming reactors can be difficult, and managing the nucleation heat effectively is still an ongoing challenge [109].

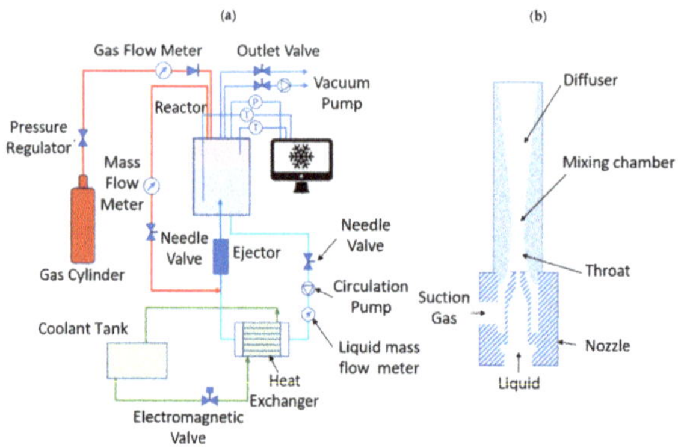

Figure 12. (**a**) Diagram of the bubble reactor; (**b**) schematic of the ejection nozzle containing premixing chamber. Reproduced with permission from *Applied Science*, 2021 [109].

Traditional hydrate reactors face two significant challenges: inefficient heat transfer and the buildup of hydrates on reactor walls. To address these issues, a novel reactor design incorporating a spiral groove has been developed [110]. This innovative design causes a swirling motion within the reactor, enhancing heat transfer and preventing hydrate accumulation. The utilization of nanobubbles significantly increases the surface area available for gas–liquid interaction, thereby accelerating hydrate formation. In essence, this advanced reactor design overcomes the limitations of conventional methods by optimizing heat transfer, minimizing hydrate buildup, and promoting efficient gas–liquid contact, eventually leading to enhanced hydrate production [111].

Ejectors and bubble-based reactors exhibit high gas-liquid contact efficiency, but their continuous operation is often disturbed by hydrate crystal accumulation in ejector throats, mixing chambers, and outlet lines. To address this, design improvements such as hydrophobic-coated internal surfaces, spiral groove nozzles, or vibrating ejector assemblies can be incorporated to minimize fouling. Using nanobubble injectors with size-controlled dispersion increases mixing uniformity by preventing blockages.

4.3. Continuous Process Reactor

In an effort to advance CO_2 hydrate technologies toward industrial application, it is important to transition from batch systems to continuous processes, which would closely mimic large-scale operations. Continuous-flow reactors have better control over reaction parameters such as flow rates, temperature gradients, and residence times, all of which influence hydrate formation kinetics and heat transfer behavior. Unlike batch reactors, continuous systems can maintain steady-state operations, improve energy efficiency, and provide hydrate production under flow-dynamic conditions that are relevant to pipeline and energy storage technologies.

Yang et al. [112] investigated conditions needed for continuous synthesis of CO_2 hydrates. They investigated CO_2 clathrate hydrate formation at increased linear fluid velocities and high gas volume fractions. For this purpose, they constructed a continuous-flow reactor (CFR) containing eleven major components which include a saturated CO_2 gas delivery system and water supply, a continuous-flow reactor, chillers, a gas–slurry separator, an accumulator, a flash reactor for hydrate slurry decomposition, a sample collection source, ThermoFisher, Waltham, MA, USA analyzer/sample collection stations, and a data acquisition system. Before starting to form the CO_2 clathrate hydrate, the heat exchange between the CFR and the ambient surroundings was determined by calorimetric experiments to assess the heat loss to the surroundings. The researchers first needed to account for heat loss from the apparatus itself. To ensure accurate heat transfer measurements within the reactor, researchers performed a preliminary calibration to understand the heat transfer characteristics under non-reactive conditions. First, this calibration was conducted by running the system with nitrogen instead of CO_2 and analyzing how variations in coolant flow rate and temperature affected heat transfer.

By simulating non-reactive conditions, they were able to evaluate the heat transfer between the coolant and the reactor environment and establish baseline parameters for subsequent experiments. These calibrations helped identify both the heat transfer efficiency and key operational factors, such as the impact of coolant flow rates and temperature regulation on reactor performance.

Figure 13 shows the reactor design, which allowed for the simulation of hydrate production conditions by providing a controlled evaluation of heat transfer efficiency during hydrate formation. By maintaining a specific range of coolant flow rates and temperatures below freezing, the reactor ensured effective heat management. This ability to mimic real-world hydrate production scenarios under controlled conditions demonstrated the reactor's suitability for studying hydrate formation, while providing flexibility in adjusting process variables [112].

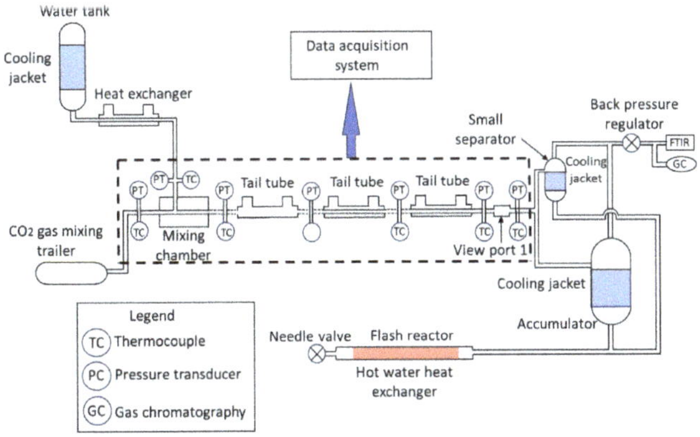

Figure 13. Continuous-flow reactor for hydrate production with integrated cooling and sampling. Reproduced with permission from *Energy and Fuels*, 2008 [112].

Continuous reactors help hydrate production under flow conditions mimicking real-world systems, but they pose challenges related to slurry handling, sedimentation, and heat balance. These can be prevented by integrating slurry-compatible positive displacement pumps, inline slurry separation chambers, and smart valve systems that respond to flow resistance. By adding modular heat exchanger segments or localized recirculation zones, one can improve thermal control and avoid cold zones by ensuring consistent hydrate growth while avoiding clogging during long operation.

4.4. Fixed-Bed Reactors

A fixed-bed reactor is a widely used type of reactor in chemical processes, where a solid catalyst or reactant bed remains stationary while a fluid (either gas or liquid) flows through it. Fixed-bed reactors are utilized to enhance the formation of CO_2 hydrates by allowing CO_2 gas to flow through a bed of water-saturated porous material under controlled temperature and pressure conditions [113].

A typical fixed-bed reactor setup for CO_2 hydrate formation, as shown in Figure 14, it consists of a cylindrical vessel packed with porous materials, such as silica gel, activated carbon or sand that is saturated with water. CO_2 gas is introduced at one end of the reactor and allowed to flow through the fixed bed. As the CO_2 gas comes into contact with the water held within the pores of the solid material, it dissolves in the water and forms CO_2 hydrates under the appropriate thermodynamic formation conditions (low temperature and high pressure). The porous material provides a large surface area for the gas–liquid interaction, which is essential for efficient hydrate formation. The fixed-bed configuration also allows for continuous gas flow, which can be useful for studying the kinetics of hydrate formation and for scaling up the process for industrial applications [114]. Table 6 summarizes various reactor types and volumes used by different researchers for CO_2 hydrate formation. It contains consolidated details on the type and size of reactors (e.g., stirred, fixed-bed, bubble reactor), the gases utilized (CO_2, N_2, H_2, CH_4), and the additives (silica gel, THF, TBAB) employed to enhance hydrate formation.

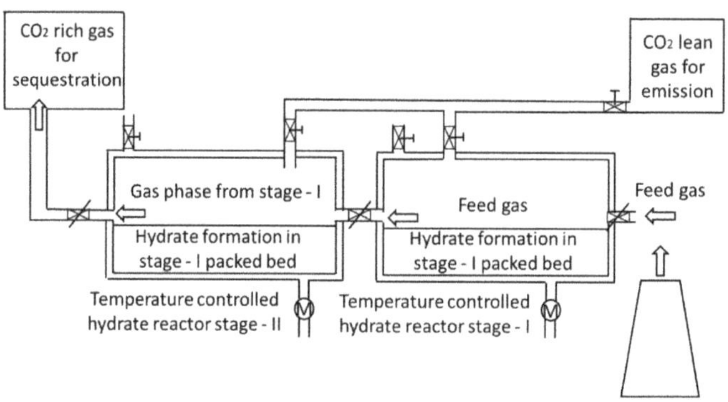

Figure 14. Fixed-bed reactor packed with porous materials for gas–liquid interaction. Reproduced with permission from *Chemical Engineering Science*, 2015 [92].

Fixed-bed reactors facilitate hydrate formation within water-saturated porous media, but scaling these systems is restricted by pore fouling, an increase in pressure drop, and thermal bottlenecks. Over time, hydrate crystal growth blocks the interstitial channels, cutting off gas flow and reducing formation rates. This can be addressed by including the use of periodically backflushed dual-flow systems, impregnated metallic foams with high thermal conductivity, and rotational bed designs that allow gentle disturbance to break early-stage blockage.

4.5. Comparisons of Different Types of CO_2 Hydrate Reactors

This section offers a short summary and comparisons of different designs of CO_2 hydrate reactors. It includes an overview of various reactor configurations and experimental conditions reported in CO_2 hydrate research. Table 7 shows the reactor types, volumes, gases used, and specific additives used to promote hydrate formation across different studies.

Table 7. Summary of reactor configurations and additives used in CO_2 hydrate research.

Authors, and Year	Reactor Type and Volume	Gases Utilized	Additive	References
Seo et al., 2001	Stirrer, 50 cm^3	CO_2, CH_4	N/A	[115]
Linga et al., 2007	Stirrer, 323 cm^3	CO_2, N_2 and CO_2, H_2	THF	[116]
Seo and Kang, 2010	Stirrer, 500 cm^3	CO_2, H_2	Silica gel	[117]
Li et al., 2010	Stirrer, 56.4 cm^3	CO_2, N_2	TBAB	[118]
Babu et al., 2013	Fixed bed, 1240 cm^3	CO_2, H_2	Silica gel	[119]
Sun et al., 2014	Stirrer, 1500 cm^3	CO_2, N_2 and CH_4, N_2 and CH_4, H_2	THF	[120]

Table 8 has the consolidated comparative analysis of the major reactor types used for CO_2 hydrate formation and their respective advantages, limitations, induction times, and scalability characteristics.

Table 8. Comparative analysis of reactor types.

Reactor type	Advantages	Disadvantages	Induction Time	Scalability	References
Stirred reactor	High gas-liquid contact area Uniform temperature distribution Lab-scale reproducibility	High energy input for mixing Impeller fouling at high hydrate fractions Limited scalability due to viscosity issues	24–261 min (depends on RPM)	Batch-limited	[121]
Bubble-Forming (Ejector) Reactors	Rapid hydrate formation via microbubbles Lower energy than stirred reactors Continuous operation feasible	Nozzle clogging Poor heat dissipation at large scales Limited data on long-term performance	180–360 min	Scale specific optimization	[108]
Continuous-Flow Reactors	Steady-state operation Suitable for industrial-scale production Integrated heat exchangers reduce energy loss	Complex pressure/temperature control Risk of hydrate plugging in pipes High capital cost	~0 min (instantaneous)	Designed particularly for industrial continuous operation	[122]
Fixed-Bed Reactors	Low energy input High surface area (porous media) Passive operation	Slow kinetics Pore blockage Difficult to regenerate porous media	50–250 min	Limited by mass and pore clogging	[114]

4.6. Challenges in Scaling CO_2 Hydrate Systems

While laboratory-scale studies have significantly advanced our understanding of CO_2 hydrate formation mechanisms and promoter performance, the conversion of these findings to industrial applications is not feasible yet, due to several engineering challenges. One of the primary issues is hydrate slurry viscosity management. As hydrate crystals form and grow within a reactor, the resulting slurry can have high viscosity and non-Newtonian behavior, particularly at solid fractions above 20–30%, resulting in increased resistance to flow, greater pumping power requirements, and potential jamming in the pipelines and heat exchangers [123]. Stirred tank reactors that are widely used at lab scale face scalability limitations due to torque overload and inefficient mixing when hydrate slurries thicken during prolonged operation [124]. Addressing this requires the design of robust mechanical systems, such as screw-driven or pipe-loop reactors, which can maintain solid suspension while minimizing energy losses and impeller fouling. Another critical hurdle is heat transfer optimization during both hydrate formation and dissociation. The exothermic nature of hydrate crystallization necessitates rapid heat removal to prevent local overheating and phase instability, while the endothermic dissociation process demands uniform heat input across the reactor. Lab-scale setups often use jacketed vessels or internal coils, but industrial-scale systems require compact, high-efficiency heat exchangers such as microchannel, plate-fin, or helically coiled systems to maintain thermal control without increasing system volume or cost [125,126]. The local heat transfer coefficient drops significantly as the solid fraction increases, emphasizing the need for advanced designs that prevent wall-layer accumulation and thermal insulation by hydrate particles. Gas–liquid–solid mass transfer limitations emerge at larger scales, making it complex to maintain the uniform dispersion of CO_2 and water. Industrial systems must incorporate dynamic flow control, optimized gas injection strategies, and multiphase CFD numerical analysis to ensure reactor homogeneity and predictable hydrate growth. The integration of real-time sensors for slurry monitoring, torque, pressure drop, and thermal gradients also becomes essential to maintain process stability. Ultimately, scaling up hydrate-based CO_2 capture and thermal energy storage systems requires a synergistic approach that combines thermodynamics in mechanical design, heat exchanger integration, and flow assurance engineering.

5. Typical Characterization Techniques Used in CO_2 Hydrates Analysis

Characterization techniques are crucial for understanding CO_2 hydrates, as they provide details of their physical properties, stability, and formation mechanisms. By analyzing the crystal structure, lattice parameters, and molecular arrangements, we can better understand how hydrates form and decompose, ensuring efficient production and storage. These techniques help assess the impact of various factors, such as salts and surfactants, on hydrate stability, leading to new applications in carbon capture, gas separation, and energy storage. This chapter provides an overview of key characterization methods used to analyze CO_2 hydrates properties such as structure, thermal, and stability properties. It discusses Fourier-transform infrared (FTIR) spectroscopy for studying crystallization mechanisms, X-ray diffraction (XRD) for crystallographic analysis and CO_2 storage capacity [127], and high-pressure differential scanning calorimetry (HP DSC) for evaluating enthalpy changes and thermodynamic behavior. Finally, the T-history method is introduced as a simplified yet effective approach for characterizing phase-change behavior. Together, these techniques enable the analysis of fundamental properties of CO_2 hydrates that make their practical applications possible in various industries.

5.1. *Fourier-Transform Infrared (FTIR) Spectroscopy*

The Fourier-transform infrared (FTIR) spectroscopy technique has been successfully used to find/analyze crystallization mechanics in many substances where crystal formation exists. One of the common studies includes the analysis of inorganic salt-based hydrate PCMs to efficiently apply them in various thermal storage applications, such as refrigeration systems. Handling and testing the formed CO_2 clathrate hydrates is difficult due to their highly unstable behavior outside their original formation conditions. In this light, it is important to develop lab instruments [128] or convert their shapes, such as making them into thin films to keep them stable until the end of testing [129,130].

In the FTIR analysis of CO_2 clathrate hydrates, specific IR bands correspond to the vibrational modes of CO_2 within the hydrate lattice, which differ from those of free CO_2.

FTIR can track these changes, providing details of formation and dissociation processes. By observing the evolution of FTIR spectra, it can be easy to monitor the kinetics of CO_2 hydrate formation by knowing the conditions under which these hydrates form. FTIR also provides the observation of hydrate dissociation by tracking the disappearance or shift of IR bands associated with the CO_2–water interactions [131].

For example, Boufares et al. measured the CO_2 consumed during the hydrate formation process by fixing the nucleation temperature ranging from 276 K to 283 K, pressure ranging from 1 to 2.5 MPa, and the concentration of CO_2 dissolved during the crystallization. As shown in Figure 15, an in situ ATR (Attenuated Total Reflection) probe is used to determine the dissolved CO_2 gas. This ATR probe is inserted through a valve from the top lid of the autoclave reactor. Then, the measurements are continuously recorded by the IR bands as CO_2 gas is injected into the reactor. This process is repeated for various temperatures, ranging from 272 K to 283 K, to measure the CO_2 absorbance by the mix.

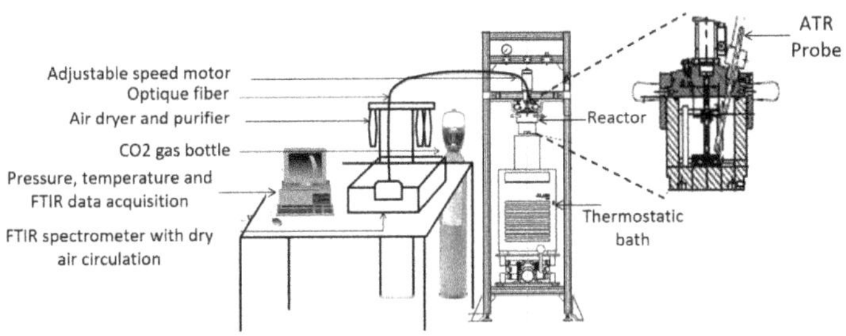

Figure 15. FTIR-based experimental setup for real-time CO_2 hydrate monitoring. Reproduced with permission from *Chemical Engineering Science*, 2018 [132].

5.2. X-Ray Diffraction (XRD)

In the analysis of gas hydrates, including CO_2 clathrate hydrates, understanding their crystallographic structure is very important. The X-ray diffraction (XRD) technique is commonly used to define/analyze the crystallographic structure analysis and the amount of CO_2 stored in the case of the CO_2 clathrate hydrate. It has an important role, as hydrates can exist in various forms, depending on factors such as water pH, pressure, and temperature. By knowing their stability, the XRD can be applied in the analysis of storage and transportation technologies for CO_2 [133].

Powder X-ray diffraction (PXRD) is a powerful technique for analyzing the crystal structure of CO_2 clathrate hydrates. PXRD works by directing X-rays at a powdered sample and analyzing the resulting diffraction pattern to determine the arrangement of atoms within the crystal lattice. This method provides some insight into the stability, formation, and decomposition of CO_2 clathrate hydrates, enabling researchers to track phase transitions, formation and decomposition kinetics, and evaluate the impact of additives on hydrate properties.

Due to the instability of CO_2 clathrate hydrates under ambient conditions, PXRD tests are conducted at high pressures (2 MPa to 10 MPa) and low temperatures (243 K to 283 K), where the hydrates remain stable. Advanced setups such as high-pressure cryogenic vessels support these experiments by providing controlled environments until the end of the test. Figure 16 shows an example of a PXRD system that allows real-time X-ray analysis during hydrate formation and decomposition.

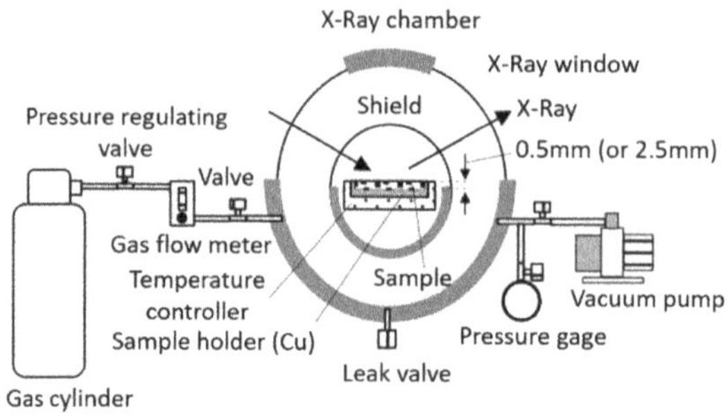

Figure 16. PXRD measurement system integrated with a CO_2 hydrate reactor. Reproduced with permission from *Fluid Phase Equilibrium*, 2016 [134].

Takeya et al. [134] showed the application of PXRD by performing temperature-dependent measurements to study CO_2 hydrates. Their setup promotes precise control of temperature and pressure, ensuring stable conditions for hydrate formation. PXRD measurements covered a 2θ range of 22.1° to 29.7°, which corresponds to the characteristic diffraction peaks of CO_2 hydrates. The fine step width of 0.02° provided high-resolution data on the crystal structure, while measurements at regular temperature intervals (10.15 K) enabled detailed tracking of structural changes under isothermal conditions. Although PXRD presents challenges such as sample preparation under high-pressure and low-temperature conditions and complex data analysis, it remains an invaluable tool for advancing our understanding of CO_2 clathrate hydrates. By revealing detailed structural information, PXRD helps researchers evaluate hydrate stability, assess the influence of additives, and optimize conditions for practical applications in gas storage, sequestration, and energy transport. This makes PXRD a critical technique for both fundamental research and industrial implementation of CO_2 hydrate technologies [134].

5.3. High Pressure Differential Scanning Calorimetry (HP-DSC)

High-pressure differential scanning calorimetry (HP DSC) is a powerful technique for characterizing the thermal behavior of CO_2 hydrates. It measures the heat flow between a sample and a reference, as the temperature changes under controlled pressure. The analysis of heat peaks observed during hydrate formation and decomposition, facilitated by High-pressure differential scanning calorimetry (HP DSC) helps determine the enthalpy change (ΔH) associated with important thermodynamic behaviors into CO_2 hydrate stability. This allows for the investigation of guest molecule exchange within hydrate cages, providing researchers with valuable data on the kinetics and thermodynamics of this phenomenon. The pressure dependence of hydrate formation and decomposition temperatures obtained through HP DSC can be utilized to evaluate CO_2 hydrate stability under varying pressure conditions [135].

The setup used for HP-DSC thermal analysis is illustrated in Figure 17. The enthalpy of dissociation of CO_2 hydrates was measured using a high-pressure micro-differential scanning calorimeter (HP-µDSC VIIa, Setaram Inc., Caluire, France). Initially, hydrates were synthesized ex situ and transferred into pre-cooled calorimetric cells under cryogenic conditions (liquid nitrogen) to prevent premature decomposition [136]. The temperature was ramped at a constant rate of 1.0 K/min, from 250 K to 305 K, under pressures up to 20 MPa. The hydrate sample mass ranged between 40 and 140 mg, and the DSC cells (volume: 0.5 cm^3) were made of Hastelloy C276. Although this conventional single-scan method enables baseline enthalpy measurement, partial water conversion and overlapping ice–hydrate transitions limit its ability to resolve metastable behaviors accurately. Researchers have been using a PVT cell equipped with a visor/window to study the phase behavior of hydrates by observing the reaction going on inside. Temperature and pressure sensors are equipped for the thermodynamic properties and pressure variations, which are all time consuming and require manual work [137].

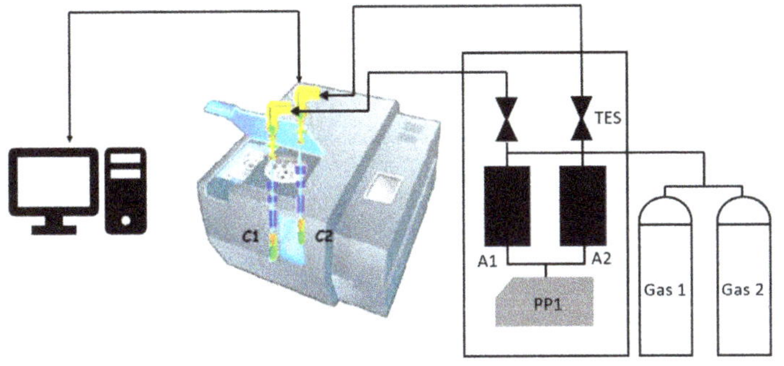

Figure 17. High-pressure differential scanning calorimetry (HP-DSC) setup for thermal analysis. Reproduced with permission from *Chemical Engineering Science*, 2008 [137].

To address these limitations, the experimental approach is compared to the multicycle calorimetric technique introduced by Robustillo et al. [138]. In the multicycle method, repeated subcooling and heating cycles (typically 20–25) are conducted under controlled pressure and temperature conditions. The heating and cooling rates vary from 0.2 to 1.0 K/min to promote a gradual increase in hydrate yield while minimizing kinetic distortions. This technique achieves >99% water-to-hydrate conversion by suppressing secondary exothermic events linked to metastable ice formation. Enthalpy values derived from this method show improved precision and do not rely on thermodynamic extrapolation through the Clausius–Clapeyron equation [138].

5.4. T-History Method

In contrast to DSC and differential thermal analysis (DTA), the T-history method provides a simpler approach for characterizing PCMs. This method was conceptualized by Yinping Zhang in 1999 [139] and refined by researchers over subsequent decades [140,141]. It provides an effective means of determining enthalpy and heat capacity. It is especially valuable for analyzing non-homogeneous samples, including those prone to supercooling. It consistently delivers superior accuracy in PCM characterization. The T-history setup for studying the phase behavior of CO_2 clathrate at high pressure is similar in concept to the setup used for other PCMs. Figure 18 shows the T-history setup constructed for testing CO_2 hydrate dissociation temperature.

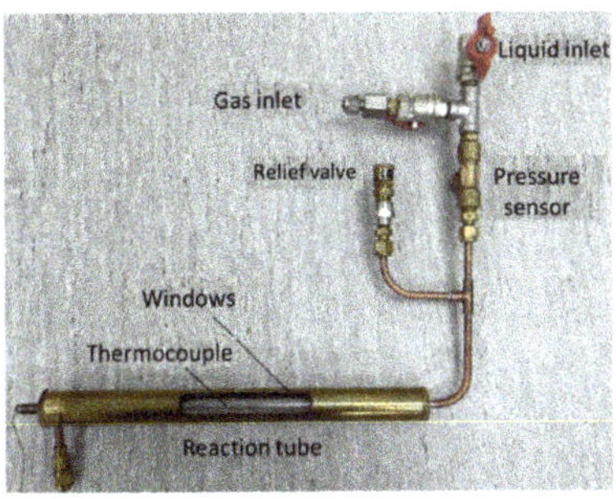

Figure 18. T-history experimental setup for dissociation analysis of CO_2 hydrates.

In this method, both the PCM and a reference material (water) are subjected to the same controlled cooling environment, and their temperature–time responses are recorded. The thermal lag during phase change is used to calculate latent heat based on a lumped thermal model. In the experimental setup, a glass tube (10 mm inner diameter, 160 mm length, 12.6 cm^3 volume) was used to hold the hydrate sample.

For sample preparations, a defined concentration of TBAB (tetra-n-butylammonium bromide) was first dissolved in deionized water to create a homogenous aqueous solution. This solution was transferred into a sealed glass tube and exposed to a regulated CO_2 gas stream under pressure. The system gradually cooled using a thermostatic bath to facilitate hydrate formation. Crystallization was confirmed through the glass on the side of the reaction tube. Once hydrate formation was complete, the sample underwent controlled thermal cycling to analyze its dissociation behavior using the T-history method. A type-K thermocouple was embedded in the sample to measure temperature changes with ± 0.1 °C accuracy, while the tube was submerged in a heat transfer fluid bath. To ensure one-dimensional heat transfer and accurate modeling, the Biot number was kept below 0.1. The phase-change enthalpy was calculated by comparing the temperature shift between sample and reference curves, accounting for specific heat, mass, and temperature differential. This method is advantageous for detecting metastable behavior such as supercooling and delayed dissociation because of its larger sample size and visual access to hydrate crystallization [127].

6. Dissociation of CO_2 Clathrate Hydrates and Advancements in Analysis of Phase-Change Enthalpy

Dissociation enthalpy is a crucial property of CO_2 clathrate hydrates, which represents the energy required to break the bonds between CO_2 molecules and the water lattice in the hydrate structure. This property is crucial for understanding the stability, formation, and potential applications in both designing thermal storage systems and analyzing the performance of CO_2 hydrate slurries used as thermo-fluids. The exact enthalpy change for CO_2 hydrate formation/dissociation depends on various factors, such as temperature, pressure, and the specific hydrate structure (I or II). A higher dissociation enthalpy indicates a more stable hydrate under given conditions [142,143].

The heat which is released or absorbed when CO_2 hydrates are formed from water and CO_2 under specific conditions of pressure and temperature is called formation enthalpy. In the reverse process, the heat required to break down CO_2 hydrates back into water and CO_2 is called dissociation enthalpy.

For hydrate formation: CO_2 + n H_2O $\rightarrow$ CO_2 n H_2O + Heat.

For hydrate dissociation: CO_2 n H_2O + Heat $\rightarrow$ CO_2 + n H_2O.

This enthalpy can be either positive or negative, depending on the process. Several experimental studies and computational simulations were used to determine the enthalpy of CO_2 hydrate formation under various conditions. Hydrates provide a compact and stable medium for storing CO_2 in solid form, and the enthalpy helps predict how much energy will be required for hydrate dissociation when the CO_2 needs to be released for sequestration or industrial use [144].

However, typical values for CO_2 hydrate dissociation range from 350 kJ/kg to 520 kJ/kg of water. This means that during dissociation, for every mole of water molecules involved in the hydrate structure, about 50–70 kJ of heat is released to the surroundings. Conversely, for hydrate formation (endothermic process), this range of values translates to the amount of heat absorbed from the surroundings per mole of water in the hydrate [145].

Once the hydrates are formed, they have to follow the phase equilibrium curve pressure and temperature conditions to stay as stable hydrates. Similarly, during the dissociation process, the hydrate conditions are below the equilibrium curve and follow the endothermic process [146]. This process basically follows three independent processes: thermal stimulation, depressurization, and chemical injection [147].

Thermal stimulation is a common approach where heat is applied to the hydrate reservoir to increase its temperature and to dissociate. As shown in Figure 19, the dissociation process happens following three stages, typically the fluid flow driving stage, the heat transfer driving stage, and the kinetic driving stage. From steps A to B, the rise in temperature is caused by heat transfer. The increase in this temperature is the main reason for the hydrate dissociation process. As the hydrates begin dissociating, this affects the pressure and temperature conditions, which can be seen as the kinetic driving force from steps B to C. This step sees the release of gas and the consumption of heat. Further heating can cause additional dissociation, but this may also lead to challenges such as increased pressure and decreased temperature due to the delayed flow of fluids (steps C–D). After this stage, if there is still a delayed transfer of dissociation between gas and water, steps D to E may appear in the process. In practice, the efficiency of this method can be limited by heat transfer issues. The water released from dissociated hydrates often forms a thin liquid film that obstructs heat transfer, and if dissociation occurs too rapidly, or if gas transfer is insufficient, it can lead to conditions that favor the reformation of hydrates [148].

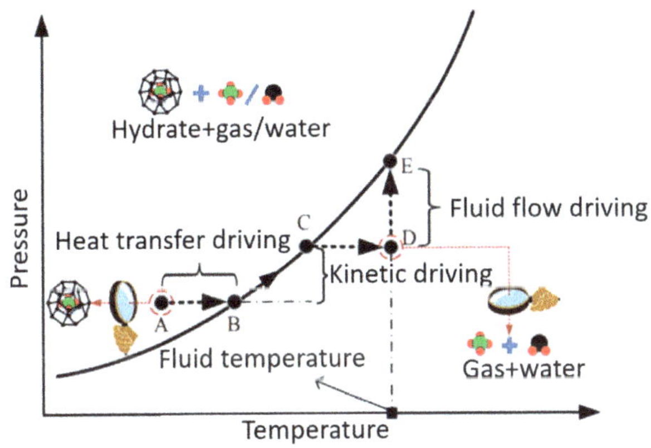

Figure 19. Thermal stimulation-induced CO_2 hydrate dissociation pathway. Reproduced with permission from Elsevier, 2024 [149].

Another method of inducing hydrate dissociation is the depressurization process, as shown in Figure 20. This occurs due to low pressure relative to the equilibrium pressure. Depressurization is another method for inducing hydrate dissociation, which involves lowering the pressure in the reservoir below the equilibrium pressure. This process starts with the release of free gas due to the reduction in pressure, which significantly lowers both the pressure and temperature in the reservoir (Step A–B). The released gas and heat then facilitate the dissociation of hydrates (Step B–C). As dissociation continues, the P-T conditions follow the equilibrium curve, and the pressure difference drives the flow of dissociated gas and water (Step C–D). As the dissociation progresses (Step D–E), the water and gas released must flow towards the outlet due to this pressure differential. If the heat supplied is insufficient to support the endothermic dissociation, the process will encounter issues, leading to step E–F [150–152].

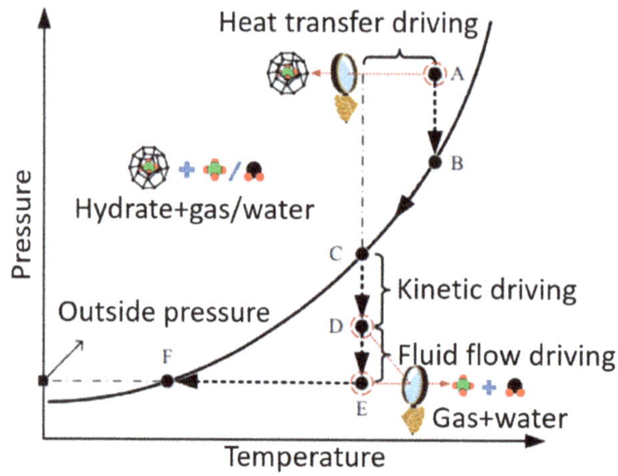

Figure 20. Depressurization-driven CO_2 hydrate dissociation stages. Reproduced with permission from Elsevier, 2024 [149].

Chemical injection is the process of adding chemicals to the reservoir that change the equilibrium conditions. This method can be effective but requires careful management to ensure that the concentration of inhibitors is optimal to destabilize the hydrates without introducing new stability issues [149,153].

Researchers have been using a PVT cell having a visor/window on the instrument to study the phase behavior of hydrates by observing the reaction going on inside. Temperature and pressure sensors are installed to record thermodynamic properties and pressure variations; however, these measurements are time-consuming and require manual work [154].

Later on, researchers started using high-pressure differential scanning calorimetry (HP-DSC), as this instrument maintains the formation and hydrate-stabilizing conditions until the test is complete. Therefore, it is considered one of the precise methods to extract equilibrium data [155–159].

For determining the heat of dissociation, and heat capacity, researchers produced hydrates within the reactor and then transferred these hydrates to a calorimetric vessel. However, the produced hydrates may become contaminated during the transfer [160,161].

Following the experimental difficulties and the possibility of sample contamination during the transfer between test instruments, as described above, there have been different approaches developed for measuring the dissociation enthalpies of CO_2 hydrates.

Xuewen et al. conducted experiments using a high-pressure tank reactor. After injecting the gas and water, the reactor was then cooled to 280 K, and CO_2 was injected to raise pressure to 3.0 MPa, with data acquisition at 20 s intervals. Hydrates formed as CO_2 was continuously injected, until the conversion was complete. Dissociation experiments were conducted by reducing pressure in steps to 2.5 MPa, 2.0 MPa, 1.5 MPa, and 1.3 MPa and by adjusting the temperature at 1.3 MPa to observe dissociation. The dissociation rate, calculated from gas flowmeter data, showed a decrease in reactor temperature due to the endothermic nature of dissociation, with stable dissociation observed at lower pressures and a significant increase in rate when the pressure was reduced further. The effect of pressure revealed two stages: stable dissociation rates at higher pressures and decreased rates at lower pressures, with significant improvement at 1.5 MPa and 1.3 MPa. Temperature effects were studied at 280 K, 282 K, and 284 K, showing increased dissociation rates of 26% and 41% at the higher temperatures, respectively, indicating that while temperature increase enabled dissociation, and pressure changes had a more substantial impact [142].

A high-pressure micro differential scanning calorimeter (HP μ-DSC) was used by Lee et al. to measure both the heat of dissociation and the dissociation temperature of the gas hydrates. This research provides an analysis of the dynamics of hydrate formation and replacement. A pressure transducer monitored the pressure within the sample cell and the formation of hydrates was achieved through a series of repeated cooling–heating cycles. For the CH_4-CO_2 replacement experiments, CH_4 was evacuated from the system, and CO_2 was subsequently injected into the sample cell. The experiments were conducted at various temperatures, generally below 273 K, to prevent the dissociation of CH_4 hydrate. Pressure was controlled and monitored throughout the experiments. CO_2 was introduced into the sample cell at pressures that were higher than those required for CH_4 hydrate dissociation but lower than those for CO_2 liquefaction. The results showed that complete conversion of water to hydrate was achieved through repeated cooling-heating cycles. The DSC method provided accurate determinations of the hydrate–liquid water–vapor equilibrium points. The CH_4-CO_2 replacement process occurred without significant dissociation or formation of new hydrates. This study also showed that the dissociation enthalpies of CH_4-CO_2 hydrates increased with the CO_2 composition, and approximately 60% of CH_4 was replaced by CO_2 in the hydrate phase [135].

Sloan and Koh et al. [16] conducted a comprehensive review of gas hydrate thermodynamics, and their work provided a detailed analysis of dissociation enthalpy using both theoretical and experimental approaches. Using the Clausius–Clapeyron equation, a fundamental thermodynamic relationship that connects the pressure–temperature (P-T) conditions of hydrate dissociation to the enthalpy change (ΔH_d), they calculated the dissociation enthalpy to be approximately 57–59 kJ/mol. This method assumes ideal behavior of the hydrate phase and a constant enthalpy change over the measured temperature range, which simplifies the analysis but requires highly accurate P-T data. This research emphasized the challenges associated with experimental measurements, such as ensuring the formation of pure CO_2 hydrates without impurities, which could skew the results. Therefore, in addition to their theoretical approach, they utilized differential scanning calorimetry (DSC) to experimentally measure the dissociation enthalpies of CO_2 hydrates under controlled pressure and temperature conditions. Their DSC experiments focused on understanding the influence of hydrate structure (structure I (sI) or structure II (sII)) and the presence of guest molecules on the enthalpy values. The results revealed that the dissociation enthalpy of CO_2 hydrates typically ranges from 50 to 60 kJ/mol, depending on the specific experimental conditions, such as pressure, temperature, and hydrate structure [162]. They demonstrated that the hydrate structure and the nature of guest molecules significantly influence the enthalpy values, with sI hydrates generally exhibiting higher dissociation enthalpies compared to sII hydrates. This dual approach—combining the Clausius–Clapeyron equation with direct DSC measurements—provided a robust and widely applicable framework for understanding CO_2 hydrate thermodynamics.

Aromada et al. (2019) [144] conducted a comprehensive study on enthalpy changes of hydrate phase transitions using a consistent thermodynamic scheme based on residual thermodynamics. This approach involved evaluating the enthalpy changes of hydrate dissociation and formation as a function of temperature, pressure, and hydration number for both CH_4 and CO_2 hydrates. It also identified the limitations of traditional methods, such as the Clausius–Clapeyron equation, which often result in inconsistent and unreliable data due to assumptions of ideal behavior and constant enthalpy over temperature ranges. By using their proposed method, as a function of pressure, temperature, and variations in hydration number, they were able to obtain more accurate and consistent enthalpy values. Their findings showed that the enthalpy change for CO_2 hydrate dissociation is approximately 10–11 kJ/mol higher than that of CH_4 hydrate within the temperature range of 273–280 K. This difference is attributed to the stronger interactions between CO_2 molecules and water in the hydrate lattice compared to CH_4, as well as the structural differences between sI (CH_4) and sI/sII (CO_2) hydrates [163].

Ma et al. (2024) [164] focused on predicting the phase equilibrium conditions and thermodynamic stability of CO_2-CH_4 mixed gas hydrates. They used theoretical models to calculate the temperature and pressure ranges for hydrate formation, which were then used to develop a multi-parameter empirical model. This model allowed for accurate predictions of hydrate formation conditions without the need for complex thermodynamic models. Their study also examined the dissociation enthalpy of CO_2-CH_4 hydrates, finding that an increase in CO_2 mole fraction led to higher dissociation enthalpy and enhanced thermodynamic stability. This is attributed to the stronger intermolecular interactions between CO_2 and water molecules compared to CH_4, as well as the structural differences between sI (CH_4-dominated) and sII (CO_2-dominated) hydrates [165]. The ability to predict hydrate formation conditions and dissociation enthalpy with high accuracy is critical for designing efficient gas separation and storage systems, especially in industries where CO_2 and CH_4 coexist, such as natural gas production and carbon capture and storage (CCS).

The enthalpy data collected from various studies shows the energy changes associated with CO_2 hydrate formation and dissociation. These measurements confirm that CO_2 hydrate formation is an exothermic process, with enthalpy values that vary depending on the condition. Understanding these energy dynamics is crucial for evaluating the feasibility of using CO_2 hydrates in different applications. Table 9 shows the different temperature and pressure conditions along with the enthalpies resulting from the addition of different single and mixed additives.

Table 9. Enthalpy measurements for CO_2 hydrate formation using different additives.

Temperature (K)	Pressure (MPa)	Additive	Enthalpy (kJ/Kg)	References
281 to 292	0.5	Cyclopentane (C_5H_{10})	500	[166]
277 to 282	1.9 to 3.4	CO_2 + Water + Ethanol	565 to 580	[11]
285 to 288	0.60 to 0.96	cyclopentane (C_5H_{10})	1833	[167]
276.46 to 278.39	0.66 to 0.91	Cyclopentanone (C_5H_8O)	677	[167]
287 to 290	0.46 to 0.81	Fluorocyclopentane (C_5H_9F)	1334	[167]

Beyond the quantification of enthalpy changes, a comprehensive understanding of the mechanisms governing the dissociation of carbon dioxide (CO_2) hydrates is crucial for the optimization of their practical applications. These applications include carbon capture and storage (CCS), cold thermal energy storage (CTES), and gas transportation systems. The breakdown of hydrates generally proceeds through three principal methods, as explained in the earlier section of this paper: thermal stimulation, depressurization, and chemical injection. Each of these pathways exhibits unique thermodynamic and kinetic properties. For comparative analysis, Table 10 displays a quantitative assessment of these dissociation mechanisms.

Table 10. Analysis of CO_2 hydrate dissociation mechanisms under varying thermal and pressure conditions.

Dissociation Mechanism	Pressure and Temperature Condition	Energy Source	Dissociation Rate	Thermal Effect	Structural Behavior	References
Depressurization	Depressurized from 3.7 MPa to 3.1 MPa at 275.6 K	10^5 to 10^7 J/m^3 (mainly the advection heat 10^3 J)	Slower gas release kinetics. Gradual dissociation over 400 min	Delayed thermal recovery	No expansion or deformation modeled	[168]
Thermal Stimulation	The bath is constant at 288.15 K (15 °C)	-	Fast release at higher temperatures	Temperature dropped from 272.15 K (−1 °C) to 269.15 K (−4 °C)	No drastic structural disruption	[169]
Combined thermal stimulation and depressurization	Depressurized from 4.5 MPa to 0.1 MPa at 288.15 K (15 °C)	-	Early dissociation onset; memory effect under disappearing water layer conditions	Max cooling 266.15 K (−7 °C) at 0.1 MPa	Sediment expanded up to 176% at lowest pressure	[169]

7. Future Heat Storage, Refrigeration, and Space Conditioning Applications

Maintaining the desired temperature within a system is crucial in many thermal management and energy-related applications. The process, which involves heat removal, is usually achieved through a refrigeration cycle. Heat transfer occurs from a higher temperature zone (heat source) to a lower temperature zone (heat sink) using a working fluid (refrigerant). The working fluid undergoes a cycle of compression, condensation, expansion, and evaporation, absorbing heat from the source during evaporation and releasing it to the sink during condensation. The heat removal capacity of a conventional refrigerant or CO_2 hydrate slurry depends on its ability to absorb the heat during a phase-change process (i.e., evaporation or thawing). This value is a crucial parameter for determining the effectiveness of a temperature control system. Latent heat is also one of the important parameters for the material we use in refrigeration and is defined as the amount of heat absorbed per unit mass of refrigerant during the liquid-to-gas phase change (evaporation), or a unit mass of CO_2 hydrate slurry during the solid-to-liquid (thawing) process. A higher latent heat value shows a greater capacity for heat removal per cycle [170,171].

While commercial air conditioning systems utilizing ice thermal storage (ITS) are well-established and widely adopted for their efficiency and environmental benefits, CO_2 clathrate hydrates present an innovative alternative with significant potential advantages. Unlike ice, CO_2 hydrates have a higher volumetric energy density, allowing for more compact and efficient thermal storage solutions. This characteristic parameter enables greater cooling capacity in a smaller footprint, which is particularly beneficial for commercial and industrial HVAC applications. CO_2 clathrate hydrates offer flexible phase-change temperatures, which can be optimized for specific cooling applications, unlike ice, which melts at a fixed temperature (0 °C). This flexibility enhances heat transfer efficiency and provides better integration with different refrigeration and air conditioning systems. Another key advantage of CO_2 hydrate-based thermal storage is its potential for simultaneous carbon capture and cooling, utilizing CO_2 as a working fluid while contributing to sustainability efforts. By replacing or supplementing ice-based thermal storage systems, CO_2 hydrates could reduce system energy consumption, enhance peak load shifting, and lower operational costs in cooling applications. A representative ice thermal energy storage configuration used in air conditioning systems is illustrated in Figure 21. Thus, integrating CO_2 clathrate hydrate thermal storage into commercial air conditioning systems could provide a more energy-efficient and environmentally friendly alternative to traditional ice-based cooling technologies [172].

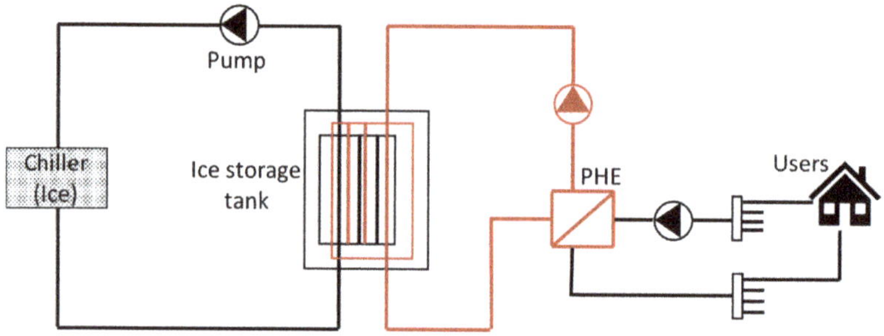

Figure 21. Ice thermal energy storage (ITS) configuration for air conditioning applications. Reproduced with permission from *Journal of Building Engineering*, 2022 [172].

Below example applications of CO_2 hydrate utilization in refrigeration systems are discussed, as adopted by different researchers.

7.1. CO_2 Hydrate Used in Air-Conditioning (AC) System

CO_2 hydrates and their slurries are considered as potential high-performance coolants and/or secondary refrigerants in next-generation energy-saving AC systems. In CO_2 hydrate applications, the latent heat ranges between 350 kJ/kg and 520 kJ/kg, with a phase-change temperature range between 278 and 285 K. For this temperature range, which is slightly above the freezing point of water, the key feature of clathrate hydrate slurry (CHS), as shown in Figure 22, is its high thermal density, which is almost 1.5 times greater than that of chilled water [170]. CHS has a thermal density ranging from 10 to 17 Mcal/m^3, providing enhanced energy storage capacity compared to chilled water. This high thermal density enables CHS to achieve the same cooling effect, with roughly half the flow rate required by chilled water, thereby reducing the required pump power and leading to significant energy savings. CHS forms at temperatures similar to chilled water (278–285 K), allowing refrigeration equipment to operate at temperatures above freezing, and requires simpler refrigeration equipment compared to systems using ice production. Its non-cohesive nature prevents freezing and blockages in pipes and indoor units, promoting direct transportation within buildings, similar to chilled water systems. Figure 22 shows the basic schematics of how hydrates can be possibly used in air conditioning systems [173].

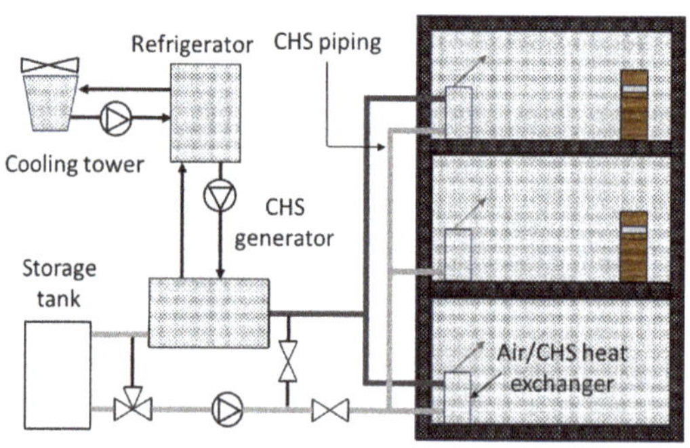

Figure 22. Clathrate hydrate slurry (CHS) application in air conditioning systems. Arrows in the figure indicate the direction of CHS and cooling fluid flow through the system components. Reproduced with permission from *JFE Technical Report*, 2004 [173].

Despite its advantages, CO_2 hydrates require high pressures (above 3 MPa) to remain stable, which complicates their use. To address this issue, thermodynamic and kinetic promoters can be added to stabilize hydrates at lower pressures. TBAB (Tetra-n-butylammonium bromide) is one of the best-known thermodynamic promoters to form CO_2 hydrates at atmospheric pressures of around 0.1 MPa and temperatures between 275–285 K [145].

A laboratory-scale hydrate-based refrigeration system was constructed to see the technical feasibility of such systems [174]. An example of this type of setup is a 5 L stainless-steel hydrate-forming reactor with a stirrer and sight glasses for visual observation and cooled using a coil-type heat exchanger connected to a chiller with a 960 W cooling capacity. The system that generates CO_2 hydrate slurry has been also validated by the food industry. As shown in Figure 23, CO_2 hydrate technology validated its effectiveness in rapid food cooling processes, such as for cooling meat from 25 °C to 8 °C without dehydration by spraying CO_2 hydrate slurry [175].

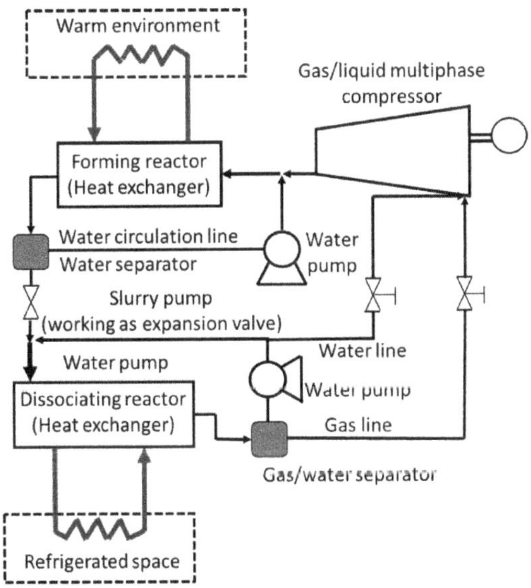

Figure 23. CO_2 hydrate slurry system for rapid cooling in the food industry. Reproduced with permission from *Applied Thermal Engineering*, 2006 [23].

For cold storage applications, the rheological properties of the CO_2 hydrate slurry are important. Studies by Jerbi et al. and others have confirmed that CO_2 hydrate slurry can flow efficiently through secondary refrigerant loops [176].

Jerbi et al. designed a refrigeration system in which CO_2 hydrate slurry was used as a two-phase secondary fluid [177]. As shown in Figure 24, there were two main parts in this system, one for the formation and storage of CO_2 hydrate slurry in a 25 L tank reactor that could resist pressures up to 4.5 MPa, and the other for gas–liquid separator (GHS) circulation and dissociation in a loop. The reactor provided a heat flux of 1–5 kW for a hydrate fraction of 10–30%. The heat exchanger in the loop allowed for the continuous decomposition of the CO_2 hydrate slurry, enabling efficient heat exchange and refrigeration performance [177].

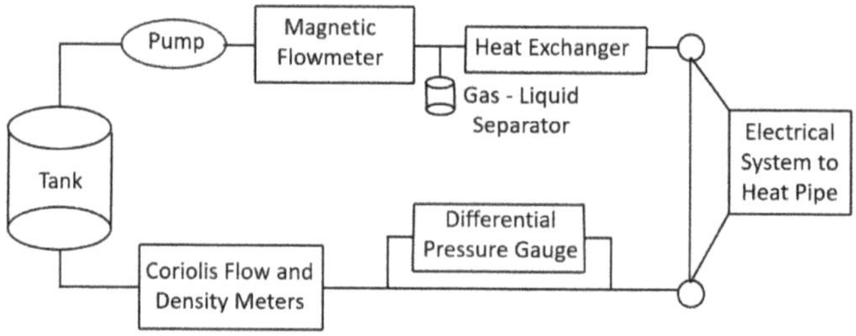

Figure 24. CO_2 hydrate slurry system operating as a two-phase secondary refrigerant. Reproduced with permission from *International Journal of Refrigeration*, 2013 [177].

The integration of CO_2 hydrate with thermodynamic promoters such as TBAB and the use of efficient system construction methods indicate the increasing feasibility of CO_2 hydrate-based refrigeration systems. These systems not only offer higher cooling capacities but also present a viable alternative to traditional refrigerants and secondary coolants, making them attractive for sustainable and energy-efficient thermal energy storage solutions.

7.2. CO_2 Hydrate Slurry as a Secondary Refrigerant System

CO_2 hydrates offer strong potential as an alternative to conventional secondary refrigerants [178]. In this system, the separated hydrate slurry is recirculated through the secondary loop to deliver cooling to end users before returning to a storage tank. At this stage, the previously separated CO_2 gas is reintroduced into the tank, which significantly reduces the overall storage volume. Studies have shown savings of up to 75%. The process also achieves a marginal coefficient of performance (COP) of up to 6, enhancing its high energy efficiency. An experimental system for CO_2 hydrate slurry-based refrigeration is illustrated in Figure 25.

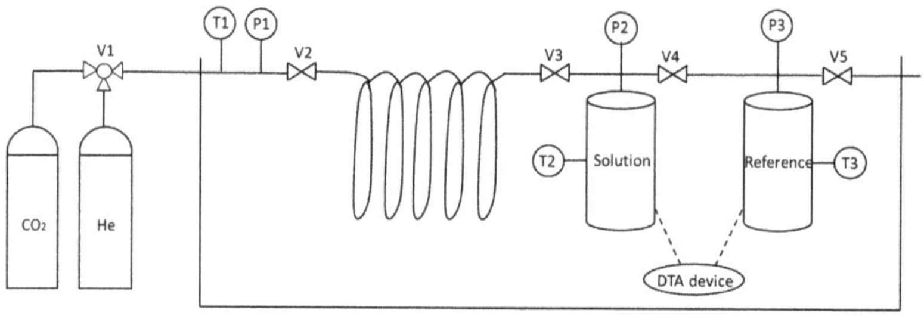

Figure 25. Experimental system for CO_2 hydrate slurry-based refrigeration. Reproduced with permission from *International Journal of Refrigeration*, 2014 [178].

The secondary refrigerant loop constructed by Oignet et al. [179], as shown in Figure 26, consists of two primary components: a reactor for forming the CO_2 hydrate slurry and a circulation loop equipped with a heating tube for heat transfer measurements. The formation reactor, a stainless-steel vessel, is designed to withstand pressures up to 3.5 MPa and has an internal volume of 26.4 L. CO_2 is injected into the reactor, where it dissolves in the liquid phase and forms hydrates when the reactor is cooled by a refrigerator fluid circulating through its jacket. A magnetic mixer is used to enhance the formation rate of CO_2 hydrate slurry by improving the mixing within the reactor. Once the hydrate slurry is formed, it is circulated through the secondary refrigerant loop, composed of stainless-steel pipes with an internal diameter of 7.74 mm. The flow rate and pressure within the loop are monitored by sensors, such as Coriolis flow and density meters, differential pressure gauges, and thermal sensors. The loop consists of a heating tube designed to measure the local convective heat transfer coefficients (CHTCs) of the hydrate slurry. By applying an electric current to the surface of the heating tube, heat is transferred from the tube's wall to the circulating slurry, and the temperature difference caused is used to calculate the local heat transfer properties. The secondary refrigerant loop in this setup helps study the behavior of CO_2 hydrate slurries under controlled conditions. The system provides valuable data on flow rates, pressure drops, and convective heat transfer, which are important properties for optimizing the performance of CO_2 hydrate slurries as secondary refrigerants. One of the key findings is that CO_2 hydrate slurries with solid fractions between 0 and 14 vol.% have convective heat transfer coefficients that are up to 2.5 times greater than those of liquid water, making them highly efficient for cold energy transport [179].

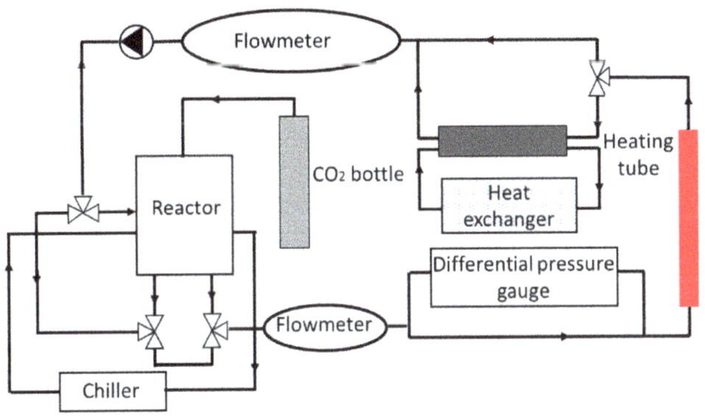

Figure 26. Secondary refrigerant loop with CO_2 hydrate formation and circulation unit. Reproduced with permission from *Applied Thermal Engineering*, 2017 [179].

While the thermal behavior was well quantified, the study did not address how the slurry behaved under flow, especially in terms of resistance and stability at higher hydrate concentrations. Fu et al. [180] measured how viscosity changed with increasing hydrate content using capillary rheometry under pressure. Their results showed a sharp rise in viscosity from around 12 mPa·s at 1.4% hydrate content to over 500 mPa·s at 17.2%. The slurry quickly moved from liquid-like to paste-like behavior, raising concerns about how such systems would perform under continuous circulation [180]. To improve flow and smooth transport, Sahu et al. [181] introduced sodium dodecyl sulfate (SDS) during hydrate formation. The presence of SDS reduced viscosity from 329 cP to 147 cP, which improved mobility while still allowing hydrate formation to proceed. This made the slurry easier to handle, without compromising its cooling capacity [181]. The particle size also influences the formation process. Hydrates formed with grains around 63 μm had the highest gas uptake and moved through the system without significant restriction. Finer particles below 10 μm, limited gas diffusion and blocked pore spaces by allowing the slurry slower to form and harder to circulate [182]. Patel et al. [183] introduced tetrahydrofuran (THF) as a promoter and observed how shear rate affected particle behavior. Below 154 s^{-1} hydrate particles started to clump, interrupting uniform flow. At 1–2 wt.% THF, the slurry remained dispersed and stable, even during high solid loading, helping to maintain consistent movement through the loop [183].

7.3. Cold Thermal Energy Storage (CTES)

In cold thermal storage, chlorofluorocarbons (CFCs) and hydrochlorofluorocarbons (HCFCs) are being phased out as refrigerants due to their severe effects on the ozone layer. Therefore, natural refrigerants such as carbon dioxide (CO_2) and ammonia have become common alternatives. Commercial refrigeration systems that use ice storage typically incur additional costs due to the energy-intensive process of freezing and maintaining ice. Systems that rely on traditional chilled water or other cooling media avoid these costs but may require larger storage capacities due to their lower energy density compared to ice [184,185].

Dealing with the problem of uneven power usage for air conditioning, implementing cold storage systems can significantly enhance energy efficiency. These systems can also ensure a consistent supply of cooling for different cooling applications. By storing cold energy during off-peak hours (such as nighttime, when electricity demand is lower), these systems can handle part of the cooling load during peak periods, reducing operational costs. In addition, refrigeration units can be designed with a smaller capacity, reducing capital costs.

A comparison of the thermophysical properties of various cold storage media, as reported by different researchers, including water, ice, a common eutectic salt (Na_2SO_4, H_2O, NaCl, NH_4Cl), TBAB hydrate, and carbon dioxide hydrate, shows the potential candidate for use in CTES [186]. Water is a common heat transfer medium but lacks the phase-change properties of ice, eutectic salts, and hydrates in terms of its energy storage capacity. Ice is highly efficient for cold storage due to its lower density of around 917 Kg/m^3 and requires a temperature of 0 °C for phase change. Eutectic salts offer moderate energy storage capacity due to its higher density of around 1490 kg/m^3 but has a lower dissociation enthalpy of 121 kJ/kg compared to ice. TBAB hydrate stores more energy than eutectic salts 193 kJ/kg but less than ice. It has a lower thermal conductivity of 0.42 W/m·K and variable viscosity of 4.8–9.6 mPa·s. CO_2 hydrate exceeds all the above in energy storage due to its high dissociation enthalpy of around 500 kJ/kg [187] and has a wide range of phase-change temperatures, ranging between 0.7 and 9.5 °C. Table 11 shows various dissociation enthalpies achieved by different media used by different researchers [188].

Table 11. Comparison of dissociation enthalpies of common cold thermal energy storage components.

Medium	Dissociation Enthalpy (kJ/kg)	References
Ice	333	[189]
Eutectic Salts	121	[186]
TBAB Hydrate	193	[190]
CO_2 Hydrate	459 to 507	[187]
CO_2 Hydrate Slurry *	370	[191]

* Changes with volume fraction.

This makes CO_2 hydrate a promising phase-change material (PCM) for air conditioning systems. The high phase equilibrium pressure of CO_2 hydrate exceeds atmospheric pressure and increases energy consumption. Numerous studies have explored ways to use CO_2 hydrate at lower pressures in air conditioning systems. The phase equilibrium pressures and dissociation enthalpies of CO_2 hydrates in the presence of various thermodynamic promoters is already discussed in Section 2 of this paper. These promoters significantly reduce the phase equilibrium pressure and increase the dissociation enthalpy, demonstrating the adaptability of CO_2 hydrate for cold storage systems.

Figure 27 illustrates a carbon dioxide hydrate-based cold thermal energy storage system that integrates with an air conditioning (AC) system. In this system, the chiller cools the water or refrigerant used to promote the formation of CO_2 hydrate within the cold storage unit during off-peak hours, when energy demand is lower. The chilled water is circulated throughout the system using a water pump, allowing the formation of CO_2 hydrate in the gas hydrate cold storage tank. This storage process is essential for capturing cold energy that can later be used during peak energy demand periods. The CO_2 hydrate serves as a phase-change material, storing a significant amount of cold energy due to its high dissociation enthalpy. When the cooling demand rises during peak hours, the stored CO_2 hydrate begins to dissociate, releasing cold energy, which is used to cool the chilled water. This chilled water is then circulated back through the AC system and to the user side, where a fan helps distribute the cooled air. In addition, the system includes a CO_2 gas storage tank, where the released carbon dioxide gas is collected. The gas compressor plays a crucial role in maintaining the necessary pressure by recompressing the CO_2 gas, ensuring efficient system operation. The system also features an expansion tank that helps regulate the thermal expansion of the water or refrigerant as the temperature changes. Various pipes and valves control the flow of gases and liquids, ensuring proper circulation and preventing backflow within the system. This design allows for more efficient energy usage by taking advantage of lower nighttime electricity rates and reducing the need for high capacity refrigeration units. By storing cold energy and using it strategically during peak demand times, the system helps lower operational costs and improve energy efficiency [192].

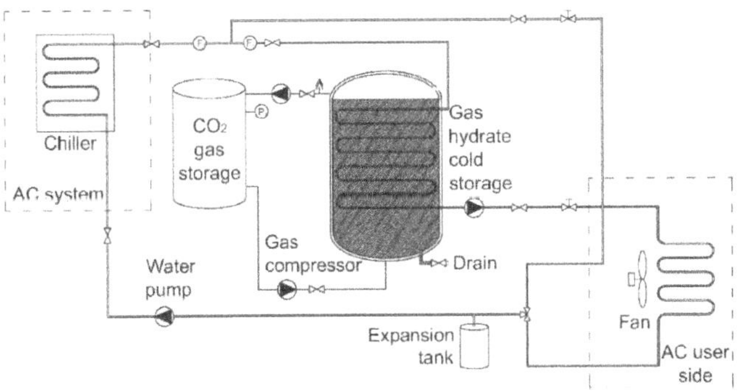

Figure 27. Integrated CO_2 hydrate-based cold thermal energy storage system with air conditioning.

Studies on methane–CO_2 replacement reactions in porous media or saline solutions could provide insights into optimizing the carbon capture process while minimizing energy consumption. Still, energy consumption remains a critical challenge for practical deployment, as the high pressures required for CO_2 hydrate formation significantly increases operational costs. Optimizing hydrate reactor designs, improving mechanical stirring systems, and enhancing the thermophysical properties of hydrates are essential for making CO_2 hydrate-based systems more competitive.

CO_2 hydrate-based systems face several key challenges that must be addressed for practical application. One major issue is the kinetics of hydrate formation, as the rate is often too slow for practical use. Additives such as surfactants and nanoparticles can enhance nucleation, but identifying cost-efficient and effective additives remains an ongoing challenge. Researchers are actively seeking catalysts to accelerate the process. Another challenge is thermal management. CO_2 hydrate formation releases a considerable amount of heat, which, if not properly managed, can lead to localized hydrate decomposition, reducing system efficiency [193]. Scaling and contamination also present significant difficulties. Industrial CO_2 streams often contain impurities such as sulfur compounds, which can inhibit hydrate formation and cause blockages in reactors. Proper pretreatment of the CO_2 stream is required to maintain efficiency and avoid fouling. In addition, ensuring the long-term stability and storage of CO_2 hydrates is difficult. Even slight variations in pressure or temperature can lead to hydrate dissociation, releasing CO_2 and negating storage benefits [194]. The material compatibility of the system components is another area of concern. The high-pressure, low-temperature conditions needed for CO_2 hydrate formation can degrade materials, increasing maintenance costs and posing safety risks. There are environmental and safety concerns related to the release of CO_2 during hydrate storage, transportation, or dissociation, which necessitate robust leak detection and safe decomposition protocols [195]. Lastly, scale-up challenges complicate the deployment of CO_2 hydrate systems. While lab-scale experiments show promise, scaling these systems to industrial levels requires precise control over pressure, temperature, and hydrate formation across large volumes. Addressing these challenges is essential for the success of CO_2 hydrate-based technologies for carbon capture and storage [196].

8. Potential Consequences of Using CO_2 Hydrates

One of the major concerns in any long-term CO_2 hydrate application, including cold thermal energy storage (CTES), is the stability of hydrates under variable conditions. Studies on geological sequestration show hydrate behavior. Zhao et al. [197] used magnetic resonance imaging (MRI) to visualize hydrate formation during simulated CO_2 leakage and found that CO_2 hydrate caps significantly reduced permeability, preventing gas migration. The hydrate layer maintained mechanical integrity under pressures up to 10 MPa and reformed quickly after temporary disturbances due to the continued availability of water and CO_2, suggesting a natural, self-repair mechanism. This behavior points to the potential for hydrate reformation to contribute to long-term stability and containment in engineered hydrate-based storage systems [197]. Gauteplass et al. [198] demonstrated that hydrate seals can form efficiently in homogeneous media and continue to function even when localized dissociation occurs. These findings highlight that under the right thermodynamic and structural conditions, hydrate systems can exhibit both sealing and self-recovery behaviors, which may benefit CTES systems exposed to cyclic operations [198]. Understanding the sealing and self-reforming behavior of CO_2 hydrates in subsurface conditions provides confidence in their durability under repeated thermal cycling in CTES systems. Such resilience is critical to maintaining system performance and preventing hydrate degradation over time.

A comparative lifecycle cost (LCC) and energy analysis between a CO_2 hydrate-based cold thermal energy storage (CTES) system and a conventional district cooling system (DCS) reveals the long-term advantages of hydrate technology despite its high-pressure formation requirements. The comparison in Table 12 shows the long-term performance of CO_2 hydrate CTES and conventional ice-based cooling systems for large-scale applications. Using standard design assumptions and updated national data, key metrics such as energy use, operating costs, and emissions were calculated and compared over a 20-year period. The analysis reflects how system efficiency, electricity rates, and carbon intensity directly influence overall performance.

Table 12. Lifecycle analysis comparing CO_2 hydrate CTES to conventional ice/PCM systems in terms of net carbon savings.

Parameter	CO_2 Hydrate CTES	Ice-Based Systems	References/Calculations
Latent Heat	507 kJ/Kg	333 kJ/Kg	[192]
Formation Pressure	Below 0.5 to up to 5 MPa	0.1 MPa	Pressure required to form
Annual Cooling Required	44,917,178 kWh	44,917,178 kWh	[199,200]
Annual Energy Consumption	6,605,795 kWh	17,967,762 kWh	[174]
Average Electricity Rate	USD 0.1276/kWh	USD 0.1276/kWh	[201]

Table 12. *Cont.*

Parameter	CO_2 Hydrate CTES	Ice-Based Systems	References/Calculations
Annual Operating Cost	USD 842,899	USD 2,292,686	Energy cost = usage × rate
Operating Cost for 20 Years	USD 16,857,989	USD 45,853,729	Annual operating cost × 20
Sequestration	Yes, as this uses CO_2 gas. Subject of the nucleation conversion rate	No	Because CO_2 hydrate is formed from the mixture of CO_2 gas and water
Environmental Effects	Some promoters are toxic, but alternatives do exist	None	Depends on the promoters used
CO_2 Emissions	0 to 0.37 Kg CO_2/kWh, subject to the reactor designs, and the CO_2 recovery rate	0.37 kg CO_2/kWh	[202]
CO_2 Emissions for 20 Years	48,750,766 kg, see above calculations performed using CO_2 emission of 0.13 CO_2/kWh	132,602,085 kg	Annual energy consumption × CO_2 emissions × 20

Table 13 shows the sensitivity analysis to assess the influence of key techno-economic and environmental parameters on the lifecycle cost (LCC) of CO_2 hydrate CTES and conventional ice-based CTES systems. Sensitivity analysis is a systematic approach used to evaluate how variations in critical input parameters affect the output of a model, thereby identifying which factors most significantly impact system performance or economic viability. In this study, sensitivity analysis was conducted to understand the robustness of lifecycle cost estimates under uncertainties in energy prices, system efficiencies, emission factors, and carbon pricing scenarios. The selected parameters show the major cost drivers and risk factors over the system lifetime. The base case values as well as the low-end and high-end scenarios for each parameter were chosen based on recent literature, regulatory data, and experimental measurements, and a Tornado figure has been plotted based on the data acquired [203].

Table 13. Sensitivity Analysis Comparison between Hydrate and Ice based CTES System.

Parameter	Base Case	Low-End Scenario	High-End Scenario	Impact on Hydrate CTES LCC (%)	Impact on Ice CTES LCC (%)	References
Electricity Rate (USD/kWh)	0.1276	0.08	0.18	±41%	±58%	[201]
CO_2 Emission Factor (kg/kWh)	0.349 (Hydrate)	0.198	0.37 (Grid)	±43%	±43%	[204]
System COP (Hydrate CTES/Ice CTES)	7.03/3.5	5.0/3.0	8.5/4.0	+58%/+50%	+50%	[205]
Hydrate Conversion Efficiency	42.7%	30%	50%	+48%/N/A	-	[206]
Carbon Pricing (USD/ton CO_2)	50	30	150	±100	±200%	[207]

A tornado diagram is a graphical representation commonly used in sensitivity analysis to illustrate the relative influence of individual input variables on a model's output. It visually ranks parameters by the magnitude of their impact, along with the most influential factors shown at the top, resembling a tornado. Each horizontal bar represents the range of variation in the output metric (in this case, lifecycle cost) resulting from the low-end and high-end changes of a single parameter while holding all other variables constant [208]. Figure 28 shows the sensitivity of lifecycle cost (LCC) for CO_2 hydrate CTES (blue) and conventional ice-based CTES (red) systems to variations in electricity rates, CO_2 emission factors, system COP, hydrate conversion efficiency, and carbon pricing.

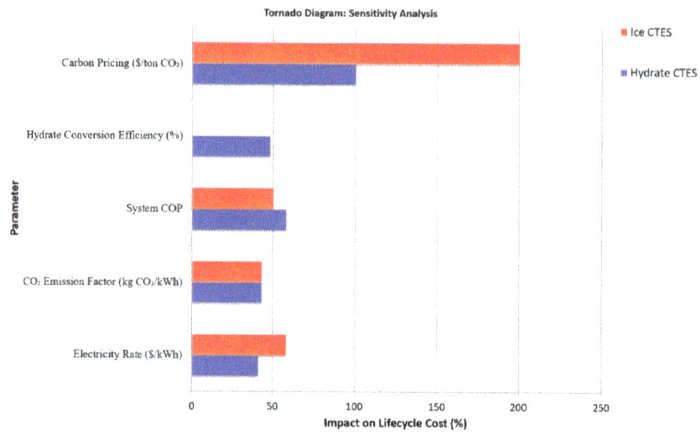

Figure 28. Tornado sensitivity analysis of hydrate CTES and ice-based CTES systems.

9. Research Gaps and Future Research Directions

CO_2 hydrates have emerged as promising materials for refrigeration applications and cold thermal energy storage (CTES) systems, offering significant advantages, such as high latent heat (500 kJ/kg) and a phase-change temperature range of 0–15 °C. Despite their potential, numerous challenges remain before CO_2 hydrate-based systems can be fully integrated into practical applications. One of the primary challenges is the need for efficient CO_2 gas storage with large absorption capacity and suitable absorption and regeneration conditions for cyclic operations. The absence of optimized CO_2 storage solutions has been a major barrier to the real-world implementation of CO_2 hydrate-based CTES systems.

Another issue is the low mass transfer during gas hydrate formation, which is caused by limited gas–water interface, hydrate blockages, and low CO_2 solubility. To address this, future designs should focus on increasing the gas–liquid interface and enhancing mass transfer, using methods such as porous media, spraying, circulation, or bubbling. Additionally, while the dissociation temperature of CO_2 hydrates is a reliable indicator of phase equilibrium, the formation temperature—affected by stochastic nucleation—is an essential property that directly impacts the efficiency of the chiller in a CTES system. This stochastic nature of hydrate formation complicates kinetic studies and poses challenges for system optimization.

Thermal hysteresis is another obstacle, where a lag between heating and cooling during phase transitions affects heat transfer between gas hydrates and the heat transfer fluid (HTF), leading to reduced system efficiency. The mechanisms behind thermal hysteresis in CO_2 hydrates and other phase-change materials (PCMs) are still unclear, and further research is needed to address this issue. Additionally, methods for determining the onset points of hydrate formation and dissociation are currently based on the visual observation of hydrate crystals, which can be inaccurate for quantitative studies. There is no unified method for such measurements, making it difficult to compare data across different studies.

CO_2 hydrate slurry has proven to be a suitable cooling carrier for air conditioning (AC) systems. However, challenges such as maintaining the mobility of the slurry, predicting its three-phase flow behavior in piping systems, and preventing hydrate dissociation due to pressure loss remain unresolved. Future research should focus on improving the rheological properties of the slurry and finding more accurate methods for predicting and managing three-phase flow in various piping configurations.

Thermodynamic promoters such as TBAB (Tetra-n-butylammonium bromide) have shown reliability in stabilizing CO_2 hydrate formation at lower pressures. However, it is hypothesized that when promoters are used, both CO_2–promoter double hydrates and promoter hydrates form. The formation of CO_2 double hydrates is often limited by mass transfer and the driving force, leading to the dominance of promoter hydrates. To enhance gas hydrate formation, it is necessary to develop better materials and systems that improve mass transfer in the presence of promoters. The long-term effects of additives such as TBAB in CTES systems are not yet fully understood. These additives may have side effects, including the potential to reduce energy storage density, impair system safety, or negatively impact other desired properties. Careful selection and further study of additives are crucial to the advancement of CO_2 hydrate-based CTES systems.

Research comparing CO_2 hydrates with conventional PCMs and water highlights key tradeoffs. CO_2 hydrates offer a higher dissociation enthalpy compared to water, making them more effective for energy storage. However, CO_2 hydrates require high pressures for formation, increasing operational costs and energy consumption. To address this, the use of thermodynamic and kinetic promoters can help lower the required formation pressure and accelerate hydrate growth. Advancements in experimental methods, such as incorporating ultrasonic waves or magnetic fields, could enhance hydrate formation efficiency without compromising stability.

Despite these challenges, CO_2 hydrates hold significant potential for revolutionizing cold thermal energy storage and carbon capture technologies. Future research efforts should focus on equipment optimization, additive development, and process integration to improve overall system efficiency. Structural improvements to heat exchangers, especially those designed to prevent hydrate layer formation on cold surfaces, will also be crucial in maintaining high heat transfer rates and preventing clogging. With further advancements, commercially viable hydrate slurries with improved flow and heat transfer characteristics can be developed and deployed. Such developments could significantly enhance energy efficiency in large-scale cooling applications, such as air conditioning and refrigeration systems, while reducing carbon emissions.

Author Contributions: S.B.A.: methodology, investigation, resources, data curation, writing—original draft preparation, and visualization; N.V.D.: conceptualization, methodology, writing—review and editing, supervision, and project administration; D.M.: writing—review and editing, and resources; J.K.: conceptualization, methodology, resources, writing—review and editing, supervision, project administration, and funding acquisition. All authors have read and agreed to the published version of the manuscript.

Funding: This work is related to the U.S. Department of Navy award N00014-23-1-2124, issued by the Office of Naval Research. The United States Government has a royalty-free license worldwide for all copyrightable material contained herein.

Conflicts of Interest: The authors declare no conflict of interest.

References

1. Lachheb, M.; Younsi, Z.; Youssef, N.; Bouadila, S. Enhancing Building Energy Efficiency and Thermal Performance with PCM-Integrated Brick Walls: A Comprehensive Review. *Build. Environ.* **2024**, *256*, 111476. [CrossRef]
2. Dincer, I. On Thermal Energy Storage Systems and Applications in Buildings. *Energy Build.* **2002**, *34*, 377–388. [CrossRef]
3. TCBU Editorial Thermal Control Magazine. 16 June 2023. Available online: https://www.thermalcontrolmagazine.com/ (accessed on 15 May 2025).
4. Franco, A.; Cillari, G. Energy Sustainability of Food Stores and Supermarkets through the Installation of PV Integrated Plants. *Energies* **2021**, *14*, 5678. [CrossRef]
5. Sharma, A.; Tyagi, V.V.; Chen, C.R.; Buddhi, D. Review on Thermal Energy Storage with Phase Change Materials and Applications. *Renew. Sustain. Energy Rev.* **2009**, *13*, 318–345. [CrossRef]
6. Jouhara, H.; Żabnieńska-Góra, A.; Khordehgah, N.; Ahmad, D.; Lipinski, T. Latent Thermal Energy Storage Technologies and Applications: A Review. *Int. J. Thermofluids* **2020**, *5–6*, 100039. [CrossRef]

7. Awan, H.T.A.; Kumar, L.; Wong, W.P.; Walvekar, R.; Khalid, M. Recent Progress and Challenges in MXene-Based Phase Change Material for Solar and Thermal Energy Applications. *Energies* **2023**, *16*, 1977. [CrossRef]
8. Lamrani, B.; Johannes, K.; Kuznik, F. Phase Change Materials Integrated into Building Walls: An Updated Review. *Renew. Sustain. Energy Rev.* **2021**, *140*, 110751. [CrossRef]
9. Nishad, S.; Krupa, I. Phase Change Materials for Thermal Energy Storage Applications in Greenhouses: A Review. *Sustain. Energy Technol. Assess.* **2022**, *52*, 102241. [CrossRef]
10. Bodner, J.; Koksharov, J.; Dammel, F.; Stephan, P. Analysis of Low-temperature Pumped Thermal Energy Storage Systems Based on a Transcritical CO_2 Charging Process. *Energy Sci. Eng.* **2023**, *11*, 3289–3306. [CrossRef]
11. Fournaison, L.; Delahaye, A.; Chatti, I.; Petitet, J.-P. CO_2 Hydrates in Refrigeration Processes. *Ind. Eng. Chem. Res.* **2004**, *43*, 6521–6526. [CrossRef]
12. Chen, H.; Han, B.; Lang, C.; Wen, M.; Fan, B.; Liu, Z. Hydrates for Cold Storage: Formation Characteristics, Stability, and Promoters. *Appl. Sci.* **2021**, *11*, 10470. [CrossRef]
13. Jäger, A.; Vinš, V.; Gernert, J.; Span, R.; Hrubý, J. Phase Equilibria with Hydrate Formation in H_2O+CO_2 Mixtures Modeled with Reference Equations of State. *Fluid Phase Equilib.* **2013**, *338*, 100–113. [CrossRef]
14. Liu, X.; Cao, Q.; Xu, D.; Luo, S.; Guo, R. Improved Methane Storage Capacity of Methane Hydrate Promoted by Vesicles from Carboxylate Surfactants and Quaternary Ammonium. *J. Nat. Gas. Sci. Eng.* **2021**, *93*, 103990. [CrossRef]
15. Izquierdo-Ruiz, F.; Otero-de-la-Roza, A.; Contreras-García, J.; Prieto-Ballesteros, O.; Recio, J. Effects of the CO_2 Guest Molecule on the SI Clathrate Hydrate Structure. *Materials* **2016**, *9*, 777. [CrossRef]
16. Sloan, E.D., Jr.; Koh, C.A. *Clathrate Hydrates of Natural Gases*; CRC Press: Boca Raton, FL, USA, 2007; ISBN 9780429129148.
17. Sun, Q.; Kang, Y.T. Review on CO_2 Hydrate Formation/Dissociation and Its Cold Energy Application. *Renew. Sustain. Energy Rev.* **2016**, *62*, 478–494. [CrossRef]
18. Sinehbaghizadeh, S.; Saptoro, A.; Mohammadi, A.H. CO_2 Hydrate Properties and Applications: A State of the Art. *Prog. Energy Combust. Sci.* **2022**, *93*, 101026. [CrossRef]
19. Ma, Z.W.; Zhang, P.; Bao, H.S.; Deng, S. Review of Fundamental Properties of CO_2 Hydrates and CO_2 Capture and Separation Using Hydration Method. *Renew. Sustain. Energy Rev.* **2016**, *53*, 1273–1302. [CrossRef]
20. Wroblewsk, S. On the Composition of the Hydrate of Carbonic Acid. *Acad. Sci.* **1882**, *94*, 954–958. Available online: https://www.chemeurope.com/en/encyclopedia/Zygmunt_Florenty_Wr%C3%B3blewski.html (accessed on 15 May 2025). (In French).
21. Ke, W.; Svartaas, T.M.; Chen, D. A Review of Gas Hydrate Nucleation Theories and Growth Models. *J. Nat. Gas. Sci. Eng.* **2019**, *61*, 169–196. [CrossRef]
22. Nguyen, N.N.; Galib, M.; Nguyen, A.V. Critical Review on Gas Hydrate Formation at Solid Surfaces and in Confined Spaces—Why and How Does Interfacial Regime Matter? *Energy Fuels* **2020**, *34*, 6751–6760. [CrossRef]
23. Farhang, F.; Nguyen, A.V.; Sewell, K.B. Fundamental Investigation of the Effects of Hydrophobic Fumed Silica on the Formation of Carbon Dioxide Gas Hydrates. *Energy Fuels* **2014**, *28*, 7025–7037. [CrossRef]
24. Sowa, B.; Maeda, N. Statistical Study of the Memory Effect in Model Natural Gas Hydrate Systems. *J. Phys. Chem. A* **2015**, *119*, 10784–10790. [CrossRef] [PubMed]
25. Uchida, T.; Yamazaki, K.; Gohara, K. Gas Nanobubbles as Nucleation Acceleration in the Gas-Hydrate Memory Effect. *J. Phys. Chem. C* **2016**, *120*, 26620–26629. [CrossRef]
26. Kou, X.; Feng, J.-C.; Li, X.-S.; Wang, Y.; Chen, Z.-Y. Memory Effect of Gas Hydrate: Influencing Factors of Hydrate Reformation and Dissociation Behaviors. *Appl. Energy* **2022**, *306*, 118015. [CrossRef]
27. Bollengier, O.; Choukroun, M.; Grasset, O.; Le Menn, E.; Bellino, G.; Morizet, Y.; Bezacier, L.; Oancea, A.; Taffin, C.; Tobie, G. Phase Equilibria in the H_2O–CO_2 System between 250–330K and 0–1.7GPa: Stability of the CO_2 Hydrates and H_2O-Ice VI at CO_2 Saturation. *Geochim. Cosmochim. Acta* **2013**, *119*, 322–339. [CrossRef]
28. Zhang, Z.; Guo, G.-J.; Wu, N.; Kusalik, P.G. Molecular Insights into Guest and Composition Dependence of Mixed Hydrate Nucleation. *J. Phys. Chem. C* **2020**, *124*, 25078–25086. [CrossRef]
29. Veluswamy, H.P.; Kumar, R.; Linga, P. Hydrogen Storage in Clathrate Hydrates: Current State of the Art and Future Directions. *Appl. Energy* **2014**, *122*, 112–132. [CrossRef]
30. Zhou, X.; Zhou, J.; Chen, P.; Wen, H.; Zang, X.; Fan, S.; Liang, D. Spectral Analysis on the Raman Peaks of CH_4, CO_2, and N_2 in the Hydrate Phase. *Energy Fuels* **2025**, *39*, 1579–1587. [CrossRef]
31. Dholabhai, P.D.; Kalogerakis, N.; Bishnoi, P.R. Equilibrium Conditions for Carbon Dioxide Hydrate Formation In Aqueous Electrolyte Solutions. *J. Chem. Eng. Data* **1993**, *38*, 650–654. [CrossRef]
32. Hammerschmidt, E.G. Formation of Gas Hydrates in Natural Gas Transmission Lines. *Ind. Eng. Chem.* **1934**, *26*, 851–855. [CrossRef]
33. Ng, H.-J.; Robinson, D.B. Hydrate Formation in Systems Containing Methane, Ethane, Propane, Carbon Dioxide or Hydrogen Sulfide in the Presence of Methanol. *Fluid Phase Equilib.* **1985**, *21*, 145–155. [CrossRef]
34. Miller, S.L.; Smythe, W.D. Carbon Dioxide Clathrate in the Martian Ice Cap. *Science* **1970**, *170*, 531–533. [CrossRef]

35. Adamson, A.W.; Jones, B.R. Physical Adsorption of Vapors on Ice. IV. Carbon Dioxide. *J. Colloid. Interface Sci.* **1971**, *37*, 831–835. [CrossRef]
36. Wendland, M.; Hasse, H.; Maurer, G. Experimental Pressure–Temperature Data on Three- and Four-Phase Equilibria of Fluid, Hydrate, and Ice Phases in the System Carbon Dioxide–Water. *J. Chem. Eng. Data* **1999**, *44*, 901–906. [CrossRef]
37. Yasuda, K.; Ohmura, R. Phase Equilibrium for Clathrate Hydrates Formed with Methane, Ethane, Propane, or Carbon Dioxide at Temperatures below the Freezing Point of Water. *J. Chem. Eng. Data* **2008**, *53*, 2182–2188. [CrossRef]
38. Kimura, H.; Fuseya, G.; Takeya, S.; Hachikubo, A. Carbon Isotope Fractionation during the Formation of CO_2 Hydrate and Equilibrium Pressures of $12CO_2$ and $13CO_2$ Hydrates. *Molecules* **2021**, *26*, 4215. [CrossRef]
39. Ohno, H.; Oyabu, I.; Iizuka, Y.; Hondoh, T.; Narita, H.; Nagao, J. Dissociation Behavior of C_2H_6 Hydrate at Temperatures below the Ice Point: Melting to Liquid Water Followed by Ice Nucleation. *J. Phys. Chem. A* **2011**, *115*, 8889–8894. [CrossRef]
40. Melnikov, V.P.; Nesterov, A.N.; Podenko, L.S.; Reshetnikov, A.M.; Shalamov, V.V. NMR Evidence of Supercooled Water Formation during Gas Hydrate Dissociation below the Melting Point of Ice. *Chem. Eng. Sci.* **2012**, *71*, 573–577. [CrossRef]
41. Longhi, J. Phase Equilibria in the System CO_2-H_2O I: New Equilibrium Relations at Low Temperatures. *Geochim. Cosmochim. Acta* **2005**, *69*, 529–539. [CrossRef]
42. Al Ghafri, S.Z.S.; Forte, E.; Maitland, G.C.; Rodriguez-Henríquez, J.J.; Trusler, J.P.M. Experimental and Modeling Study of the Phase Behavior of (Methane + CO_2 + Water) Mixtures. *J. Phys. Chem. B* **2014**, *118*, 14461–14478. [CrossRef]
43. Larson, S.D. Phase Studies of the Two-Component Carbon-Dioxide—Water System Involving the Carbon-Dioxide Hydrate. Ph.D. Thesis, University of Illinois Urbana-Champaign, Champaign, IL, USA, 1955.
44. Fan, S.-S.; Guo, T.-M. Hydrate Formation of CO_2-Rich Binary and Quaternary Gas Mixtures in Aqueous Sodium Chloride Solutions. *J. Chem. Eng. Data* **1999**, *44*, 829–832. [CrossRef]
45. Semenov, A.; Mendgaziev, R.; Stoporev, A.; Istomin, V.; Tulegenov, T.; Yarakhmedov, M.; Novikov, A.; Vinokurov, V. Direct Measurement of the Four-Phase Equilibrium Coexistence Vapor–Aqueous Solution–Ice–Gas Hydrate in Water–Carbon Dioxide System. *Int. J. Mol. Sci.* **2023**, *24*, 9321. [CrossRef]
46. Babu, P.; Yao, M.; Datta, S.; Kumar, R.; Linga, P. Thermodynamic and Kinetic Verification of Tetra-*n*-Butyl Ammonium Nitrate ($TBANO_3$) as a Promoter for the Clathrate Process Applicable to Precombustion Carbon Dioxide Capture. *Environ. Sci. Technol.* **2014**, *48*, 3550–3558. [CrossRef]
47. Choi, J.W.; Chung, J.T.; Kang, Y.T. CO_2 Hydrate Formation at Atmospheric Pressure Using High Efficiency Absorbent and Surfactants. *Energy* **2014**, *78*, 869–876. [CrossRef]
48. Nashed, O.; Partoon, B.; Lal, B.; Sabil, K.M.; Shariff, A.M. Investigation of Functionalized Carbon Nanotubes' Performance on Carbon Dioxide Hydrate Formation. *Energy* **2019**, *174*, 602–610. [CrossRef]
49. Gambelli, A.M.; Stornelli, G.; Di Schino, A.; Rossi, F. Methane and Carbon Dioxide Hydrate Formation and Dissociation in Presence of a Pure Quartz Porous Framework Impregnated with CuSn12 Metallic Powder: An Experimental Report. *Materials* **2021**, *14*, 5115. [CrossRef]
50. Lv, X.; Lu, D.; Liu, Y.; Zhou, S.; Zuo, J.; Jin, H.; Shi, B.; Li, E. Study on Methane Hydrate Formation in Gas–Water Systems with a New Compound Promoter. *RSC Adv.* **2019**, *9*, 33506–33518. [CrossRef]
51. Gross, A.; Pisipati, A.; Mahajan, D. Investigating the Effect of Carbon Dioxide to Propane Ratio on Incubation Time and Gas Hydrate Conversion Yield. In Proceedings of the AIChE Annual Meeting 2024, San Diego, CA, USA, 27–31 October 2024.
52. Lee, Y.; Kim, H.; Lee, W.; Kang, D.W.; Lee, J.W.; Ahn, Y.-H. Thermodynamic and Kinetic Properties of CO_2 Hydrates and Their Applications in CO_2 Capture and Separation. *J. Environ. Chem. Eng.* **2023**, *11*, 110933. [CrossRef]
53. Majid, A.A.A.; Worley, J.; Koh, C.A. Thermodynamic and Kinetic Promoters for Gas Hydrate Technological Applications. *Energy Fuels* **2021**, *35*, 19288–19301. [CrossRef]
54. Circone, S.; Stern, L.A.; Kirby, S.H.; Durham, W.B.; Chakoumakos, B.C.; Rawn, C.J.; Rondinone, A.J.; Ishii, Y. CO_2 Hydrate: Synthesis, Composition, Structure, Dissociation Behavior, and a Comparison to Structure I CH_4 Hydrate. *J. Phys. Chem. B* **2003**, *107*, 5529–5539. [CrossRef]
55. Zhou, X.; Long, Z.; He, Y.; Shen, X.; Liang, D. Phase Equilibria and the Crystallographic Properties of TBAB–CO_2 Semiclathrate Hydrates. *J. Chem. Eng. Data* **2018**, *63*, 1249–1255. [CrossRef]
56. Arjmandi, M.; Chapoy, A.; Tohidi, B. Equilibrium Data of Hydrogen, Methane, Nitrogen, Carbon Dioxide, and Natural Gas in Semi-Clathrate Hydrates of Tetrabutyl Ammonium Bromide. *J. Chem. Eng. Data* **2007**, *52*, 2153–2158. [CrossRef]
57. Herslund, P.J.; Daraboina, N.; Thomsen, K.; Abildskov, J.; von Solms, N. Measuring and Modelling of the Combined Thermodynamic Promoting Effect of Tetrahydrofuran and Cyclopentane on Carbon Dioxide Hydrates. *Fluid Phase Equilib.* **2014**, *381*, 20–27. [CrossRef]
58. Zhang, J.S.; Lee, J.W. Equilibrium of Hydrogen + Cyclopentane and Carbon Dioxide + Cyclopentane Binary Hydrates. *J. Chem. Eng. Data* **2009**, *54*, 659–661. [CrossRef]
59. Lee, Y.-J.; Kawamura, T.; Yamamoto, Y.; Yoon, J.-H. Phase Equilibrium Studies of Tetrahydrofuran (THF) + CH_4, THF + CO_2, CH_4 + CO_2, and THF + CO_2 + CH_4 Hydrates. *J. Chem. Eng. Data* **2012**, *57*, 3543–3548. [CrossRef]

60. Delahaye, A.; Fournaison, L.; Marinhas, S.; Chatti, I.; Petitet, J.-P.; Dalmazzone, D.; Fürst, W. Effect of THF on Equilibrium Pressure and Dissociation Enthalpy of CO_2 Hydrates Applied to Secondary Refrigeration. *Ind. Eng. Chem. Res.* **2006**, *45*, 391–397. [CrossRef]
61. Adisasmito, S.; Sloan, E.D. Hydrates of Hydrocarbon Gases Containing Carbon Dioxide. *J. Chem. Eng. Data* **1992**, *37*, 343–349. [CrossRef]
62. He, Y.; Sun, M.-T.; Chen, C.; Zhang, G.-D.; Chao, K.; Lin, Y.; Wang, F. Surfactant-Based Promotion to Gas Hydrate Formation for Energy Storage. *J. Mater. Chem. A Mater.* **2019**, *7*, 21634–21661. [CrossRef]
63. Rehman, A.N.; Bavoh, C.B.; Pendyala, R.; Lal, B. Research Advances, Maturation, and Challenges of Hydrate-Based CO_2 Sequestration in Porous Media. *ACS Sustain. Chem. Eng.* **2021**, *9*, 15075–15108. [CrossRef]
64. Wang, P.; Teng, Y.; Zhu, J.; Bao, W.; Han, S.; Li, Y.; Zhao, Y.; Xie, H. Review on the Synergistic Effect between Metal–Organic Frameworks and Gas Hydrates for CH_4 Storage and CO_2 Separation Applications. *Renew. Sustain. Energy Rev.* **2022**, *167*, 112807. [CrossRef]
65. Kang, S.-P.; Lee, J.-W. Kinetic Behaviors of CO_2 Hydrates in Porous Media and Effect of Kinetic Promoter on the Formation Kinetics. *Chem. Eng. Sci.* **2010**, *65*, 1840–1845. [CrossRef]
66. Molokitina, N.S.; Nesterov, A.N.; Podenko, L.S.; Reshetnikov, A.M. Carbon Dioxide Hydrate Formation with SDS: Further Insights into Mechanism of Gas Hydrate Growth in the Presence of Surfactant. *Fuel* **2019**, *235*, 1400–1411. [CrossRef]
67. Kumar, A.; Sakpal, T.; Linga, P.; Kumar, R. Influence of Contact Medium and Surfactants on Carbon Dioxide Clathrate Hydrate Kinetics. *Fuel* **2013**, *105*, 664–671. [CrossRef]
68. Arora, A.; Kumar, A.; Bhattacharjee, G.; Kumar, P.; Balomajumder, C. Effect of Different Fixed Bed Media on the Performance of Sodium Dodecyl Sulfate for Hydrate Based CO_2 Capture. *Mater. Des.* **2016**, *90*, 1186–1191. [CrossRef]
69. Okutani, K.; Kuwabara, Y.; Mori, Y.H. Surfactant Effects on Hydrate Formation in an Unstirred Gas/Liquid System: An Experimental Study Using Methane and Sodium Alkyl Sulfates. *Chem. Eng. Sci.* **2008**, *63*, 183–194. [CrossRef]
70. Nesterov, A.N.; Reshetnikov, A.M.; Manakov, A.Y.U.; Adamova, T.P. Synergistic Effect of Combination of Surfactant and Oxide Powder on Enhancement of Gas Hydrates Nucleation. *J. Energy Chem.* **2017**, *26*, 808–814. [CrossRef]
71. Liu, Y.; Chen, B.; Chen, Y.; Zhang, S.; Guo, W.; Cai, Y.; Tan, B.; Wang, W. Methane Storage in a Hydrated Form as Promoted by Leucines for Possible Application to Natural Gas Transportation and Storage. *Energy Technol.* **2015**, *3*, 815–819. [CrossRef]
72. Prasad, P.S.R.; Sai Kiran, B. Clathrate Hydrates of Greenhouse Gases in the Presence of Natural Amino Acids: Storage, Transportation and Separation Applications. *Sci. Rep.* **2018**, *8*, 8560. [CrossRef]
73. Cho, S.-G.; Kim, K.; Sa, J.-H. Promotion Effects of Hydrophobic Amino Acids on CO_2 Hydrate Formation Kinetics under Isochoric/Isobaric and Stirred/Nonstirred Conditions. *Energy Fuels* **2024**, *38*, 526–535. [CrossRef]
74. Das, S.K.; Choi, S.U.S.; Patel, H.E. Heat Transfer in Nanofluids—A Review. *Heat. Transf. Eng.* **2006**, *27*, 3–19. [CrossRef]
75. Wei, B.; Zou, C.; Li, X. Experimental Investigation on Stability and Thermal Conductivity of Diathermic Oil Based TiO_2 Nanofluids. *Int. J. Heat Mass Transf.* **2017**, *104*, 537–543. [CrossRef]
76. Bai, Y.; Lu, H.; Ma, F.; He, Y.; Wang, F. Carbon Nanotube-Based Nanopromoters for Gas Hydrate Formation. *J. Nat. Gas. Sci. Eng.* **2021**, *94*, 104109. [CrossRef]
77. Wang, Y.; Niu, A.; Jiao, W.; Chen, J.; Zhang, P.; Li, J. Effects of the Presence of Promoters Sodium Dodecyl Sulfate, Nanographite, and Tetra-*n*-Butylammonium Bromide on the Formation of CO_2 Hydrates. *Energy Fuels* **2022**, *36*, 10269–10277. [CrossRef]
78. Said, S.; Govindaraj, V.; Herri, J.-M.; Ouabbas, Y.; Khodja, M.; Belloum, M.; Sangwai, J.S.; Nagarajan, R. A Study on the Influence of Nanofluids on Gas Hydrate Formation Kinetics and Their Potential: Application to the CO_2 Capture Process. *J. Nat. Gas. Sci. Eng.* **2016**, *32*, 95–108. [CrossRef]
79. Mohammadi, M.; Haghtalab, A.; Fakhroueian, Z. Experimental Study and Thermodynamic Modeling of CO_2 Gas Hydrate Formation in Presence of Zinc Oxide Nanoparticles. *J. Chem. Thermodyn.* **2016**, *96*, 24–33. [CrossRef]
80. Mohammadi, A.; Manteghian, M.; Haghtalab, A.; Mohammadi, A.H.; Rahmati-Abkenar, M. Kinetic Study of Carbon Dioxide Hydrate Formation in Presence of Silver Nanoparticles and SDS. *Chem. Eng. J.* **2014**, *237*, 387–395. [CrossRef]
81. Jiao, L.; Wan, R.; Wang, Z. Experimental Investigation of CO_2 Hydrate Dissociation in Silica Nanoparticle System with Different Thermal Conductivity. *Int. J. Thermophys.* **2021**, *42*, 170. [CrossRef]
82. Li, X.-S.; Xia, Z.-M.; Chen, Z.-Y.; Yan, K.-F.; Li, G.; Wu, H.-J. Gas Hydrate Formation Process for Capture of Carbon Dioxide from Fuel Gas Mixture. *Ind. Eng. Chem. Res.* **2010**, *49*, 11614–11619. [CrossRef]
83. Ma, S.; Tian, X.; Liu, Z.; Wu, Z.; Li, G.; Guan, X.; Zheng, J.; Yang, M. Formation and Decomposition Characteristics of CO_2+ TBAB Hydrate for a Safer CO_2 Storage. *Energy* **2024**, *307*, 132801. [CrossRef]
84. Lirio CFda, S.; Pessoa, F.L.P.; Uller, A.M.C. Storage Capacity of Carbon Dioxide Hydrates in the Presence of Sodium Dodecyl Sulfate (SDS) and Tetrahydrofuran (THF). *Chem. Eng. Sci.* **2013**, *96*, 118–123. [CrossRef]
85. Torré, J.-P.; Haillot, D.; Rigal, S.; de Souza Lima, R.; Dicharry, C.; Bedecarrats, J.-P. 1,3 Dioxolane versus Tetrahydrofuran as Promoters for CO_2-Hydrate Formation: Thermodynamics Properties, and Kinetics in Presence of Sodium Dodecyl Sulfate. *Chem. Eng. Sci.* **2015**, *126*, 688–697. [CrossRef]

86. Zhang, M.; Sun, B.; Liu, S.; Chen, L.; Gao, Y.; Wang, Z. Review on Cooperative Effect of Compound Additives on CO_2 Hydrate Formation: Recent Advances and Future Directions. *Energy Fuels* **2023**, *37*, 5667–5688. [CrossRef]
87. Prasad, P.S.R.; Kiran, B.S. Are the Amino Acids Thermodynamic Inhibitors or Kinetic Promoters for Carbon Dioxide Hydrates? *J. Nat. Gas Sci. Eng.* **2018**, *52*, 461–466. [CrossRef]
88. Bhuyan, A.; Ahmaruzzaman, M. Metal-Organic Frameworks: A New Generation Potential Material for Aqueous Environmental Remediation. *Inorg. Chem. Commun.* **2022**, *140*, 109436. [CrossRef]
89. Cao, Y.; Yang, P.; Zhao, R.; Wang, F. Recent Progress in Enzyme Immobilization to Metal–Organic Frameworks to Enhance the CO_2 Conversion Efficiency. *Molecules* **2025**, *30*, 251. [CrossRef]
90. Saikia, T.; Al-Jaberi, J.; Al Hamad, J.S.; Helal, A.; Sultan, A. Application of Metal Organic Frameworks for the Inhibition of CO_2 Hydrates in Gas Dominated Pipelines. *Gas. Sci. Eng.* **2023**, *110*, 204879. [CrossRef]
91. Kazemi, A.; Pordsari, M.A.; Tamtaji, M.; Manteghi, F.; Ghaemi, A.; Rohani, S.; Goddard, W.A. Environmentally Friendly Synthesis and Morphology Engineering of Mixed-Metal MOF for Outstanding CO_2 Capture Efficiency. *Chem. Eng. J.* **2025**, *505*, 158951. [CrossRef]
92. Pandey, J.; von Solms, N. Metal–Organic Frameworks and Gas Hydrate Synergy: A Pandora's Box of Unanswered Questions and Revelations. *Energies* **2022**, *16*, 111. [CrossRef]
93. Wang, X.; Shao, S.; Yang, G.; Yan, Q.; Yuan, H.; Chen, C.; Wang, F. Experimental Verification and 4E Analysis of a Hydrate-Based CO_2 Capture System for Flue Gas Decarbonization. *Carbon Neutrality* **2025**, *4*, 11. [CrossRef]
94. Phan, A.; Striolo, A. Chemical Promoter Performance for CO_2 Hydrate Growth: A Molecular Perspective. *Energy Fuels* **2023**, *37*, 6002–6011. [CrossRef]
95. Pang, W.; Ge, Y.; Chen, M.; Zhang, X.; Wen, H.; Fu, Q.; Lei, X.; Li, Q.; Zhou, S. Large-Scale Experimental Investigation of Hydrate-Based Carbon Dioxide Sequestration. *Energies* **2024**, *17*, 3103. [CrossRef]
96. Minnesota Department of Health (MDH). *Tetrahydrofuran Toxicological Summary Sheet: Health-Based Guidance for Water*; Minnesota Department of Health: St. Paul, MN, USA, 2018.
97. New Jersey Department of Health and Senior Services (NJDHSS). *Hazardous Substance Fact Sheet: Tetrahydrofuran (THF)*; New Jersey Department of Health and Senior Services: Trenton, NJ, USA, 2004.
98. Veluswamy, H.P.; Chin, W.I.; Linga, P. Clathrate Hydrates for Hydrogen Storage: The Impact of Tetrahydrofuran, Tetra-n-Butylammonium Bromide and Cyclopentane as Promoters on the Macroscopic Kinetics. *Int. J. Hydrogen Energy* **2014**, *39*, 16234–16243. [CrossRef]
99. Yodpetch, V.; Inkong, K.; Veluswamy, H.P.; Kulprathipanja, S.; Rangsunvigit, P.; Linga, P. Investigation on the Amino Acid-Assisted CO_2 Hydrates: A Promising Step Toward Hydrate-Based Decarbonization. *ACS Sustain. Chem. Eng.* **2023**, *11*, 2797–2809. [CrossRef]
100. Pivezhani, F.; Roosta, H.; Dashti, A.; Mazloumi, S.H. Investigation of CO_2 Hydrate Formation Conditions for Determining the Optimum CO_2 Storage Rate and Energy: Modeling and Experimental Study. *Energy* **2016**, *113*, 215–226. [CrossRef]
101. Partoon, B.; Malik, S.N.A.; Azemi, M.H.; Sabil, K.M. Experimental Investigations on the Potential of SDS as Low-dosage Promoter for Carbon Dioxide Hydrate Formation. *Asia Pac. J. Chem. Eng.* **2013**, *8*, 916–921. [CrossRef]
102. Clarke, M.A.; Bishnoi, P.R. Determination of the Intrinsic Kinetics of CO_2 Gas Hydrate Formation Using in Situ Particle Size Analysis. *Chem. Eng. Sci.* **2005**, *60*, 695–709. [CrossRef]
103. Partoon, B.; Javanmardi, J. Effect of Mixed Thermodynamic and Kinetic Hydrate Promoters on Methane Hydrate Phase Boundary and Formation Kinetics. *J. Chem. Eng. Data* **2013**, *58*, 501–509. [CrossRef]
104. Martín, M.; Montes, F.J.; Galán, M.A. Bubbling Process in Stirred Tank Reactors II: Agitator Effect on the Mass Transfer Rates. *Chem. Eng. Sci.* **2008**, *63*, 3223–3234. [CrossRef]
105. Hould, N.; Elanany, M.S.; Aleisa, R.M.; Al-Majnouni, K.A.; Al-Malki, A.; Abba, I. Evaluating Polymeric Inhibitors of Ethane Clathrate Hydrates. *J. Nat. Gas Sci. Eng.* **2015**, *24*, 543–549. [CrossRef]
106. Linga, P.; Kumar, R.; Lee, J.D.; Ripmeester, J.; Englezos, P. A New Apparatus to Enhance the Rate of Gas Hydrate Formation: Application to Capture of Carbon Dioxide. *Int. J. Greenh. Gas Control* **2010**, *4*, 630–637. [CrossRef]
107. Takahashi, M.; Kawamura, T.; Yamamoto, Y.; Ohnari, H.; Himuro, S.; Shakutsui, H. Effect of Shrinking Microbubble on Gas Hydrate Formation. *J. Phys. Chem. B* **2003**, *107*, 2171–2173. [CrossRef]
108. Tang, L.-G.; Li, X.-S.; Feng, Z.-P.; Lin, Y.-L.; Fan, S.-S. Natural Gas Hydrate Formation in an Ejector Loop Reactor: Preliminary Study. *Ind. Eng. Chem. Res.* **2006**, *45*, 7934–7940. [CrossRef]
109. Ghaani, M.R.; Schicks, J.M.; English, N.J. A Review of Reactor Designs for Hydrogen Storage in Clathrate Hydrates. *Appl. Sci.* **2021**, *11*, 469. [CrossRef]
110. Xin, Y.; Zhang, J.; He, Y.; Wang, C. Modelling and Experimental Study of Hydrate Formation Kinetics of Natural Gas-water-surfactant System in a Multi-tube Bubble Column Reactor. *Can. J. Chem. Eng.* **2019**, *97*, 2765–2776. [CrossRef]
111. Ghaani, M.R.; Kusalik, P.G.; English, N.J. Massive Generation of Metastable Bulk Nanobubbles in Water by External Electric Fields. *Sci. Adv.* **2020**, *6*, eaaz0094. [CrossRef]

112. Yang, D.; Le, L.A.; Martinez, R.J.; Currier, R.P.; Spencer, D.F.; Deppe, G. Heat Transfer During CO_2 Hydrate Formation in a Continuous Flow Reactor. *Energy Fuels* **2008**, *22*, 2649–2659. [CrossRef]
113. Linga, P.; Haligva, C.; Nam, S.C.; Ripmeester, J.A.; Englezos, P. Gas Hydrate Formation in a Variable Volume Bed of Silica Sand Particles. *Energy Fuels* **2009**, *23*, 5496–5507. [CrossRef]
114. Kumar, A.; Sakpal, T.; Linga, P.; Kumar, R. Enhanced Carbon Dioxide Hydrate Formation Kinetics in a Fixed Bed Reactor Filled with Metallic Packing. *Chem. Eng. Sci.* **2015**, *122*, 78–85. [CrossRef]
115. Seo, Y.-T.; Lee, H.; Yoon, J.-H. Hydrate Phase Equilibria of the Carbon Dioxide, Methane, and Water System. *J. Chem. Eng. Data* **2001**, *46*, 381–384. [CrossRef]
116. Linga, P.; Kumar, R.; Englezos, P. Gas Hydrate Formation from Hydrogen/Carbon Dioxide and Nitrogen/Carbon Dioxide Gas Mixtures. *Chem. Eng. Sci.* **2007**, *62*, 4268–4276. [CrossRef]
117. Seo, Y.; Kang, S.-P. Enhancing CO_2 Separation for Pre-Combustion Capture with Hydrate Formation in Silica Gel Pore Structure. *Chem. Eng. J.* **2010**, *161*, 308–312. [CrossRef]
118. Li, X.-S.; Xu, C.-G.; Chen, Z.-Y.; Wu, H.-J. Tetra-n-Butyl Ammonium Bromide Semi-Clathrate Hydrate Process for Post-Combustion Capture of Carbon Dioxide in the Presence of Dodecyl Trimethyl Ammonium Chloride. *Energy* **2010**, *35*, 3902–3908. [CrossRef]
119. Babu, P.; Kumar, R.; Linga, P. Pre-Combustion Capture of Carbon Dioxide in a Fixed Bed Reactor Using the Clathrate Hydrate Process. *Energy* **2013**, *50*, 364–373. [CrossRef]
120. Sun, Q.; Chen, G.; Guo, X.; Liu, A. Experiments on the Continuous Separation of Gas Mixtures via Dissolution and Hydrate Formation in the Presence of THF. *Fluid Phase Equilib.* **2014**, *361*, 250–256. [CrossRef]
121. Li, A.; Jiang, L.; Tang, S. An Experimental Study on Carbon Dioxide Hydrate Formation Using a Gas-Inducing Agitated Reactor. *Energy* **2017**, *134*, 629–637. [CrossRef]
122. Lozada Garcia, N.; SDamaceno, D.; Fardelone, L.C.; Ferreira de Mattos Silvares, A.; Ferreira Del Pintor, J.P.; Teixeira Mendes, A.H.; Costa, M.F.; Lopes, J.C.B.; Won Park, S.; Nunhez, J.R. Characteristics of Continuous CO_2 Hydrate Formation Process Using a NetMIX Reactor. *Chem. Eng. Sci.* **2023**, *280*, 119023. [CrossRef]
123. Shi, B.-H.; Chai, S.; Wang, L.-Y.; Lv, X.; Liu, H.-S.; Wu, H.-H.; Wang, W.; Yu, D.; Gong, J. Viscosity Investigation of Natural Gas Hydrate Slurries with Anti-Agglomerants Additives. *Fuel* **2016**, *185*, 323–338. [CrossRef]
124. Kumar, A.; Khatri, D.; Lee, J.D.; Kumar, R. Crystallization Kinetics for Carbon Dioxide Gas Hydrate in Fixed Bed and Stirred Tank Reactor. *Korean J. Chem. Eng.* **2016**, *33*, 1922–1930. [CrossRef]
125. Wu, Y.; Zhi, C.; Wang, Z.; Chen, Y.; Wang, C.; Chen, Q.; Tan, G.; Ming, T. Enhanced Thermal and Mechanical Performance of 3D Architected Micro-Channel Heat Exchangers. *Heliyon* **2023**, *9*, e13902. [CrossRef]
126. Berrouk, A.S.; Jiang, P.; Safiyullah, F.; Basha, M. CFD Modelling of Hydrate Slurry Flow in a Pipeline Based on Euler-Euler Approach. *Progress. Comput. Fluid. Dyn. An. Int. J.* **2020**, *20*, 156. [CrossRef]
127. Wang, X.; Dennis, M. Phase Equilibrium and Formation Behaviour of CO_2-TBAB Semi-Clathrate Hydrate at Low Pressures for Cold Storage Air Conditioning Applications. *Chem. Eng. Sci.* **2016**, *155*, 294–305. [CrossRef]
128. Oancea, A.; Grasset, O.; Le Menn, E.; Bollengier, O.; Bezacier, L.; Le Mouélic, S.; Tobie, G. Laboratory Infrared Reflection Spectrum of Carbon Dioxide Clathrate Hydrates for Astrophysical Remote Sensing Applications. *Icarus* **2012**, *221*, 900–910. [CrossRef]
129. Golikov, O.; Yerezhep, D.; Yerlanov, T. A Semi-Automatic Examination of CO_2 Structures in Thin Films at Low Temperatures. *Int. J. Math. Phys.* **2023**, *14*, 53–60. [CrossRef]
130. Douïeb, S.; Fradette, L.; Bertrand, F.; Haut, B. Impact of the Fluid Flow Conditions on the Formation Rate of Carbon Dioxide Hydrates in a Semi-batch Stirred Tank Reactor. *AIChE J.* **2015**, *61*, 4387–4401. [CrossRef]
131. Uchida, T.; Sum, A.K. IR and R Aman Spectroscopy of Clathrate Hydrates. In *Clathrate Hydrates*; Wiley: Hoboken, NJ, USA, 2022; pp. 569–629.
132. Boufares, A.; Provost, E.; Dalmazzone, D.; Osswald, V.; Clain, P.; Delahaye, A.; Fournaison, L. Kinetic Study of CO_2 Hydrates Crystallization: Characterization Using FTIR/ATR Spectroscopy and Contribution Modeling of Equilibrium/Non-Equilibrium Phase-Behavior. *Chem. Eng. Sci.* **2018**, *192*, 371–379. [CrossRef]
133. Takeya, S.; Kida, M.; Minami, H.; Sakagami, H.; Hachikubo, A.; Takahashi, N.; Shoji, H.; Soloviev, V.; Wallmann, K.; Biebow, N.; et al. Structure and Thermal Expansion of Natural Gas Clathrate Hydrates. *Chem. Eng. Sci.* **2006**, *61*, 2670–2674. [CrossRef]
134. Takeya, S.; Muromachi, S.; Yamamoto, Y.; Umeda, H.; Matsuo, S. Preservation of CO_2 Hydrate under Different Atmospheric Conditions. *Fluid Phase Equilib.* **2016**, *413*, 137–141. [CrossRef]
135. Lee, S.; Lee, Y.; Lee, J.; Lee, H.; Seo, Y. Experimental Verification of Methane–Carbon Dioxide Replacement in Natural Gas Hydrates Using a Differential Scanning Calorimeter. *Environ. Sci. Technol.* **2013**, *47*, 13184–13190. [CrossRef]
136. Burla, S.K.; Pinnelli, P.S.R. Experimental Evidence on the Prolonged Stability of CO_2 Hydrates in the Self-Preservation Region. *Case Stud. Chem. Environ. Eng.* **2023**, *7*, 100335. [CrossRef]
137. Gupta, A.; Lachance, J.; Sloan, E.D.; Koh, C.A. Measurements of Methane Hydrate Heat of Dissociation Using High Pressure Differential Scanning Calorimetry. *Chem. Eng. Sci.* **2008**, *63*, 5848–5853. [CrossRef]

138. Robustillo, M.D.; de Menezes, D.E.S.; de Alcântara Pessôa Filho, P. Experimental Determination of Dissociation Temperatures and Enthalpies of Methane, Ethane and Carbon Dioxide Hydrates up to 90 MPa by Using a Multicycle Calorimetric Procedure. *Fuel* **2022**, *312*, 122896. [CrossRef]
139. Yinping, Z.; Yi, J.; Yi, J. A Simple Method, the -History Method, of Determining the Heat of Fusion, Specific Heat and Thermal Conductivity of Phase-Change Materials. *Meas. Sci. Technol.* **1999**, *10*, 201–205. [CrossRef]
140. Hong, H.; Kim, S.K.; Kim, Y.-S. Accuracy Improvement of T-History Method for Measuring Heat of Fusion of Various Materials. *Int. J. Refrig.* **2004**, *27*, 360–366. [CrossRef]
141. Kapica, P.; Karwacki, J.; Bykuć, S. The Use of the T-History Method to Estimate Thermal Capacity and Latent Heat for RT15 and RT18 Materials. *E3S Web Conf.* **2018**, *70*, 01008. [CrossRef]
142. Cao, X.; Yang, K.; Xia, W.; Tang, G.; Bian, J. Dissociation Experiment and Dissociation Rate Model of CO_2 Hydrate. *Nat. Gas Ind. B* **2021**, *8*, 607–614. [CrossRef]
143. Honda, K.; Fujikawa, R.; Ma, X.; Yamamoto, N.; Fujiwara, K.; Kaneko, A.; Abe, Y. The Formation and Growth Model of a CO_2 Hydrate Layer Based on Molecular Dynamics. *AIChE J.* **2022**, *68*, e17406. [CrossRef]
144. Aromada, S.A.; Kvamme, B.; Wei, N.; Saeidi, N. Enthalpies of Hydrate Formation and Dissociation from Residual Thermodynamics. *Energies* **2019**, *12*, 4726. [CrossRef]
145. Hu, Q.; Xiao, X. Formation Methods and Applications of Carbon Dioxide Hydrate: An Overview. *Carbon Capture Sci. Technol.* **2023**, *7*, 100113. [CrossRef]
146. Sun, S.; Zhao, J.; Yu, D. Dissociation Enthalpy of Methane Hydrate in Salt Solution. *Fluid Phase Equilib.* **2018**, *456*, 92–97. [CrossRef]
147. Kondori, J.; Zendehboudi, S.; Hossain, M.E. A Review on Simulation of Methane Production from Gas Hydrate Reservoirs: Molecular Dynamics Prospective. *J. Pet. Sci. Eng.* **2017**, *159*, 754–772. [CrossRef]
148. Ma, S.; Zheng, J.; Tang, D.; Lv, X.; Li, Q.; Yang, M. Experimental Investigation on the Decomposition Characteristics of Natural Gas Hydrates in South China Sea Sediments by a Micro-Differential Scanning Calorimeter. *Appl. Energy* **2019**, *254*, 113653. [CrossRef]
149. Yang, M.; Zhao, J.; Zheng, J.; Song, Y. Hydrate Reformation Characteristics in Natural Gas Hydrate Dissociation Process: A Review. *Appl. Energy* **2019**, *256*, 113878. [CrossRef]
150. Jamaluddin, A.K.M.; Kalogerakis, N.; Bishnoi, P.R. Modelling of Decomposition of a Synthetic Core of Methane Gas Hydrate by Coupling Intrinsic Kinetics with Heat Transfer Rates. *Can. J. Chem. Eng.* **1989**, *67*, 948–954. [CrossRef]
151. Yin, Z.; Khurana, M.; Tan, H.K.; Linga, P. A Review of Gas Hydrate Growth Kinetic Models. *Chem. Eng. J.* **2018**, *342*, 9–29. [CrossRef]
152. Yin, Z.; Chong, Z.R.; Tan, H.K.; Linga, P. Review of Gas Hydrate Dissociation Kinetic Models for Energy Recovery. *J. Nat. Gas. Sci. Eng.* **2016**, *35*, 1362–1387. [CrossRef]
153. Khurana, M.; Yin, Z.; Linga, P. A Review of Clathrate Hydrate Nucleation. *ACS Sustain. Chem. Eng.* **2017**, *5*, 11176–11203. [CrossRef]
154. Dalmazzone, D.; Sales Silva, L.P.; Delahaye, A.; Fournaison, L. Calorimetric Characterization of Clathrate and Semiclathrate Hydrates. In *Gas Hydrates 1*; Wiley: Hoboken, NJ, USA, 2017; pp. 145–176.
155. Handa, Y.P. Compositions, Enthalpies of Dissociation, and Heat Capacities in the Range 85 to 270 K for Clathrate Hydrates of Methane, Ethane, and Propane, and Enthalpy of Dissociation of Isobutane Hydrate, as Determined by a Heat-Flow Calorimeter. *J. Chem. Thermodyn.* **1986**, *18*, 915–921. [CrossRef]
156. Lievois, J.S.; Perkins, R.; Martin, R.J.; Kobayashi, R. Development of an Automated, High Pressure Heat Flux Calorimeter and Its Application to Measure the Heat of Dissociation and Hydrate Numbers of Methane Hydrate. *Fluid Phase Equilib.* **1990**, *59*, 73–97. [CrossRef]
157. Kang, S.-P.; Lee, H.; Ryu, B.-J. Enthalpies of Dissociation of Clathrate Hydrates of Carbon Dioxide, Nitrogen, (Carbon Dioxide+ Nitrogen), and (Carbon Dioxide + Nitrogen+ Tetrahydrofuran). *J. Chem. Thermodyn.* **2001**, *33*, 513–521. [CrossRef]
158. de Menezes, D.É.S.; Sum, A.K.; Desmedt, A.; de Alcântara Pessôa Filho, P.; Fuentes, M.D.R. Coexistence of SI and SII in Methane-Propane Hydrate Former Systems at High Pressures. *Chem. Eng. Sci.* **2019**, *208*, 115149. [CrossRef]
159. de Menezes, D.É.S.; de Alcântara Pessôa Filho, P.; Fuentes, M.D.R. Use of 1-Butyl-3-Methylimidazolium-Based Ionic Liquids as Methane Hydrate Inhibitors at High Pressure Conditions. *Chem. Eng. Sci.* **2020**, *212*, 115323. [CrossRef]
160. Handa, Y.P.; Hawkins, R.E.; Murray, J.J. Calibration and Testing of a Tian-Calvet Heat-Flow Calorimeter Enthalpies of Fusion and Heat Capacities for Ice and Tetrahydrofuran Hydrate in the Range 85 to 270 K. *J. Chem. Thermodyn.* **1984**, *16*, 623–632. [CrossRef]
161. Handa, Y.P.; Yamamuro, O.; Oguni, M.; Suga, H. Low-Temperature Heat Capacities of Xenon and Krypton Clathrate Hydrates. *J. Chem. Thermodyn.* **1989**, *21*, 1249–1262. [CrossRef]
162. Anderson, R.; Chapoy, A.; Tohidi, B. Phase Relations and Binary Clathrate Hydrate Formation in the System H_2–THF–H_2O. *Langmuir* **2007**, *23*, 3440–3444. [CrossRef]
163. Servio, P.; Englezos, P. Effect of Temperature and Pressure on the Solubility of Carbon Dioxide in Water in the Presence of Gas Hydrate. *Fluid Phase Equilib.* **2001**, *190*, 127–134. [CrossRef]

164. Ma, H.; Liu, J.; Zhang, Y.; Li, J.; Kan, J.; Li, N. Prediction of Phase Equilibrium Conditions and Thermodynamic Stability of CO_2-CH_4 Gas Hydrate. *Appl. Sci.* **2024**, *14*, 2320. [CrossRef]
165. Skovborg, P.; Rasmussen, P. A Mass Transport Limited Model for the Growth of Methane and Ethane Gas Hydrates. *Chem. Eng. Sci.* **1994**, *49*, 1131–1143. [CrossRef]
166. Chami, N.; Bendjenni, S.; Clain, P.; Osswald, V.; Delahaye, A.; Fournaison, L.; Dalmazzone, D. Thermodynamic Characterization of Mixed Gas Hydrates in the Presence of Cyclopentane as Guest Molecule for an Application in Secondary Refrigeration. *Chem. Eng. Sci.* **2021**, *244*, 116790. [CrossRef]
167. Matsumoto, Y.; Makino, T.; Sugahara, T.; Ohgaki, K. Phase Equilibrium Relations for Binary Mixed Hydrate Systems Composed of Carbon Dioxide and Cyclopentane Derivatives. *Fluid Phase Equilib.* **2014**, *362*, 379–382. [CrossRef]
168. Zhu, M.; Zhang, H.; Shi, Y.; Zhou, J.; Fu, L. Changes in Heat and Energy During Depressurization-Induced Natural Gas Hydrate Dissociation in Porous Media. *Processes* **2025**, *13*, 1023. [CrossRef]
169. Cheng, C.; Wang, F.; Zhang, J.; Qi, T.; Xu, P.; Zheng, J.; Zhao, J.; Zhang, H.; Xiao, B.; Li, L.; et al. Gas–Liquid–Solid Migration Characteristics of Gas Hydrate Sediments in Depressurization Combined with Thermal Stimulation Dissociation. *ACS Omega* **2019**, *4*, 17547–17555. [CrossRef] [PubMed]
170. Sari, O.; Hu, J.; Eicher, S.; Egolf, P.W.; Homsy, P. Thermo-Physical and Flow Properties of CO_2 Hydrate Slurry. In Proceedings of the International Refrigeration and Air Conditioning Conference, West Lafayette, IN, USA, 14–17 July 2008.
171. Karthikeyan, A.; Aakhash Sivan, V.; Maher Khaliq, A.; Anderson, A. Performance Improvement of Vapour Compression Refrigeration System Using Different Phase Changing Materials. *Mater. Today Proc.* **2021**, *44*, 3540–3543. [CrossRef]
172. Zhang, W.; Jin, X.; Hong, W. The Application and Development of District Cooling System in China: A Review. *J. Build. Eng.* **2022**, *50*, 104166. [CrossRef]
173. Hidemasa, O.; Shingo, T. Air-Conditioning System Using Clathrate Hydrate Slurry. *JFE Tech. Rep.* **2004**, *3*, 1–5.
174. Xie, N.; Tan, C.; Yang, S.; Liu, Z. Conceptual Design and Analysis of a Novel CO_2 Hydrate-Based Refrigeration System with Cold Energy Storage. *ACS Sustain. Chem. Eng.* **2019**, *7*, 1502–1511. [CrossRef]
175. Ogawa, T.; Ito, T.; Watanabe, K.; Tahara, K.; Hiraoka, R.; Ochiai, J.; Ohmura, R.; Mori, Y.H. Development of a Novel Hydrate-Based Refrigeration System: A Preliminary Overview. *Appl. Therm. Eng.* **2006**, *26*, 2157–2167. [CrossRef]
176. Jerbi, S.; Delahaye, A.; Oignet, J.; Fournaison, L.; Haberschill, P. Rheological Properties of CO_2 Hydrate Slurry Produced in a Stirred Tank Reactor and a Secondary Refrigeration Loop. *Int. J. Refrig.* **2013**, *36*, 1294–1301. [CrossRef]
177. Jerbi, S.; Delahaye, A.; Fournaison, L.; Haberschill, P. Design of a New Circulation Loop and Heat Transfer of CO_2 Hydrate Slurry. In Proceedings of the 9th IIR Conference on PCMs and Slurries for Refrigeration and Air Conditioning, Sofia, Bulgaria, 29 September–1 October 2010.
178. Youssef, Z.; Hanu, L.; Kappels, T.; Delahaye, A.; Fournaison, L.; Zambrana, C.; Pollerberg, C. Experimental Study of Single CO_2 and Mixed CO_2+ TBAB Hydrate Formation and Dissociation in Oil-in-Water Emulsion. *Int. J. Refrig.* **2014**, *46*, 207–218. [CrossRef]
179. Oignet, J.; Hoang, H.M.; Osswald, V.; Delahaye, A.; Fournaison, L.; Haberschill, P. Experimental Study of Convective Heat Transfer Coefficients of CO_2 Hydrate Slurries in a Secondary Refrigeration Loop. *Appl. Therm. Eng.* **2017**, *118*, 630–637. [CrossRef]
180. Fu, W.; Wei, W.; Wang, H.; Huang, B.; Wang, Z. Study on the Rheology of CO_2 Hydrate Slurry by Using the Capillary Method. *J. Mar. Sci. Eng.* **2022**, *10*, 1224. [CrossRef]
181. Sahu, C.; Kumar Prasad, S.; Kumar, R.; Sangwai, J.S. High-Pressure Rheological Signatures of CO_2 Hydrate Slurries Formed from Gaseous and Liquid CO_2 Relevant for Refrigeration, Pipeline Transportation, Carbon Capture, and Geological Sequestration. *Sep. Purif. Technol.* **2023**, *309*, 123087. [CrossRef]
182. Zhang, X.; Li, P.; Shan, T.; Liu, Q.; Li, J.; Huang, T.; Wu, Q.; Zhang, P. Experimental Study on the Influence of Particle Size and Grain Grading on the CO_2 Hydrate Formation and Storage Process in Porous Media. *Energy* **2024**, *305*, 132328. [CrossRef]
183. Patel, S.K.; Kumar, Y.; Kumar, S.; Lal, B.; Sangwai, J. CO_2 Hydrate Slurry Flow Characteristics Using Tetrahydrofuran as a Promoter for Carbon Capture, Transportation, and Sequestration. *Energy Fuels* **2025**, *39*, 7362–7372. [CrossRef]
184. Bachu, S. Sequestration of CO_2 in Geological Media: Criteria and Approach for Site Selection in Response to Climate Change. *Energy Convers. Manag.* **2000**, *41*, 953–970. [CrossRef]
185. Marinhas, S.; Delahaye, A.; Fournaison, L.; Dalmazzone, D.; Fürst, W.; Petitet, J.-P. Modelling of the Available Latent Heat of a CO_2 Hydrate Slurry in an Experimental Loop Applied to Secondary Refrigeration. *Chem. Eng. Process. Process Intensif.* **2006**, *45*, 184–192. [CrossRef]
186. Li, G.; Hwang, Y.; Radermacher, R. Review of Cold Storage Materials for Air Conditioning Application. *Int. J. Refrig.* **2012**, *35*, 2053–2077. [CrossRef]
187. Kim, S.; Lee, S.H.; Kang, Y.T. Characteristics of CO_2 Hydrate Formation/Dissociation in H_2O+THF Aqueous Solution and Estimation of CO_2 Emission Reduction by District Cooling Application. *Energy* **2017**, *120*, 362–373. [CrossRef]
188. Lin, W.; Delahaye, A.; Fournaison, L. Phase Equilibrium and Dissociation Enthalpy for Semi-Clathrate Hydrate of CO_2+TBAB. *Fluid Phase Equilib.* **2008**, *264*, 220–227. [CrossRef]

189. Janajreh, I.; Zhang, H.; El Kadi, K.; Ghaffour, N. Freeze Desalination: Current Research Development and Future Prospects. *Water Res.* **2023**, *229*, 119389. [CrossRef]
190. Oyama, H.; Shimada, W.; Ebinuma, T.; Kamata, Y.; Takeya, S.; Uchida, T.; Nagao, J.; Narita, H. Phase Diagram, Latent Heat, and Specific Heat of TBAB Semiclathrate Hydrate Crystals. *Fluid Phase Equilib.* **2005**, *234*, 131–135. [CrossRef]
191. Cao, X.; Yang, K.; Bian, J. Investigation of CO_2 Hydrate Slurry Flow Characteristics with Particle Dissociation for Carbon Storage and Transportation. *Process Saf. Environ. Prot.* **2021**, *152*, 427–440. [CrossRef]
192. Wang, X.; Zhang, F.; Lipiński, W. Carbon Dioxide Hydrates for Cold Thermal Energy Storage: A Review. *Sol. Energy* **2020**, *211*, 11–30. [CrossRef]
193. Zhang, L.; Xu, S.; Li, X.; Zhang, Y.; Yang, R.; Ouyang, Q.; Ren, S. Reaction Kinetic Characteristics and Model of Methane Hydrate Formation in Porous Media. *Energy Fuels* **2017**, *31*, 8548–8559. [CrossRef]
194. Viswanadhan, S.K.; Singh, A.; Veluswamy, H.P. Hydrate-Based Gas Separation (HBGS) Technology Review: Status, Challenges and Way Forward. *Gas Sci. Eng.* **2024**, *131*, 205465. [CrossRef]
195. Aminnaji, M.; Qureshi, M.F.; Dashti, H.; Hase, A.; Mosalanejad, A.; Jahanbakhsh, A.; Babaei, M.; Amiri, A.; Maroto-Valer, M. CO_2 Gas Hydrate for Carbon Capture and Storage Applications—Part 1. *Energy* **2024**, *300*, 131579. [CrossRef]
196. Englezos, P.; Lee, J.D. Gas Hydrates: A Cleaner Source of Energy and Opportunity for Innovative Technologies. *Korean J. Chem. Eng.* **2005**, *22*, 671–681. [CrossRef]
197. Zhao, G.; Zheng, J.; Gong, G.; Chen, B.; Yang, M.; Song, Y. Formation Characteristics and Leakage Termination Effects of CO_2 Hydrate Cap in Case of Geological Sequestration Leakage. *Appl. Energy* **2023**, *351*, 121896. [CrossRef]
198. Gauteplass, J.; Almenningen, S.; Ersland, G.; Barth, T.; Yang, J.; Chapoy, A. Multiscale Investigation of CO_2 Hydrate Self-Sealing Potential for Carbon Geo-Sequestration. *Chem. Eng. J.* **2020**, *381*, 122646. [CrossRef]
199. Warmup Inc. What Is the Recommended BTU per Square Foot of Heating? Available online: https://www.warmup.com/blog/btu-per-square-foot-heating-rule-of-thumb (accessed on 15 May 2025).
200. Vaughn, M. *ASHRAE Handbook Online*; ASHRAE: Atlanta, GA, USA, 2016.
201. U.S. Energy Information Administration. *Massachusetts State Profile and Energy Estimates*. Available online: https://www.eia.gov/state/?sid=MA (accessed on 15 May 2025).
202. Ember and Energy Institute. Carbon Intensity of Electricity Generation, 2000 to 2023. Available online: https://ember-energy.org/app/uploads/2024/11/Global-Electricity-Review-2023.pdf (accessed on 15 May 2025).
203. Zhu, L.; Zhang, J.; Gao, Y.; Tian, W.; Yan, Z.; Ye, X.; Sun, Y.; Wu, C. Uncertainty and Sensitivity Analysis of Cooling and Heating Loads for Building Energy Planning. *J. Build. Eng.* **2022**, *45*, 103440. [CrossRef]
204. U.S. Environmental Protection Agency (EPA). *Emission Factors for Greenhouse Gas Inventories: GHG Emission Factors Hub*; U.S. EPA: Washington, DC, USA, 2025; Update 2025.
205. Choi, S.; Park, J.; Kang, Y.T. Experimental Investigation on CO_2 Hydrate Formation/Dissociation for Cold Thermal Energy Harvest and Transportation Applications. *Appl. Energy* **2019**, *242*, 1358–1368. [CrossRef]
206. Wang, M.-L.; Sun, Y.-F.; Chen, H.-N.; Zhong, J.-R.; Ren, L.-L.; Wang, M.; Rao, D.; Liu, Y.-Q.; Hao, Y.-B.; Liu, B.; et al. A New Approach to High Conversion CO_2 Hydrate Sequestration by CO_2/Water Emulsion Injection into Marine Sediments. *Chem. Eng. J.* **2025**, *503*, 158375. [CrossRef]
207. World Bank. Group State and Trends of Carbon Pricing Dashboard. 2024. Available online: https://carbonpricingdashboard.worldbank.org/ (accessed on 15 May 2025).
208. Abdellatif, M.; Azizmohammadi, S.; Stiedl, G.; Pichler, M.; Ott, H. Sensitivity Analysis of the Methanation Process in Underground Hydrogen Storage: A Case Study in Upper Austria. *Int. J. Hydrogen Energy* **2025**, *105*, 1164–1177. [CrossRef]

energies

MDPI

Article

Reliability Analysis and Economic Evaluation of Thermal Reflective Insulators

Davide Borelli *, Alessandro Cavalletti, Paolo Cavalletti and Luca Antonio Tagliafico

Department of Mechanical, Energy, Management and Transport Engineering, Division of Thermal Energy and Environmental Conditioning, University of Genoa, Via all'Opera Pia 15/A, 16145 Genoa, Italy
* Correspondence: davide.borelli@unige.it; Tel.: +39-010-33-52572

Abstract: High-performance thermal insulators allow a dramatic reduction in the thickness of coatings, thanks to their low thermal conductivity. This study provides an overview about thermal insulation materials, with regards to heat reflective insulators in particular. Then, the numerical investigation method adopted to compute the thermal resistance associated with reflective insulators is introduced. This method has been used in turn to check the accuracy of the declared, measured performance of different, heat-reflective materials on the market. Many manufacturers of reflective insulators were available to provide information and a good agreement between the declared and expected thermal resistance has been found. The choice of a non-experimental approach is meant to check the validity of an already performed test on a reflective insulator using a predictive approach instead of standard, additional testing. Then, the insulation of five typical walls at three different sites in Italy has been simulated, showing that most of heat-reflective materials cannot achieve the maximum required transmittance. Interstitial condensation is likely to occur in specific cases, also because of the aluminum layers inside. The economic analyses showed comparable costs for both heat reflective and traditional insulators, and their cost effectiveness needs to be evaluated case by case.

Keywords: thermal-reflective insulation; radiant barriers; interstitial condensation; cost effectiveness; transmittance numerical evaluation; reflective foils

1. Introduction

The current political crisis has caused a steep increase in oil price, returning to costs comparable to the those in 1973 (the year of the Yom-Kippur war). Since then, European countries have been implementing concrete strategies and policies to improve their energy efficiency and reduce the energy dependence on fossil fuels [1]. For these reasons, the decrease in the energy-need in buildings, through the reduction in their thermal losses, represents one of the most cost-effective actions to cope with such an issue.

Insulation techniques can actually act either jointly or singularly on the three possible ways of heat transfer: conduction, convection, and radiation. More generally, the total equivalent resistance of a material (R_{tot}) can be estimated as the sum of the conductivity resistance for solid conduction (R_{solid}), gas conduction (R_{gas}), radiation (R_{rad}), and eventually convection R_{conv} (1):

$$R_{tot} = R_{solid} + R_{gas} + R_{rad} + R_{conv} \tag{1}$$

The higher each component, the higher the total resistance. Equation (1) frames all the possible strategies to reduce thermal losses. The parameter R_{solid} depends on the material thickness and on its thermal conductivity, defined as a measure of the ability of a substance to transfer heat by conduction. The measurement of this parameter can be carried out by means of either the guarded hot plate or the flow-meter methods [2,3].

A brief overview of the insulators available on the market and actually adopted in energy efficiency for buildings is presented.

Traditional thermal insulation materials (Table 1) try to reduce as far as possible the term associated to conduction [4–6]. A minimum limit value for this group of insulators can be found in the conductivity of motionless air (about 0.025 W/m K) [7].

Table 1. Types and average thermal performance of some traditional thermal insulation materials.

Insulation Type	Origin	Name	λ [W/mK]
Fibrous structure	Natural	Coconut fiber	0.050
	Natural	Sheep wool	0.037–0.064
	Synthetic	Rock wool	0.035–0.040
	Synthetic	Glass wool	0.032–0.067
Cellular structure	Natural	Expanded granulated cork	0.040–0.060
	Synthetic	EPS polystyrene	0.035–0.044
	Synthetic	XPS polystyrene	0.034–0.038
	Synthetic	XPS polystyrene + graphite	0.025–0.030
Porous structure	Natural	Pumice	0.110
	Natural	Vermiculite	0.077–0.082
	Synthetic	Cellular cement	0.070

The family of **Vacuum insulated panels (VIPs)**, belonging to innovative insulators, rely on a layer with a vacuum inside where heat can only propagate by means of radiation. This strategy can lead to very low equivalent conductivities (about 0.01 W/mK) with a weak point where every cut or perforation seriously damages the panel, at least doubling its conductivity [8]. Many works are available in the literature about the topic, starting from general reviews [9,10], and going more in detail with specific studies: e.g., influence of water inside the material [11], variation of thermal conductivity with different materials filling the VIP [12], and future development of VIPs [13]. Usually, the most relevant flaw in these insulators consists of the aluminum layers put inside of the VIPs (or similar materials) which almost block the passage of water vapor with a consequent risk of interstitial condensation.

Gas-filled panels (GFPs) represent another family of insulators exploiting a principle similar to the one for VIPs, where the void is replaced by gases with a conductivity lower than air. Regardless, the VIP panels still seem to be preferable to the GFP for durability reasons [14,15].

On the other hand, **aerogel materials** belong to a group where the improved conductivity is reached by the replacement of the liquid part of a gel with gases during a supercritical drying process. Aerogels lower the limit conductivity for traditional insulators (0.015 W/mK against 0.025 W/mK for motionless air) [16] thanks to the so-called Knudsen effect, which is a reduction in thermal conductivity in gases when the size of the containing cavity is almost equal to the mean free path [17]. A further improvement in the performance of aerogels can be reached with carbon black, which suppresses the radiative transfer as well [18].

Nanotechnology-based materials try to exploit the same Knudsen effect adopted in aerogels with a dramatic decrease in the size of the pores and therefore in conductivity. Such technology could overcome the issue of excessively high vapor resistance highlighted for VIPs and any perforation would not decrease its performance. The Knudsen effect in very small pores (order of magnitude of nanometers) can be associated to the fact that gas molecules are more likely to hit the walls of the pores instead of other molecules. In this context, the sizes of the pores (compared to the mean free path of the gas molecules) determine the reduction in conductivity. In actual fact, an excessive decrease in the cavities size might cause a relevant increase in the radiation heat transfer, since they tend to be lower than the wavelength of infrared radiations (about three orders of magnitude, 10 nm as reference pore size against 10 μn for infrared radiations) [19,20]. The contribution of radiation heat transfer against the conduction/convection ones still has to be exhaustively estimated and tested [20]. Regardless, if the conductive/convective mechanisms still remain dominant, the conductivity of the nanotechnology-based insulator can reach very low values (about 0.004 W/mK) with the consequent use of very thin layers to insulate buildings. This group is composed by mortars, resins, plasters, and paints containing nanoparticles, usually hollow glass or ceramic microspheres of variable size (between 20 and 120 μm of diameter), which are bonded by a mixture of synthetic rubber and other polymers granting the mechanical properties, flexibility, and water resistance as well. Their effective composition is under industrial secrecy and therefore different explanations are provided by the different manufacturers, leaving room for discussion about the reliability of the insulating mechanism and on the accuracy of the models available in the literature [21–24]. Moreover, the technical information about conductivity for nanotechnology materials is, most of the time, contradictory, and very different values can be found on the market, differing even of two orders of magnitude, e.g., from 0.001 W/mK [21] up to 0.14 W/mK [22]. No CE marking, neither compulsory nor voluntary, is available for nanotechnology-based materials since there is neither a harmonized standard nor a manufacturer's voluntary adoption of ETA (European Technical Assessment). The performance needs to be tested in authorized laboratories within one of the member countries of the European community [25] to adopt the declared conductivity reported in the technical sheet. Moreover, the evaluation has to be conducted following a statistical approach (i.e., a single test is not sufficient), following the methodology reported in UNI ISO 10456 [26].

Ultimately, **heat-reflective materials** can be found: The thermal insulation system is made of one or more low-emissivity reflective surfaces, in contact with one or more cavities filled with air. They work with a range of wavelengths between about 5 μm and 50 μm (infra-red field) because the radiation heat exchange occurs between bodies at room temperature. The reference standard for materials exploiting the contribution of either reflective or low-emissivity surfaces in contact with an air cavity can be found in [27]. One of the most effective applications of reflective materials can be found in [28], where the simple application of a reflective panel between a radiator and the perimetral wall (common pattern in residential applications) leads to an energy saving of 3.50% in the worst insulation conditions. Several studies were developed in the past regarding heat reflective insulators and were mainly divided into two groups: numerical approaches, with the aim to be implemented into simulation software [29–42], and experimental approaches to characterize their properties [43–49]. A research review about heat-reflective materials [50] estimates that about 59% of the studies were concerned with experimental approaches (either field measurement or indoor test rigs) whereas only 41% regarded numerical simulations (e.g., building performance or CFD).

To the best of the authors' knowledge, no relevant study about the reliability of the declared performance of reflective material has been conducted. In addition, most of scientific literature does not account for the compatibility of the heat-reflective materials with the required thermal resistance. The economic aspects are lacking as well, neglecting the cost effectiveness of these materials compared to other traditional insulating solutions. The present study aims to cover these topics, by means of the following steps:

- Firstly, the thermal resistance of the insulator has been computed by means of the theoretical approach contained in the UNI EN ISO 6946:2018 standard [51]. The results have been compared with the values reported in the declared transmittance data of different heat reflecting materials available on the market. A substantial good agreement with most of materials has been found. The names of the different brands considered are kept anonymous. The numerical approach has been intentionally chosen instead of experimental tests in order to test the reliability of the measured values declared by the producers by means of a predictive numerical approach.
- The most effective position for heat reflective insulators when considering cavity, brick and stone walls has been investigated. Indeed, the thermal resistance of such insulators is influenced by their position in a multilayer wall.
- Then, the insulation of the main and most common walls in Italy with heat-reflective materials has been examined. In particular, cavity, stone, and brick walls were considered, assuming three different locations in Italy (Genoa—climatic zone D, Milan—

climatic zone B, and Naples—climatic zone C). The limits set by the Italian laws in terms of transmittance and interstitial condensation have been checked.

- A simplified economic analysis is proposed to identify the insulating material which grants a required performance (in terms of thermal transmittance) at the lowest possible price. Clearly, the conclusions on this topic do depend on the market price fluctuations of thermal insulation materials, which have been experiencing a steep increase since the "Superbonus" [52] tax incentives came into force. In fact, it is preliminary estimated that an increase in the prices for building insulation of about 35% has occurred in the past couple of years. Although the analyses refer to a period former to 2020, the results still provide interesting conclusions because the increase in the prices has affected almost homogenously the group of heat reflective thermal insulation materials.

2. Materials and Methods

This section proposes a first recall on the numerical method used to assess the thermal resistance of reflective insulators. Then, the walls used in the simulations are presented, recalling also for the mandatory energetic efficiency Italian limits (i.e., total thermal resistance of walls according to the different Italian climatic zones).

2.1. Thermal Resistance in Cavity Walls—Reflective Insulators

The thermal resistance associated to a cavity filled with air can be carried out by means of the approach reported in the UNI EN ISO 6946:2018 standard [51] and previous numerical studies [2,3,26,27,50,53] which can be applied both to air voids (i.e., the width/length are comparable to the thickness) and to air layers (i.e., the width/length are more than 10 times the thickness along the heat flow direction). In particular, the adopted approach applies to unventilated airspaces with length and width which both exceed 10 times the thickness. Even airspaces with openings which allow no airflow through the layer and do not have an extended area (500 mm^2 per meter of length for vertical air layers or 500 mm^2 per square meter of surface area for horizontal air layers) can be considered as unventilated. Assuming that the heat exchange between two opposite emitting surfaces with an interposed air layer depends on the mean surface temperature, the thermal resistance can be expressed as:

$$R_g = 1/(h_a + h_r) \tag{2}$$

where:

R_g is the thermal resistance of the air cavity, [m^2K/W];

h_a is the convective heat transfer coefficient, [W/m^2K];

h_r is the radiative coefficient, [W/m^2K].

In turn, h_a can be determined according to the direction of the heat flow and the temperature difference (Tables 2 and 3). Indeed, the convective heat transfer is assumed to be either constant or function of the width of the air gap with low temperature differences (<5 K), then it depends on temperature as the difference increases (>5 K).

Table 2. Values for the convective heat transfer coefficient (h_a) according to the direction of the heat flow for temperature difference lower than 5 K [51].

Direction of the Heat Flow	h_a [W/m^2K] [1]
Horizontal	1.25
Upwards	1.95
Downwards	$0.12d^{0.44}$

[1] if the values are lower than $0.025/d$, where d is the thickness of the air gap in the direction of the heat flow, then $0.025/d$ is assumed.

Table 3. Values for the convective heat transfer coefficient (h_a) according to the direction of the heat flow for temperature difference higher than 5 K [51].

Direction of the Heat Flow	h_a [W/m^{2}K] [1]
Horizontal	$0.73\ T^{1/3}$
Upwards	$1.14\ T^{1/3}$
Downwards	$0.09\ T^{0.187} d^{-0.44}$

[1] if the values are lower than $0.025/d$, where d is the thickness of the air gap in the direction of the heat flow, then $0.025/d$ is assumed.

On the other hand, the radiative coefficient (h_r) can be represented as:

$$h_r = E\ h_{r0} \tag{3}$$

with:

h_{r0} radiative coefficient for a black body surface [W/m^{2}K], as reported in Equation (4);

E surface-to-surface emittance, formulated in Equation (5).

The radiative coefficient for a black body surface has been expressed assuming unit values for the view factors, since the airspace is usually large and thin

$$h_{r0} = 4\ \sigma\ T_m^{\ 3} \tag{4}$$

where:

σ is the Stefan-Boltzmann constant (5.67×10^{-8} W/m^{2}K^4);

T_m is the average temperature of the surface and of its surroundings, assumed to be at 282 K for the case study. Therefore, the radiative coefficient for the black body surface is 5.1 W/m^{2}K.

Finally, the emittance is given by:

$$E = 1/(1/\varepsilon_1 + 1/\varepsilon_2 - 1) \tag{5}$$

where ε_1 and ε_2 are the hemispherical emissivities of the surfaces bounding the airspace. These values should account for the effects of deterioration and dust as well.

The so-estimated thermal resistance of a cavity can be accounted in the global transmittance for a multilayer wall by means of Equation (6):

$$U_{tot} = 1/(R_{si} + \sum s_j/\lambda_j + R_{gk} + R_{se}), \tag{6}$$

where:

R_{si} and R_{se} are the surface resistances determined according to the direction of the heat flow (m^{2}K/W) [51];

s_j and λ_j represent respectively the thickness (m) and the conductivities (W/m K) of the jth layer;

R_{gk} is the thermal resistance of the kth cavity (m^{2}K/W).

The values for the materials were taken from the UNI/TR 11552:2014 Italian standard [54].

2.2. *Type Walls Adopted in the Simulations and Computation of the Global Transmittance*

The carried-out simulations apply the reflective insulation to different kinds of walls which represent some of the most common construction typologies in the Italian building stock. Eight configurations have been evaluated, including brick, cavity, and stone walls. The results for all the simulations can be found in Appendix A, whereas three relevant cases are analyzed in detail as follows:

- cavity walls (Figure 1a): typical of the post-war buildings with reinforced concrete structure. Usually, two layers of bricks (respectively 8 and 12 cm) are separated by an air cavity which can be from 4 cm up to 16–18 cm wide. The contribution of the air cavity to the global thermal performance of the wall shows a very weak dependence on the thickness. Indeed, the thermal resistance of unventilated air cavities at least 25 mm wide, with high emissivity surfaces, are associated to constant thermal resistances [51]. The total transmittance associated to such walls (without reflective insulation) is usually between 1 and 1.2 W/m^{2}K. The reflective insulator shall be simulated within the air cavity in the most effective position (i.e., either in the middle of the air cavity or on one of the two faces between bricks). The average temperature of the cavity surfaces is assumed to be at 10 °C, representative of a winter steady-state condition common in Italian climatic zones from A to E;

- brick walls with a low insulation layer (built between 1976 and 1990) (Figure 1b): the wall is entirely made of bricks and an insufficient external insulation was applied for instance to avoid interferences; for instance, the reduction in the width of the balconies. The global initial transmittance of the wall is assumed to be 0.39 W/m^{2}K;
- stone walls (Figure 1c): such typology is common, especially in historical buildings, which are in turn subjected to architectural constraints and therefore a classical external coating is not feasible. The global transmittance of these walls depends on the total thickness, and an average of 2.33 W/m^{2}K could be considered representative of this kind of walls.

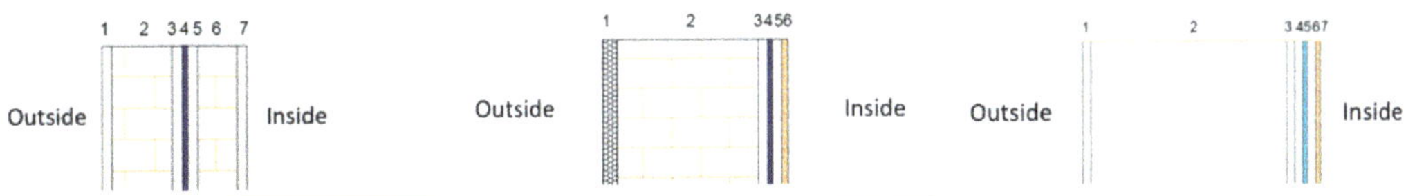

	Stratigraphy	s [m]	λ [W/mK]
1	Lime-cement plaster	0.02	0.7
2	Hollow brick 12 cm	0.12	0.387
3	Air cavity	0.02	-
4	Thermoreflective insulation	-	-
5	Air cavity	0.02	-
6	Hollow brick 8 cm	0.08	0.387
7	Lime plaster	0.02	0.7

(a)

	Stratigraphy	s [m]	λ [W/mK]
1	Expanded sintered polystyrene EPS	0.03	0.04
2	Alveolated brick 30 cm	0.3	0.18
3	Air cavity	0.02	-
4	Thermoreflective insulation	-	-
5	Air cavity	0.02	-
6	Drywall	0.012	0.2

(b)

	Stratigraphy	s [m]	λ [W/mK]
1	Lime cement plaster	0.02	0.9
2	Stone wall	0.54	2.4
3	Lime plaster	0.02	0.7
4	Air cavity	0.02	-
5	Thermoreflective insulation	-	-
6	Air cavity	0.02	-
7	Drywall	0.012	0.2

(c)

Figure 1. Kinds of walls investigated in the present work to simulate the contribution of the reflective insulator with the assumed conductivities for the different layers. (**a**) Cavity wall; (**b**) Brick wall; (**c**) Stone wall.

2.3. Maximum Legislative Limits for the Total Trasmittance

The simulations were performed assuming three different climatic zones among the Italian zones: E (e.g., Milan), C (e.g., Naples) and D (e.g., Genoa). The other climatic zones were preliminary neglected because they are associated to a smaller part of the Italian territory. Table 4 recalls the legislative limits according to the different climatic zones.

Table 4. Reference value for thermal transmittance of vertical opaque structures, to the exterior, and unheated spaces or against the ground [55].

Climatic Zone	U_{ref} [W/m^{2}K]
A-B	0.43
C	0.34
D	0.29
E	0.26
F	0.24

2.4. Economic Analysis–Cost Effectiveness

The "cost-effectiveness analysis" allows the comparison of several alternatives according to their cost and a measured (not economic) useful effect (e.g., thermal transmittance for the present study) [56]. The cost-effectiveness index for the i-th alternative can be expressed by means of Equation (7):

$$CE_i = C_i / E_i, \quad (7)$$

where:

C_i is the cost of the *i*th alternative [€/m^2], for our case the cost of both the materials and the installation.

E_i is the related benefit, i.e., the thermal resistance for the present study [m^{2}K/W].

As a consequence, the unit measure for the cost effectiveness (CE) depends on the related benefit. In this case CE can be measured in [€W/Km^4].

As regards to the costs for heat-reflective materials, most of them were declared by the producers, except for material 1, where the manufacturer did not reply to the question and therefore was excluded from this part of the study. On the other hand, the cost of installation, when not declared, was estimated basing on the regional price list carried out by Regione Liguria. The installation cost obtained by means of the regional price list leads to values comparable to the declared ones.

The cost-effectiveness approach was carried out twice: at first, it was applied to the heat-reflective materials under study, and then it was applied for a comparison among different thermal insulation materials.

3. Results

The present section presents the main results obtained following the approach illustrated above.

Firstly, the thermal resistance is computed following the numerical approach illustrated before, comparing the results with the declared, tested performance of the heat-reflective materials on the market.

Secondly, the most effective position of the reflective insulator in multilayer walls (especially cavity walls) is investigated to justify the chosen stratigraphy adopted in the following simulations. This issue is not trivial because the best, ideal position may not be always feasible or the same for different, existing walls.

In conclusion, the reflective insulators formerly verified are applied to the chosen walls to enquire their convenience from the point of view of both thermal performance and cost effectiveness.

3.1. Comparison between Declared and Computed Thermal Resistence for Heat Reflective Insulators

Table 5 presents the main results about the calculation performed to compute the expected total thermal resistance for each of the eight, heat reflective insulators considered, according to the numerical approach illustrated above. The materials are kept anonymous, but the samples here examined represent quite a complete range of heat-reflective materials available on the market.

Table 5. Calculated thermal transmittance for eight heat reflective thermal insulation materials–main results.

Mat. N.	ε_1 [-]	ε_2 [-]	E [-]	h_r [W/m^{2}K]	Air Cavity Thermal Resistance [W/m^{2}K]	Core Thermal Resistance [W/m^{2}K]	Total Thermal Resistance [W/m^{2}K]
1	0.06	0.9	0.060	0.304	0.644	0.285	1.572
2	0.05	0.9	0.050	0.254	0.665	2.761	4.091
3	0.05	0.9	0.050	0.254	0.665	0.968	2.298
4	0.02	0.9	0.020	0.102	0.740	1.521	3.000
5	0.05	0.9	0.050	0.254	0.665	0.200	1.530
6	0.05	0.9	0.050	0.254	0.665	0.420	1.750
7	0.04	0.9	0.040	0.203	0.688	0.417	1.793
8	0.12	0.9	0.118	0.604	0.539	1.252	2.331

The total thermal transmittance referred to in Table 5 is associated to a horizontal heat flow of two air gaps 20 mm thick each, divided by the reflective insulation (Figure 2). The information required (i.e., emissivity, conductivity, and thickness) was taken from the data sheets of the materials to ensure the comparability of the results whereas the convective heat transfer coefficient has been assumed according to Table 2.

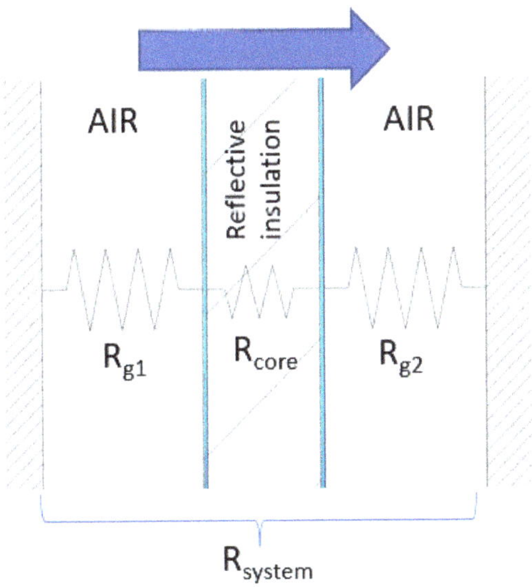

Figure 2. Scheme of the system of thermal resistances considered to compare the calculated results with the declared ones. The resistances associated to the air gaps (R_{g1} and R_{g2}) also consider the reflective contribution. R_{core} represents instead the conductive thermal resistance associated to the core material itself.

Then, a comparison between the computed and measured thermal resistances declared by the producers is provided in Table 6.

In general, predicted and declared values are in good agreement, with a variation below 5%. Only one case differs (about 13%-material 8) and the manufacturer, when informed about this result, admitted that the value needed to be certified again.

Table 6. Comparison between declared and calculated thermal resistance for eight heat-reflective materials available on the market. The trademarks are kept anonymous.

Material Number	Calculated Thermal Resistance [m^2K/W]	Declared Measured Thermal Resistance [m^2K/W]	Variation [%]
1	1.57	1.61	2
2	4.09	4.05	−1
3	2.30	2.28	−1
4	3.00	3.00	0
5	1.53	1.52	−1
6	1.75	1.68	−4
7	1.79	1.86	4
8	2.33	2.64	13

3.2. Best Position of the Heat Reflective, Insulating Layer

The effectiveness of the reflective insulating layer changes according to its position inside a multilayer element. Although common insulators (acting mainly on conductivity) work better (or equally) when applied from the outside, heat-reflective materials need to be installed on the inner side of the wall for winter insulation because their effectiveness depends on the direction of the heat flow. Clearly, such a pattern might act negatively during summer, as it prevents the heat inside to flow through the wall instead of reflecting the heat coming from the outside. The risk of interstitial condensation recalled in the introductory section may be influenced in addition by the position of the insulator.

In the simulations carried out, the reflective materials have been assumed as applied on the inner side of the stone and brick walls by means of frames and pins in aluminum or wood. This pattern leads to two benefits: firstly, the insulator can work on both sides because it is not in direct touch with the wall. Secondly, the issue of interstitial condensation loses importance if the supporting frame is provided with openings on the top and on the bottom to allow ventilation. Unluckily, this feature is not very common in ordinary cavity walls. Among the examined types, the cavity wall deserves an additional remark, because the air gap between the two layers of bricks represents one of the most effective positions for the heat reflective material. This configuration avoids the increase in the total thickness of the wall, unlike the cases of solid multilayer walls (e.g., brick or stone walls). Figure 3 shows the effect of the different positions of the heat reflective layer (i.e., adherent to the layer of bricks; on the inner side of the wall; in the middle of the air cavity) on the total wall transmittance.

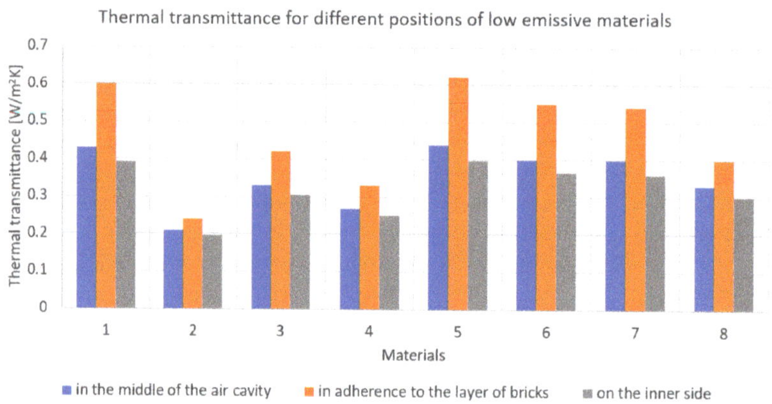

Figure 3. Variation of the thermal transmittances obtained for a cavity wall, positioning the heat reflective layer either in the middle of the air cavity (blue, no contact) or in the air cavity but in contact with the layer of the bricks (orange) or on the inner side of the wall (grey).

Figure 3 illustrates the obtained transmittance for a cavity wall applying singularly the seven heat-reflective materials considered (horizontal axis) in three different conditions: in the middle of the air cavity (blue columns), in adherence to the layer of bricks within the cavity (orange column), and on the inner side of the wall (grey columns). Clearly, the highest transmittance is always reached in the case of insulator inside the cavity but in contact with one of the two layers of bricks. This result is due to the loss of the heat reflective contribution of the insulator side in contact with the bricks.

On the other hand, the cases of insulation in the middle of the air cavity are quite similar to the ones with insulation on the inner side of the wall. The slight improvement in this last case (grey columns) is due to the additional thermal resistance provided by the additional drywall, necessary to protect the heat reflective insulating layer from shocks.

3.3. Application of the Heat Reflective Thermal Insulation Materials to the Case Study Walls

In this section, the eight heat-reflective materials previously compared with the numerical results are applied to the walls described in Section 2 (Materials and Methods). A comparison with the limits imposed by the law for three Italian climatic zones (i.e., maximum transmittance) is performed, with a brief recall to the issue of interstitial condensation.

- Cavity wall: Figures 4 and 5 show, respectively, the application in the middle of the air cavity and in touch with one of the layers of bricks. Only one of the eight heat-reflective materials leads to a total transmittance within the limit for the climatic zones associated to Genoa (i.e., D) and Milan (i.e., E). As far as the climatic zone C (Naples) is considered, just four materials are eligible to respect the threshold. Clearly, the case in Figure 5 with respect to Figure 4 slightly worsens due the installation of the insulator in touch with the layer of bricks (and the consequent partial loss of the reflective insulation on one side).

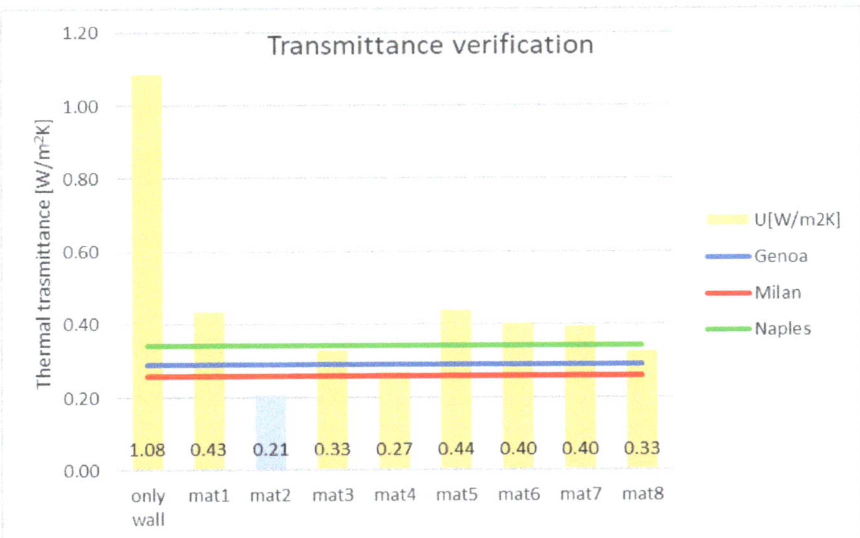

Location	U_{ref} [W/m²K]
Genoa	0.29
Milan	0.26
Naples	0.34

Figure 4. Thermal transmittance in cavity walls with different heat reflective insulating layers (application in the middle of the air cavity). The red, blue, and green lines represent the threshold for different cities.

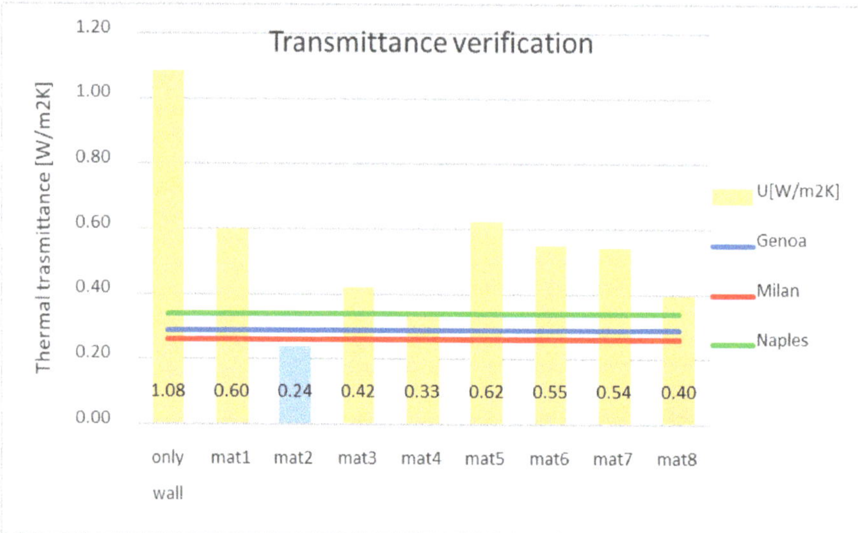

Location	U_{ref} [W/m^{2}K]
Genoa	0.29
Milan	0.26
Naples	0.34

Figure 5. Thermal transmittance in cavity walls with different heat reflective insulating layers (in touch with one of the layers of bricks). The red, blue, and green lines represent the threshold for different cities.

- Brick wall: the contribution provided by means of the heat reflective insulator to the existing/reduced insulating coating allows the limits to be respected independently from the material (Figure 6).

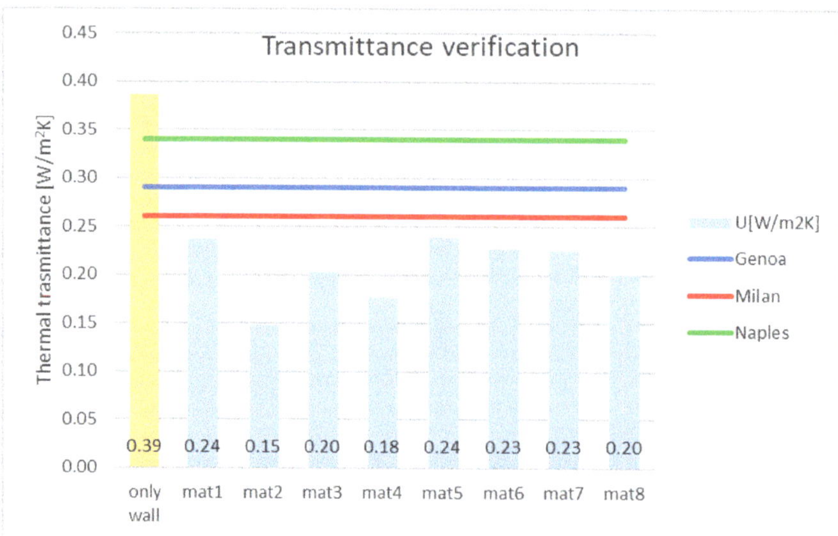

Location	U_{ref} [W/m^{2}K]
Genoa	0.29
Milan	0.26
Naples	0.34

Figure 6. Thermal transmittance in brick walls with different heat reflective insulating layers. The red, blue, and green lines represent the threshold for different cities.

- Stone wall (Figure 7): only one material grants the limit transmittance for climatic zones D and E (Genova and Milan, respectively). On the other hand, only two reflective insulators grant the limits for climatic zone C (Naples).

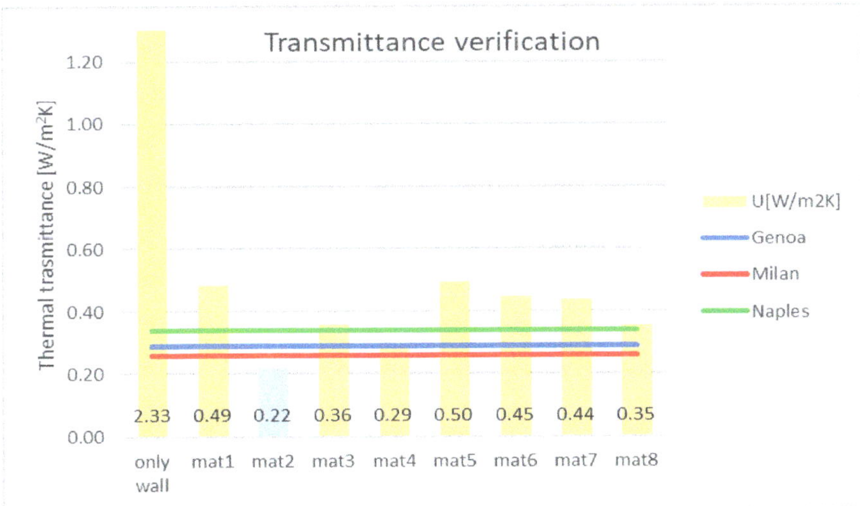

Location	U_{ref} [W/m^{2}K]
Genoa	0.29
Milan	0.26
Naples	0.34

Figure 7. Thermal transmittance in stone walls with different heat reflective insulating layers. The red, blue, and green lines represent the threshold for different cities.

More generally, all the configurations concerning the inner insulation determined an increase in the wall thickness of about 6 cm, and accounted for finishes such as plasters and drywalls. This in turn implies a not negligible reduction in the useful surface of the flats in existing buildings. Only the case of heat reflective insulator inside the cavity wall could overcome such an issue, but it clearly represents a solution suitable only for new constructions.

On the other hand, assuming an expanded polyurethane with an average conductivity of 0.023 W/mK, 6 cm thick (the same thickness of the heat reflective insulating package) provides a thermal resistance of about 2.6 m^{2}K/W which is equal or higher than the declared one of 6 out of 8 heat-reflective materials investigated (Table 6). In addition, the heat reflective insulators are usually provided with at least an aluminum layer which acts on the reflection mechanism, but it has the side effect of a very high vapor resistance (order of magnitude 300.000).

As a consequence, problems about interstitial condensation might arise when adopting heat reflective insulators, whereas other solutions (e.g., expanded polyurethane) are very likely to bypass the problem (a vapor resistance of about 100 for such insulators). In the case studies considered above, no condensation occurred, except for the stone wall, where this issue might become relevant also according to the standard use of the building, the climatic zone, and the vapor resistance of the heat reflective material.

Indeed, each stratigraphy was tested by means of a commercial program implementing Glaser's method. For reasons of synthesis, the full results are omitted in the present work, and the case in which condensation occurred is reported on only (Figure 8). In any case, the issue of interstitial condensation shall be the topic of further, specific studies.

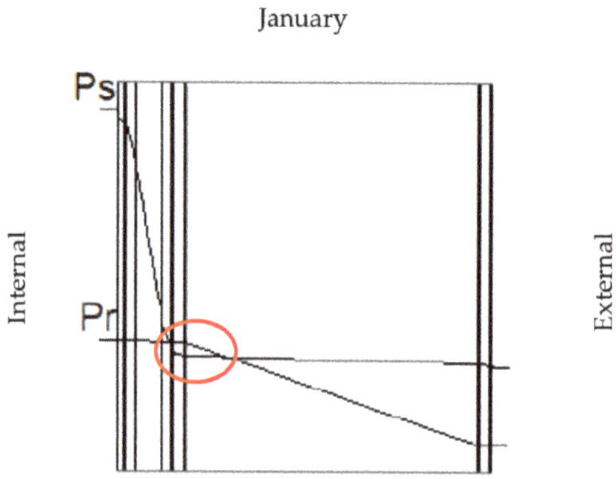

Figure 8. Extract from the diagram of monthly pressures referred to the month of January for wall type 5.1 (see also Annex A for further information). Condensation occurs in the region where the lines representing, respectively, the saturation (P_s) and relative (P_r) pressure intersect (red circle).

3.4. Cost-Effectiveness Analysis

3.4.1. Heat-Reflective Materials

Table 7 and Figure 9 present the costs for each material and its installation, with the correspondent cost-effectiveness index. The installation cost accounts for the case of internal insulation with a plasterboard covering:

Table 7. Material, installation, and total cost, with the cost-effectiveness index for the heat-reflective materials.

Material Number	Material Cost [€/m^2]	Installation Cost [€/m^2]	Total Cost [€/m^2]	CE_i [€W/m^{4}K]
2	30.00	49.00	79.00	19.31
3	11.80	48.32	60.12	26.16
4	18.00	42.07	60.07	20.02
5	4.46	47.01	51.47	33.64
6	15.50	48.32	63.82	36.47
7	11.40	48.32	59.72	33.31
8	17.60	40.00	57.60	24.71

The cost for the materials varies from a minimum of 4.46 €/m^2 (material number 5) up to a maximum of 30 €/m^2 (material number 2) whereas the costs for installation are within the range between 40 and 49 €/m^2. The most convenient heat reflective insulators among the ones under study is material number 2. Besides the highest material cost, the relevant contribution to the total thermal resistance makes it the most eligible material (4 m^{2}K/W against an average of 2 m^{2}K/W for other heat reflective insulators, see Table 6 for more information). On the other hand, cases number 5, 6, and 7 have to be discarded due to their poor convenience.

Figure 10 plots the total cost against the obtainable thermal resistance, superposing a second-degree polynomial approximation. The yellow region identifies the minimum of the plot where a thermal resistance between about 2.3 and 2.6 m^{2}K/W corresponds to a minimum total cost between about 55 and 62 €/m^2. Such a region can be conceived as a starting point to compare heat reflective insulators to most diffused ones such as EPS. Indeed, the value of the range between 2.3 and 2.6 m^{2}K/W can be obtained with 6 cm thick EPS with a maximum thermal conductivity of 0.023 W/mK. Such thickness is comparable to the required one by the heat reflective insulators, including the air cavity, the frames, and the finishes. Even from an economical point of view, the total costs can be considered as comparable.

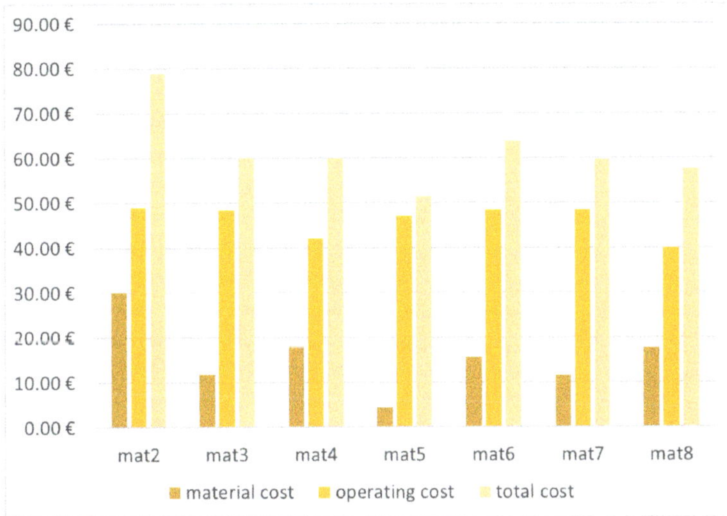

Figure 9. Graphical representation of costs for the heat-reflective materials reported in Table 7.

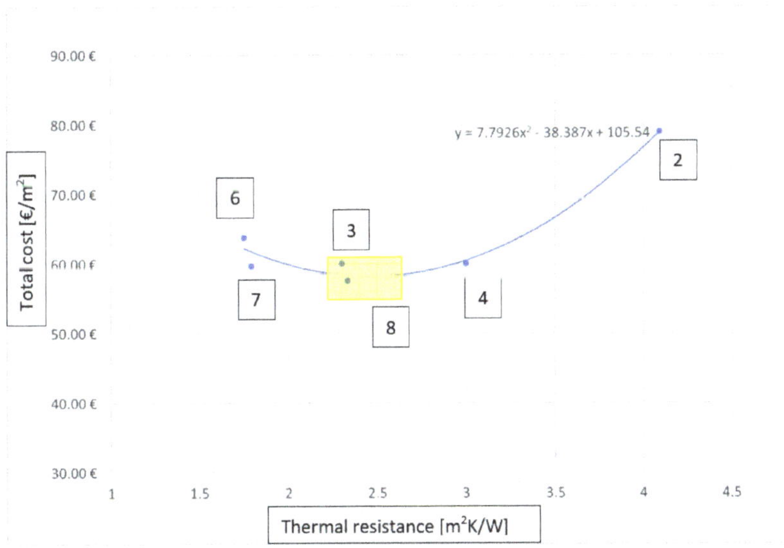

Figure 10. Total cost of the heat-reflective materials (numbered in the squares) with the correspondent thermal resistance. Approximation of the results with a second-degree polynomial equation.

From this starting point, the materials falling on the left of the yellow region have to be discarded because they provide a lower thermal resistance and require a higher total cost.

On the contrary, the materials on the right (e.g., 4 and 2) present a higher total cost (up to 25% higher than the minimum obtained cost) in front of a thermal resistance which is almost doubled, resulting in the most convenient ones, as already outlined by the cost-effectiveness index (Table 7).

The polynomial approximation provides a useful tool to assess the cost effectiveness of other thermal insulators not investigated in the study by entering the graph with the total cost and the thermal resistance.

3.4.2. Traditional and Heat Reflective Thermal Insulation Materials–Cost Comparison

This paragraph describes the cost comparison between traditional and reflective thermal insulation materials by means of the cost-effectiveness approach.

The information about heat-reflective materials is the average of the costs reported in the previous Table 7, whereas the costs for traditional materials in Table 8 have been estimated referring to the regional price lists and the manufacturers' data.

Table 8. Costs of traditional and heat reflective thermal insulation materials.

Insulating Material	Material Cost [€/m^2]	Total Cost (Including Installation) [€/m^2]
Rock wool panels	12.00	47.50
Calcium silicates panels	22.50	125.00
Wood fiber panels	30.00	65.00
EPS panels	11.50	37.50
Glass wool	3.75	55.00
Heat reflective panels (average)	15.54	61.69

Figure 11 shows that the total cost for heat reflective insulation is on average comparable to a traditional insulating solution. Therefore, the cost effectiveness of heat-reflective materials needs to be computed case by case and in accordance with the required thermal resistance.

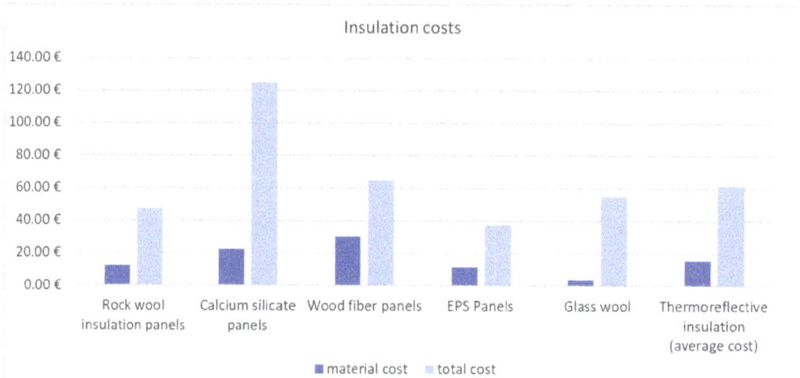

Figure 11. Total cost for insulation with different materials of the heat-reflective materials with the correspondent thermal resistance. Approximation of the results with a second-degree polynomial equation.

4. Conclusions and Future Developments

The energy efficiency strategy concerning insulating coatings in existing buildings introduces many issues (e.g., limited space for insulating coating and maximum limit transmittance) which require the use of innovative insulators, instead of traditional ones.

This present work provides a general overview about the available thermal insulation materials, with a specific insight into heat reflective ones. Their theorical thermal resistance was computed by means of the presented numerical approach and was used to verify the accuracy of the measured performances declared by the producers. A very good agreement with different anonymized brands has been found. Only one out of eight investigated materials, representing the main heat reflective insulators on the market, presented a variation higher than 5% between declared and estimated performance.

Then, the application of the heat reflective thermal insulation materials to different walls in the most common Italian climatic zones (C, D, and E) was simulated. According to the results, most of the reflective insulators cannot satisfy the required limits in terms of transmittance themselves, almost independently from the climatic zone. The use of reflective materials only as an integration into a thinner external coating is the most effective strategy (Figures 5–8). Indeed, all considered heat-reflective materials lead to transmittances below the limits for all considered climatic zones. Concerning the external integration insulator, the required thickness (about 3 cm) reduces the impact of the insulation on the external façade, whereas the maximum conductivity (about 0.04 W/mK) is highly compatible with most traditional and cheap insulators. Regardless, all considered reflective materials present an intrinsic flaw concerned with the risk of interstitial condensations due to the aluminum layers inside. The reported case studies have shown only one case (stone wall) where interstitial condensation occurred.

A cost-effectiveness analysis has been carried out on the investigated materials, highlighting that the most eligible heat reflective insulator in terms of costs and performance is not always the cheapest one. The results are not relevantly influenced by the steep increase in prices for building materials occurred in Italy in the past months because the rise affected the cost of the insulators in a very similar way. The cost-effectiveness analyses also provided a comparison with the traditional insulators, showing that some heat-reflective materials (30% of the enquired ones, two out of seven materials) are never convenient due to their high cost compared to the low, provided thermal resistance. About 40% of the materials (three out of seven) resulted in being comparable to traditional insulating solutions from the points of view of the total cost, thermal resistance, and required thickness. Only 30% of the investigated materials (two out of seven) almost doubled the thermal resistance at an increase in the total price of about 25%. As a consequence, the adoption of the heat reflective insulators is not always preferable to other classic solutions and needs to be evaluated according to the specific case study.

In future developments, the remaining Italian climatic zones and other kinds of walls should be included in the simulations. Indeed, these solutions would result as being highly effective in buildings where the external outline cannot be modified, (e.g., ancient mansions, castles, or heritage-protected buildings). Another very common case is represented by buildings where there is not enough space to install an external coating and, therefore, internal insulation might result mandatory. In addition, innovative strategies to cope with the interstitial condensation should be investigated because the heat reflective mechanism is typically based on material with a very high vapor resistance (i.e., aluminum layers inside the materials). An instance might be represented by ventilation inside the cavities whereas a more interesting approach concerns the coupling of internal heat reflective insulators to mechanized ventilation systems.

Author Contributions: Conceptualization, P.C. and D.B.; methodology, P.C., D.B., A.C. and L.A.T.; software, A.C.; validation, D.B., P.C. and A.C.; writing—original draft preparation, A.C.; writing—review and editing, D.B. and P.C. and L.A.T.; supervision, L.A.T., D.B. and P.C. All authors have read and agreed to the published version of the manuscript.

Funding: This research received no external funding.

Conflicts of Interest: The authors declare no conflict of interest.

Appendix A

This appendix illustrates the eight cases analyzed reporting for each: the stratigraphy, the thermal transmittance reached by means of different heat-reflective layers and a graph where the transmittances are compared to the Italian limits for the three main climatic zones, as done previously in the paper. Please note that the numbers in each stratigraphy always start from the external one.

Wall type 1

	Stratigraphy	s [m]	λ [W/mK]
1	lime-cement plaster	0.02	0.9
2	hollow brick 12 cm	0.12	0.387
3	air cavity	0.02	-
4	Thermoreflective insulation	-	-
5	air cavity	0.02	-
6	alveolated brick 20 cm	0.2	0.18
7	lime plaster	0.02	0.7

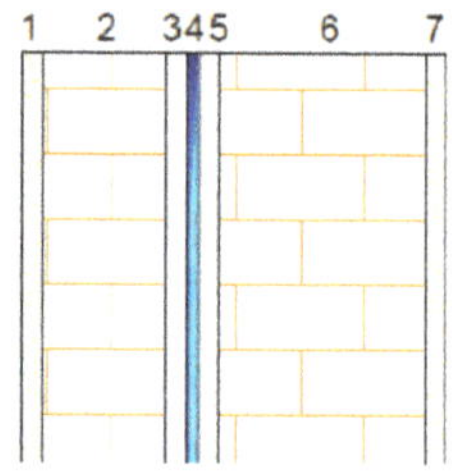

Figure A1. Wall type 1: stratigraphy with thicknesses and conductivities for the different layers. The numbers start from the outside.

Table A1. Wall type 1: thermal transmittance reached with different heat-reflective materials.

Material Number	Calculated Thermal Resistance [W/m^{2}K]
Only wall	0.546
1	0.311
2	0.174
3	0.254
4	0.215
5	0.315
6	0.295
7	0.291
8	0.252

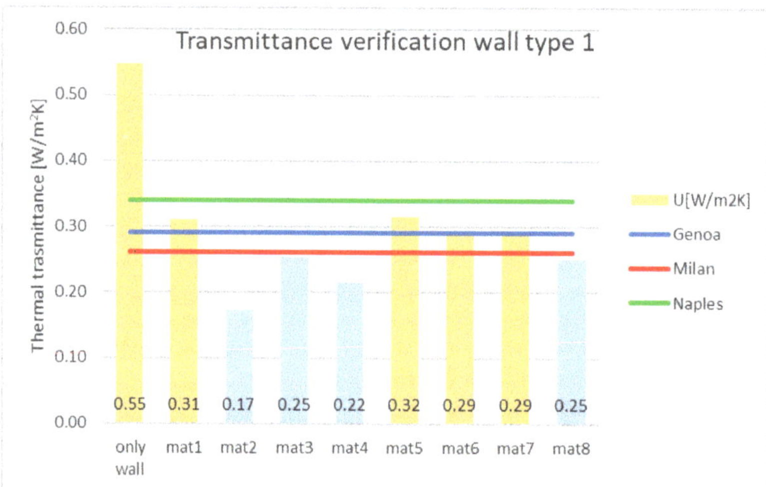

Figure A2. Wall type 1: transmittance check for the three main Italian climatic zones. In blue are the heat-reflective materials which respect all the limits.

Wall type 2

	Stratigraphy	s [m]	λ [W/mK]
1	lime-cement plaster	0.02	0.7
2	hollow brick 8 cm	0.08	0.387
3	air cavity	0.02	-
4	Thermoreflective insulation	-	-
5	air cavity	0.02	-
6	alveolated brick 30 cm	0.3	0.18
7	lime plaster	0.02	0.7

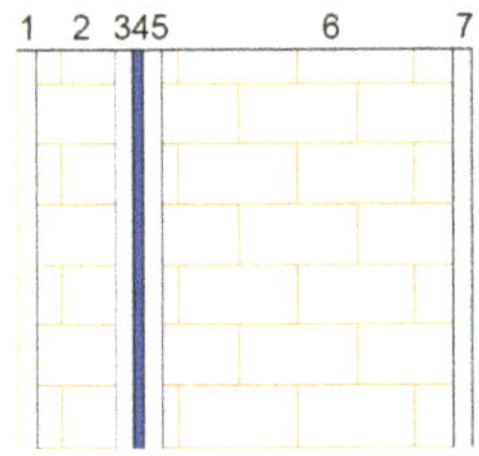

Figure A3. Wall type 2: stratigraphy with thicknesses and conductivities for the different layers. The numbers start from the outside.

Table A2. Wall type 2: thermal transmittance reached with different heat-reflective materials.

Material Number	Calculated Thermal Resistance [W/m^{2}K]
Only wall	0.437
1	0.272
2	0.162
3	0.228
4	0.196
5	0.276
6	0.260
7	0.257
8	0.226

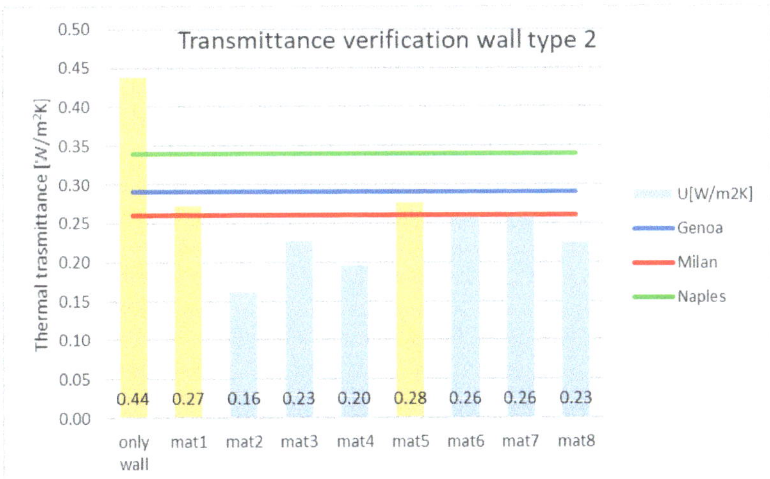

Figure A4. Wall type 2: transmittance check for the three main Italian climatic zones. In blue are the heat-reflective materials which respect all the limits.

Wall type 3.1

Stratigraphy		s [m]	λ [W/mK]
1	lime plaster	0.02	0.7
2	hollow brick 8 cm	0.08	0.387
3	air cavity	0.02	-
4	Thermoreflective insulation	-	-
5	air cavity	0.02	-
6	hollow brick 12 cm	0.12	0.387
7	lime-cement plaster	0.02	0.9

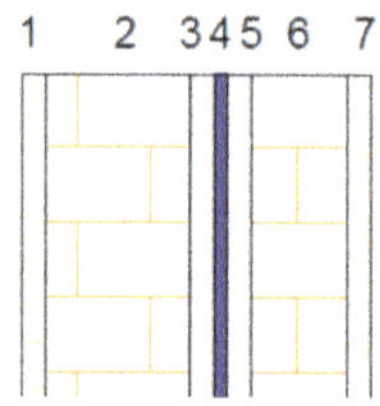

Figure A5. Wall type 3.1: stratigraphy with thicknesses and conductivities for the different layers. The numbers start from the outside.

Table A3. Wall type 3.1: thermal transmittance reached with different heat-reflective materials.

Material Number	Calculated Thermal Resistance [W/m²K]
Only wall	1.087
1	0.433
2	0.207
3	0.329
4	0.268
5	0.441
6	0.402
7	0.395
8	0.326

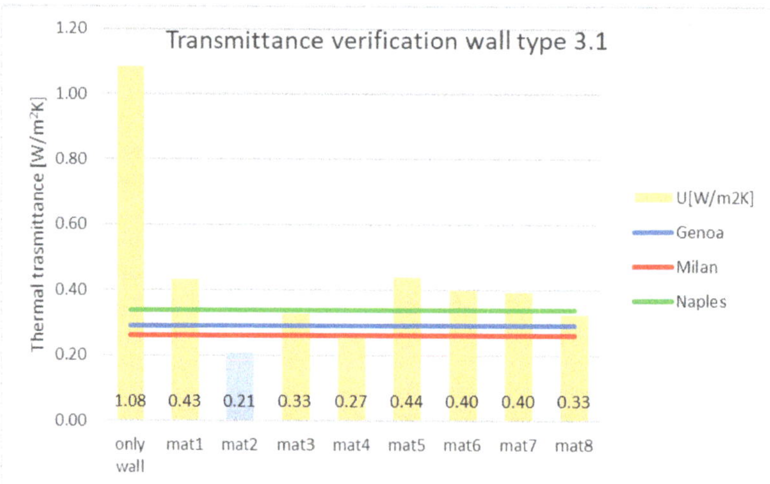

Figure A6. Wall type 3.1: transmittance check for the three main Italian climatic zones. In blue are the heat-reflective materials which respect all the limits.

Wall type 3.2

	Stratigraphy	s [m]	λ [W/mK]
1	lime plaster	0.02	0.7
2	hollow brick 8 cm	0.08	0.387
3	Thermoreflective insulation	-	-
4	air cavity	0.04	-
5	hollow brick 12 cm	0.12	0.387
6	lime-cement plaster	0.02	0.9

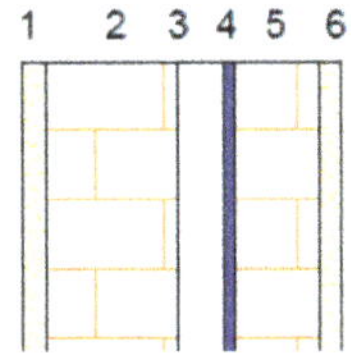

Figure A7. Wall type 3.2: stratigraphy with thicknesses and conductivities for the different layers. The numbers start from the outside.

Table A4. Wall type 3.2: thermal transmittance reached with different heat reflective materials.

Material Number	Calculated Thermal Resistance [W/m²K]
Only wall	1.087
1	0.600
2	0.240
3	0.422
4	0.334
5	0.624
6	0.549
7	0.543
8	0.395

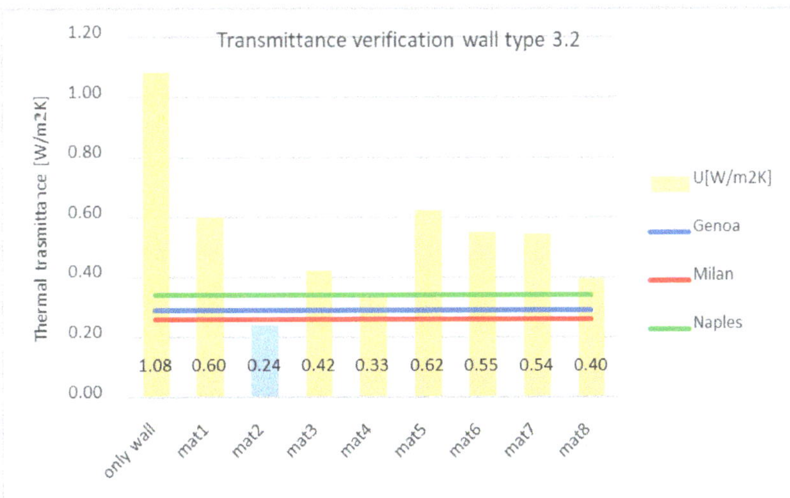

Figure A8. Wall type 3.2: transmittance check for the three main Italian climatic zones. In are blue the heat-reflective materials which respect all the limits.

Wall type 3.3

	Stratigraphy	s [m]	λ [W/mK]
1	drywall	0.012	0.2
2	air cavity	0.02	-
3	Thermoreflective insulation	-	-
4	air cavity	0.02	-
5	lime plaster	0.02	0.7
6	hollow brick 8 cm	0.08	0.387
7	air cavity	0.16	
8	hollow brick 12 cm	0.12	0.387
9	lime-cement plaster	0.02	0.9

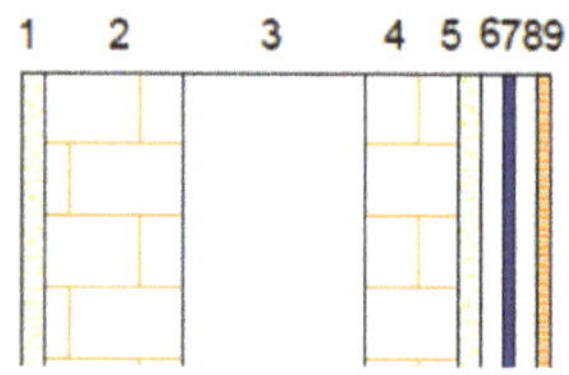

Figure A9. Wall type 3.3: stratigraphy with thicknesses and conductivities for the different layers. The numbers start from the outside.

Table A5. Wall type 3.3: thermal transmittance reached with different heat-reflective materials.

Material Number	Calculated Thermal Resistance [W/m^{2}K]
Only wall	1.087
1	0.392
2	0.197
3	0.305
4	0.251
5	0.398
6	0.366
7	0.360
8	0.302

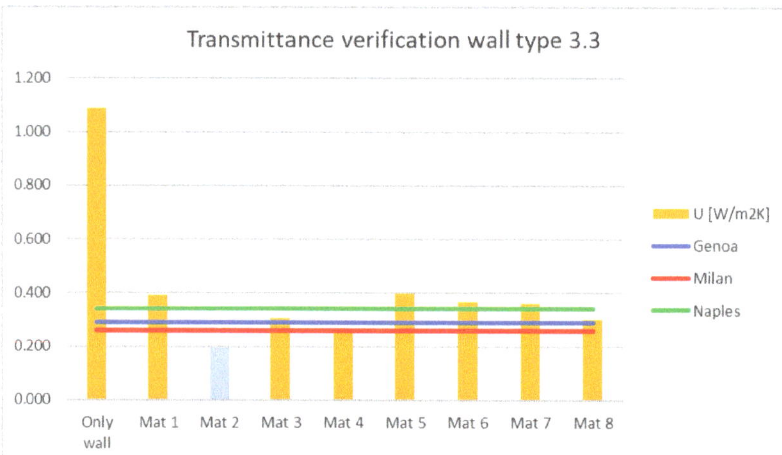

Figure A10. Wall type 3.3: transmittance check for the three main Italian climatic zones. In are blue the heat-reflective materials which respect all the limits.

Wall type 4

	Stratigraphy	s [m]	λ [W/mK]
1	Expanded sintered polystyrene EPS	0.03	0.04
2	Alveolated brick 30 cm	0.3	0.18
3	Air cavity	0.02	-
4	Thermoreflective insulation	-	-
5	Air cavity	0.02	-
6	Drywall	0.012	0.2

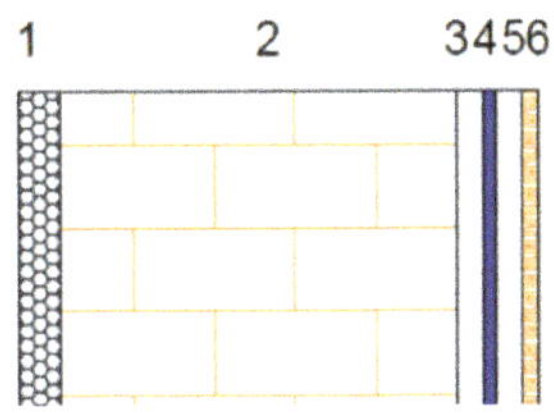

Figure A11. Wall type 4: stratigraphy with thicknesses and conductivities for the different layers. The numbers start from the outside.

Table A6. Wall type 4: thermal transmittance reached with different heat-reflective materials.

Material Number	Calculated Thermal Resistance [W/m^{2}K]
Only wall	0.386
1	0.237
2	0.148
3	0.202
4	0.177
5	0.239
6	0.227
7	0.225
8	0.201

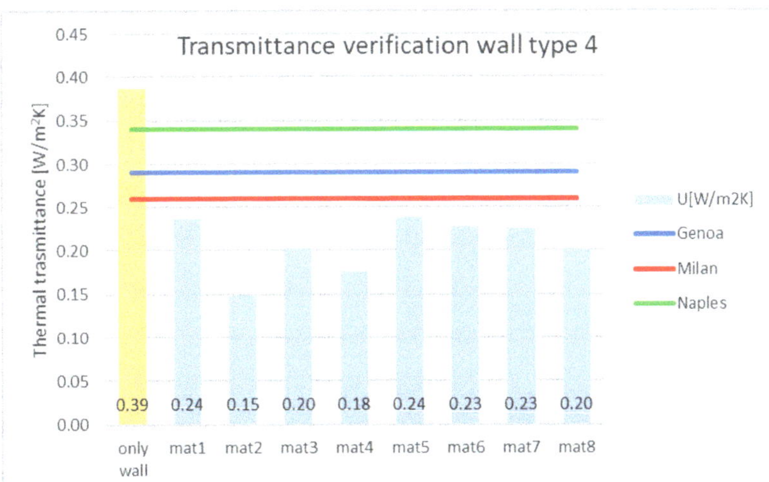

Figure A12. Wall type 4: transmittance check for the three main Italian climatic zones. In are blue the heat-reflective materials which respect all the limits.

Wall type 5.1

Stratigraphy		s [m]	λ [W/mK]
1	Stone wall	0.54	2.4
2	air cavity	0.02	-
3	Thermoreflective insulation	-	-
4	air cavity	0.02	-
5	Drywall	0.012	0.2

Figure A13. Wall type 5.1: stratigraphy with thicknesses and conductivities for the different layers. The numbers start from the outside.

Table A7. Wall type 5.1: thermal transmittance reached with different heat-reflective materials.

Material Number	Calculated Thermal Resistance [W/m^{2}K]
Only wall	2.326
1	0.485
2	0.218
3	0.359
4	0.287
5	0.495
6	0.447
7	0.438
8	0.355

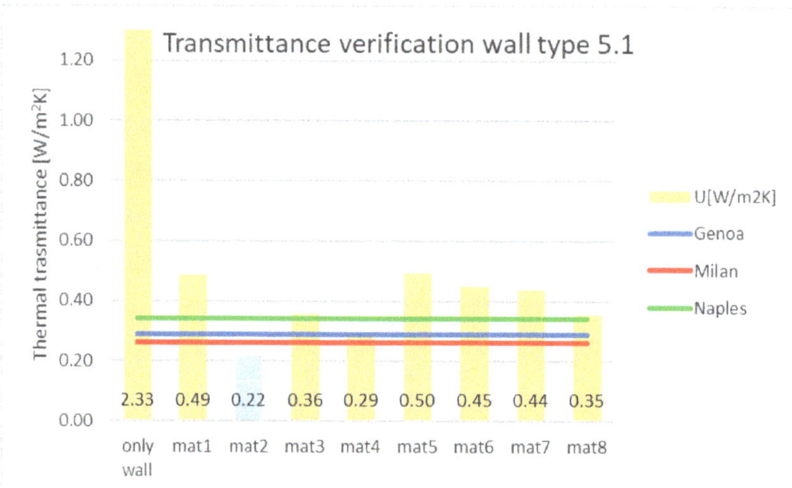

Figure A14. Wall type 5.1: transmittance check for the three main Italian climatic zones. In are blue the heat-reflective materials which respect all the limits.

Wall type 5.2

Stratigraphy		s [m]	λ [W/mK]
1	ceramic tiles	0.01	1.2
2	air cavity	0.02	-
3	Thermoreflective insulation	-	-
4	air cavity	0.02	-
5	lime-cement plaster	0.02	0.9
6	stone wall	0.5	2.4
7	lime plaster	0.02	0.7

Figure A15. Wall type 5.2: stratigraphy with thicknesses and conductivities for the different layers. The numbers start from the outside.

Table A8. Wall type 5.2: thermal transmittance reached with different heat-reflective materials.

Material Number	Calculated Thermal Resistance [W/m^{2}K]
Only wall	2.326
1	0.737
2	0.259
3	0.485
4	0.372
5	0.773
6	0.661
7	0.652
8	0.450

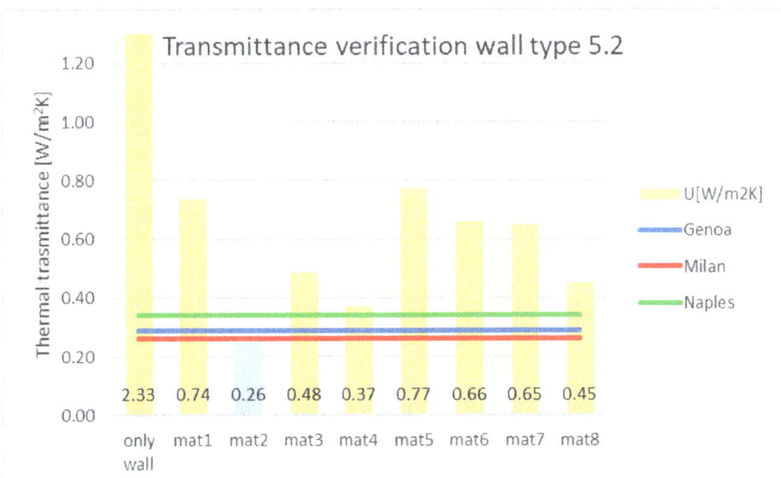

Figure A16. Wall type 5.2: transmittance check for the three main Italian climatic zones. In are blue the heat-reflective materials which respect all the limits.

References

1. Crude Oil Prices—70 Year Historical Chart. Available online: https://www.macrotrends.net/1369/crude-oil-price-history-chart (accessed on 5 July 2022).
2. *UNI EN 12664:2002*; Thermal Performance of Building Materials and Products—Determination of Thermal Resistance by Means of Guarded Hot Plate and Heat Flow Meter Methods—Dry and Moist Products of Medium and Low Thermal Resistance. British Standards Institution: London, UK, 2001.
3. *UNI EN 12667:2002*; Thermal Performance of Building Materials and Products—Determination of Thermal Resistance by Means of Guarded Hot Plate and Heat Flow Meter Methods—Products of High and Medium Thermal Resistance. British Standards Institution: London, UK, 2001.

4. Al-Homoud, M.S. Performance characteristics and practical applications of common building thermal insulation materials. *Build. Environ.* **2005**, *40*, 353–366. [CrossRef]
5. Papadopoulos, A.M. State of the art in thermal insulation materials and aims for future developments. *Energy Build.* **2005**, *37*, 77–86. [CrossRef]
6. Jelle, B.P.; Gustavsen, A.; Baetens, R. The path to the high-performance thermal building insulation materials and solutions of tomorrow. *Build. Phys. J.* **2010**, *34*, 99–123. [CrossRef]
7. Simmler, H.; Brunner, S.; Heinemann, U.; Schwab, H.; Kumaran, K.; Mukhopadhyaya, P.; Quénard, D.; Sallée, H.; Noller, K.; Kücükpinar-Niarchos, E.; et al. Vacuum Insulation Panels—Study on VIP-Components and Panels for Service Life Prediction of Vip in Building Applications (Subtask A). In *Technical Report Number SFOE-250129*; Swiss Federal Office of Energy: Berne, Switzerland, 2005. Available online: https://www.osti.gov/etdeweb/biblio/21131463 (accessed on 1 September 2022).
8. Wang, X.; Walliman, N.; Ogden, R.; Kendrick, C. VIP and their applications in buildings: A review. *Proc. Inst. Civ. Eng. -Constr. Mater.* **2007**, *160*, 145–153. [CrossRef]
9. Baetens, R.; Jelle, B.P.; Thue, J.V.; Tenpierik, M.J.; Grynning, S.; Uvsløkk, S.; Gustavsen, A. Vacuum insulation panels for building applications: A review and beyond. *Energy Build.* **2010**, *42*, 147–172. [CrossRef]
10. Alam, M.; Singh, H.; Limbachiya, M.C. Vacuum Insulation Panels (VIPs) for building construction industry–A review of the contemporary developments and future directions. *Appl. Energy* **2011**, *88*, 3592–3602. [CrossRef]
11. Beck, A.; Frank OBinder, M. Influence of water content on the thermal conductivity of vacuum panels with fumed silica kernels. In Proceedings of the 8th International Vacuum Insulation Symposium, Würzburg, Germany, 18–19 September 2007; pp. 1–9.
12. Kwon, J.S.; Jang, C.H.; Jung, H.; Song, T.H. Effective thermal conductivity of various filling materials for vacuum insulation panels. *Heat Mass Transf. Int. J.* **2009**, *52*, 5525–5532. [CrossRef]
13. Frick, J. From Dewars to VIPs-one century of progress in vacuum insulation technology. In Proceedings of the 7th International Vacuum Insulation Symposium, Dübendorf, Switzerland, 28–29 September 2005.
14. Mills, G.L.; Zeller, C.M. The performance of gas filled multilayer insulation. *AIP Conf. Proc. Am. Inst. Phys.* **2008**, *985*, 1475–1482.
15. Baetens, R.; Jelle, B.P.; Gustavsen, A. Phase change materials for building applications: A state-of-the-art review. *Energy Build.* **2010**, *42*, 1361–1368. [CrossRef]
16. Cuce, E.; Cuce, P.M.; Wood, C.J.; Riffat, S.B. Toward aerogel based thermal superinsulation in buildings: A comprehensive review. *Renew. Sustain. Energy Rev.* **2014**, *34*, 273–299. [CrossRef]
17. Lee, J.H.; Park, S.J. Recent advances in preparations and applications of carbon aerogels: A review. *Carbon* **2020**, *163*, 1–18. [CrossRef]
18. Baetens, R.; Jelle, B.P.; Gustavsen, A. Aerogel insulation for building applications: A state-of-the-art review. *Energy Build.* **2011**, *43*, 761–769. [CrossRef]
19. Mulet, J.P.; Joulain, K.; Carminati, R.; Greffet, J.J. Enhanced radiative heat transfer at nanometric distances. *Microscale Thermophys. Eng.* **2002**, *6*, 209–222. [CrossRef]
20. Joulain, K.; Mulet, J.P.; Marquier, F.; Carminati, R.; Greffet, J.J. Surface electromagnetic waves thermally excited: Radiative heat transfer, coherence properties and casimir forces revisited in the near field. *Surf. Sci. Rep.* **2005**, *57*, 59–112. [CrossRef]
21. Bozsaky, D. Recent studies on thermodynamic processes in nano-ceramic thermal insulation coatings. *Pollack Period.* **2019**, *14*, 107–116. [CrossRef]
22. Yu, H.; Zhang, H.; Zhao, J.; Liu, J.; Xia, X.; Wu, X. Thermal conductivity of micro/nano-porous polymers: Prediction models and applications. *Front. Phys.* **2022**, *163*, 1–18. [CrossRef]
23. Orbán, J. Use of nanotechnology in building industry, Part I and II. *Magy. Építéstechnika* **2012**, *50*, 40–43.
24. Chukhlanov, V.Y.; Trifonova, T.A.; Selivanov, O.G.; Ilina, M.E.; Chukhlanova, N.V. Thinfilm coatings based on hollow inorganic microspheres and polyacrylic binder. *Appl. Eng. Res. Int. J.* **2017**, *12*, 1194–1199.
25. Italian Ministry of Industry, Trade and Handicraft. *Italian Ministerial Decree 2 April 1998-Methods for Certifying the Characteristics and Energy Performance of Buildings and the Systems Connected to Them*; Italian Ministry of Industry, Trade and Handicraft: Rome, Italy, 1998. (In Italian)
26. *UNI EN ISO 10456:2008*; Building Materials and Products—Hygrothermal Properties—Tabulated Design Values and Procedures for Determining Declared and Design Thermal Values. International Organization for Standardization: Geneva, Switzerland, 2008.
27. *UNI EN 16012:2015*; Thermal Insulation for Buildings—Reflective Insulation Products—Determination of the Declared Thermal Performance. International Organization for Standardization: Geneva, Switzerland, 2015.
28. Baldinelli, G.; Asdrubali, F. Reflecting panels for radiators in residential buildings: Theoretical analysis of energy performance. In Proceedings of the 2008 Second International Conference on Thermal Issues in Emerging Technologie, Cairo, Egypt, 17–20 December 2008; pp. 449–454.
29. Mavromatidis, L.E.; Bykalyuk, A.; Lequay, H. Development of polynomial regression models for composite dynamic envelopes' thermal performance forecasting. *Appl. Energy* **2013**, *104*, 379–391. [CrossRef]
30. Mavromatidis, L.E.; Mankibi, M.; Michel, P.; Bykalyuk, A.; Santamouris, M. Guidelines to study numerically and experimentally reflective insulation systems as applied to buildings. *Adv. Build. Energy Res.* **2012**, *6*, 1–34. [CrossRef]
31. Troppova, E.; Tippner, J.; Švehlík, M. Numerical and experimental study of conjugate heat transfer in a horizontal air cavity. *Build. Simul.* **2017**, *11*, 339–346. [CrossRef]

32. Mavromatidis, L.E.; Bykalyuk, A.; Mankibi, M.; Michel, P.; Santamouris, M. Numerical investigation of a wall's optimum multilayer thermal insulation position. In Proceedings of the Building Simulation 2011: 12th Conference of International Building Performance Simulation Association, Sydney, Australia, 14–16 November 2011; pp. 1425–1438.
33. Medina, M.A.; O'Neal, D.L.; Turner, W.D. A transient heat and mass transfer model of residential attics used to simulate radiant barrier retrofits. Part I. Development. *ASME J. Sol. Energy Eng.* **1998**, *120*, 32–38. [CrossRef]
34. Mavromatidis, L.E.; Michel, P.; Mankibi, M. Study on transient heat transfer through multilayer thermal insulation: Numerical analysis and experimental investigation. *Build. Simul.* **2010**, *3*, 279–294. [CrossRef]
35. Medina, M.A. Effects of shingle absorptivity, radiant barrier emissivity, attic ventilation flowrate and roof slope on the performance of radiant barriers. *Int. J. Energy Res.* **2000**, *24*, 665–678. [CrossRef]
36. Saber, H.H. Thermal resistance of enclosed reflective airspaces in building applications. In Proceedings of the BEST4 Conference, Kansas City, MI, USA, 12–15 April 2015; pp. 1–16.
37. Close, K.B. Development and validation of numerical models for evaluation of foam-vacuum insulation panel composite boards, including edge effects. *Energies* **2018**, *11*, 2228. [CrossRef]
38. Mavromatidis, L.E.; Bykalyuk, A.; Mankibi, M.; Michel, P.; Santamouris, M. Numerical estimation of air gaps' influence on the insulating performance of multilayer thermal insulation. *Build. Environ.* **2012**, *49*, 227–237. [CrossRef]
39. Stovall, T.; Shrestha, S.; Arimilli, R.V.; Yarbrough, D.W.; Pearson, T. Analysis in support of the radiant barrier fact sheet 2010 update. In Proceedings of the Thermal Performance of the Exterior Envelopes of Whole Buildings XI International Conference, Clearwater Beach, FL, USA, 5–9 December 2010; pp. 1–9.
40. Medina, M.A.; O'Neal, D.L.; Turner, W.D. A Transient Heat and Mass Transfer Model of Residential Attics Used to Simulate Radiant Barrier Retrofits, Part II: Validation and Simulations. *ASME J. Sol. Energy Eng.* **1998**, *120*, 39–44. [CrossRef]
41. Miranville, F.; Lauret, P.; Medina, M.; Bigot, D. A Simplified Model for Radiative Transfer in Building Enclosures With Low Emissivity Walls: Development and Application to Radiant Barrier Insulation. *ASME. J. Sol. Energy Eng.* **2011**, *133*, 2120–2132. [CrossRef]
42. Winiarski, D.W.; O'Neal, D.L. A quasi-steady-state model of attic heat transfer with radiant barriers. *Energy Build.* **1996**, *24*, 183–194. [CrossRef]
43. Medina, M.A. On the performance of radiant barriers in combination with different attic insulation levels. *Energy Build.* **2000**, *33*, 31–40. [CrossRef]
44. Michels, C.; Lamberts, R.; Güths, S. Theoretical/experimental comparison of heat flux reduction in roofs achieved through the use of reflective thermal insulators. *Energy Build.* **2008**, *40*, 438–444. [CrossRef]
45. Chiu, W.K.; Richards, C.J.; Jaluria, Y. Experimental and numerical study of conjugate heat transfer in a horizontal channel heated from below. *J. Heat Transf.* **2001**, *123*, 688–697. [CrossRef]
46. Aviram, D.P.; Fried, A.N.; Roberts, J.J. Thermal properties of a variable cavity wall. *Build. Environ.* **2001**, *36*, 1057–1072. [CrossRef]
47. Roels, S.; Deurinck, M. The effect of a reflective underlay on the global thermal behaviour of pitched roofs. *Build. Environ.* **2011**, *46*, 134–143. [CrossRef]
48. D'Orazio, M.; Di Perna, C.; Di Giuseppe, E.; Morodo, M. Thermal performance of an insulated roof with reflective insulation: Field tests under hot climatic conditions. *J. Build. Phys.* **2013**, *36*, 229–246. [CrossRef]
49. Pásztory, Z.; Horváth, T.; Glass, S.; Zelinka, S. Experimental investigation of the influence of temperature on thermal conductivity of multilayer reflective thermal insulation. *Energy Build.* **2018**, *174*, 26–30. [CrossRef]
50. Ashhar, M.; Haw, L. Recent research and development on the use of reflective technology in buildings—A review. *J. Build. Eng.* **2022**, *45*, 2352–2361. [CrossRef]
51. *UNI EN ISO 6946:2018*; Building Components and Building Elements—Thermal Resistance and Thermal Transmittance—Calculation Methods. International Organization for Standardization: Geneva, Switzerland, 2018.
52. Italian Council of Ministers. *Italian Law Decree 19 May 05 2020 n. 34—Urgent Measures in the Field of Health, Support for Work and the Economy, as Well as Social Policies Related to the Epidemiological Emergency from COVID-19*; Italian Council of Ministers: Rome, Italy, 2020. (In Italian)
53. *EN 15976:2011*; Flexible Sheets for Waterproofing—Determination of Emissivity. European Committee for Standardization: Bruxelles, Belgium, 2011.
54. *UNI/TR 11552:2014*; Abacus of the Structures Constituting the Opaque Envelope of the Buildings—Thermophysical Parameters. Italian National Unification Body: Rome, Italy, 2014.
55. Ministry of Economic Development. *Italian Interministerial Decree 26 June 2015—Application of Methods for Calculating Energy Performance and Definition of Prescriptions and Minimum Requirements for Buildings*; Ministry of Economic Development: Rome, Italy, 2015. (In Italian)
56. Tuominen, P.; Reda, F.; Dawoud, W.; Elboshy, B.; Elshafei, G.; Negm, A. Economic appraisal of energy efficiency in buildings using cost effectiveness assessment. *Procedia Econ. Financ.* **2015**, *21*, 422–430. [CrossRef]

energies

MDPI

Article

Life Cycle Analysis of Innovative Technologies: Cold Formed Steel System and Cross Laminated Timber

Ornella Iuorio [1,*], Antonio Gigante [2] and Rosa Francesca De Masi [2]

[1] School of Civil Engineering, University of Leeds, Leeds LS2 9JT, UK
[2] DING—Department of Engineering, University of Sannio, 82100 Benevento, Italy
* Correspondence: o.iuorio@leeds.ac.uk; Tel.: +44-0113-343-2294

Abstract: Reducing the embodied and operational energy of buildings is a key priority for construction and real estate sectors. It is essential to prioritize materials and construction technologies with low carbon footprints for the design of new buildings. Off-site constructions systems are claimed to have the potential to deliver a low carbon build environment, but at present there are a lack of data about their real environmental impacts. This paper sheds lights on the environmental performance of two offsite technologies: cold formed steel and cross laminated timber. Specifically, the environmental impacts of a CFS technology are discussed according to six standard impact categories, which includes the global warming potential and the total use of primary energy. The study is based on a detailed cradle to gate life cycle analysis of a real case study, and discusses the impacts of both structural and non-structural components of CFS constructions. As a useful frame of reference, this work compares the environmental impacts of 1 m^2 of walls and floors of CFS technology with those of cross laminated timber, which is spreading as innovative off-site technology for the development of nearly zero energy buildings, and a conventional reinforced masonry technology, which is largely adopted in the Italian construction sector. The paper concludes with the necessity to optimize structural systems to reduce the overall embodied carbon impacts.

Keywords: life cycle analysis; cold formed steel; cross laminate timber; net zero; embodied carbon; greenhouse gases

1. Introduction

The built environment is accountable for 42% of the EU's total energy consumption and for about 35% of the greenhouse gas emissions [1]. There is an utmost urgency to reduce the carbon footprint of both existing constructions and new buildings. Numerous European policies and legislations have been recently emanated to advance and support the environmental sustainability of the construction sector. They promote the adoption of life cycle assessment used from the early stage of the design process to critically evaluate and optimize the use of material and guide towards the adoption of construction systems, processes and materials having lower environmental impacts, minimized waste and minimized water consumption. The trend is to move towards mandatory reporting of carbon emissions in the build environment, along with limiting embodied carbon (EC) emissions in projects.

The life cycle assessment (LCA) of a building includes embodied energy (EE) and operational energy. Operational energy typically has a larger impact on the total life cycle energy of a building over long lifetimes. However, as the operational efficiency of new buildings is improved, the relative significance of the embodied impacts of construction materials and processes increases [2]. Therefore, for buildings having short life-time span, the impact of embodied carbon over their full life-cycle impact is extremely significant [3]. In recognition of this, significant attention is now being paid to the quantification and reduction of the embodied carbon impacts of construction products. In the most complete form, the term 'embodied carbon' (EC) refers to the lifecycle greenhouse gas emissions, referred to as carbon dioxide equivalents (CO_2e), which arise through the manufacture and transportation of construction materials and components, during the construction process, the maintenance and the end-of-life of a construction, including demolition, reuse and recycling.

However, the embodied carbon evaluation can be restricted to pre-defined system boundaries. For example, in [4], the EE and EC contributions are calculated as the energy consumed and emissions triggered by mining, processing and manufacturing of materials ("cradle to gate"), while [5] evaluated the embodied emissions, considering a 'cradle-to-site' approach, including the materials transport to construction site. Studies have also started to quantify the impact of recycling and reuse. For instance, Peng et al. [6] have found that if the construction material is recycled, the total embodied carbon saving potential from 2018 to 2060 in China would be of 92.26 Mt CO_2e. Among the construction systems, the prefab sector is specifically embracing the challenges of producing energy-efficient dwellings by lowering both EC and operational energy. Prefab constructions, also known as offsite, have the capacity to consistently produce high energy efficient offsite buildings [7]. However, there is a lack of scientific studies that can demonstrate this. Therefore, this study looks at two very promising off-site technologies, i.e., cold formed steel (CFS) and cross laminate timber (CLT). They are the focus of this study as they are dry construction systems, currently having exponential grow thanks to their capacity of delivering high structural performance in seismic areas while also having low carbon footprint. This paper proposes a detailed comparison between three different technologies: cold formed steel (CFS), cross laminate (CLT) and a traditional reinforced masonry system. Specifically, starting from a real case study, which is a school building in Southern Italy, the environmental impacts in a cradle-to-gate perspective of the CFS technology are analysed. The embodied carbon emissions are calculated by using real project data, with the aim to develop useful data for future designers and researchers. Then, the most environmentally intensive components, such as the walls and roof, are compared with the other two technologies: CLT, which is finding wide application in nearly zero energy building, and a conventional reinforced masonry technology. The proposed comparison allows the evaluation of the difference in terms of carbon footprint between two innovative solutions, which designers could select for improving the global sustainability of the building sector.

Therefore, Section 2 describes the case studies with a detailed description of the structural and technological systems adopted in the examined CFS, CLT and traditional systems; Section 3 describes the methodological approach adopted for the cradle-to-gate life cycle assessment of the investigated CFS school and for the comparison between walls and roof of the CFS buildings, the NZEB building developed for the same climatic zone, and having CLT as primary structure, and those of a traditional construction widespread in the same region. Section 4 shows and discusses the obtained results, and Section 5 indicates the necessary steps that the construction sector should take to reduce its environmental impacts.

2. Case Studies

This section presents the investigated case studies. Note that, while the CFS school is discussed in detail, so that a comprehensive understanding of the system and of the environmental impacts of all structural and non-structural components is possible, instead, only the strictly necessary information are given for the comparison with the CLT case study and the traditional masonry building.

2.1. CFS System

The "Foundation and Primary Stage School" (British Force School, Figure 1) is a nursery and primary school realized in 2009 for the British Armed Forces in Naples, Italy [8]. The school covers an area of about 3000 m^2 and has a usable area of about 2600 m^2. This school was a pioneer in adopting the CFS system in Italy, to respond the customer's requirements of fast delivery while enabling high seismic performance and high energy efficiency. The building is located in an area of medium seismic intensity, a grade 2 seismic zone according to the Italian classification [9], and was designed according to an elastic approach with coefficient of structure equal to 1. It was also designed and built to achieve the best energy class. The CFS system adopted for the British school is based on stick-built construction, which entails CFS components being formed and cut to size in the factory, then transported on site where they were assembled with mechanical fasteners. The decision to adopt such a level of prefabrication was dictated by difficult site access, that limited the possibility of using heavy tracks and bringing large prefab components. Figure 1 shows the construction process of CFS components, from CFS walls erection (Figure 1a) to wall bracing and roof erection (Figure 1b) and the final result (Figure 1c). The opaque elements of the building made with a CSF structure are described in detail in the following sub-paragraphs.

Figure 1. CFS school construction process: (**a**) CFS structural walls, (**b**) Sheathing of CFS walls, and construction of the roof, (**c**) Finishing and final result.

Structural and Technological System

The building is set on a continuous slab foundation, which is composed of reinforced concrete beams 800 mm height and with the width ranging between 500 and 1500 mm. All the spaces among the beams are filled with recycled plastic modular elements, having sizes in plan 58 × 58 mm and 700 mm high, and a 100 mm thick reinforced concrete slab. This solution was chosen to realize a perfect planar surface, which facilitated the introduction of all the required cables and pipes for the water, electrical and mechanical systems. The foundation sets also the basis for the ground floor, as shown in Figure 2.

The walls were realized and assembled on site with 150 × 50 × 10 × 1.5 mm studs (C lipped channel), which are spaced at 600 mm and connected at the ends to 152 × 40 × 1.5 mm U section wall tracks. All CFS profiles are made by zinc coated and dip-hot galvanized S320 steel. To understand the environmental sustainability of CFS profiles, it is important to notice that they can be manufactured to precise measurements, having minimal job site scrap and consequent minimal waste, which can certainly then be recycled. These characteristics make CFS system particularly appropriate for a sustainable management of building sites [10]. CFS walls were braced with 9 mm thick type 3 OSB panels (ISO 6308, 1980) on both sides of the walls. S/HD10B metal-to-metal connectors [11] and HIT-RE 500 with HIS adhesive-bonded anchors were placed as hold-down connections at the corners of resisting walls and 8 mm mechanical connection were instead used as shear anchors along the walls. The wall stratigraphy is reported in Figure 3 and aimed at the eco-efficiency of the building in its life-cycle, starting from the use of eco-compatible materials, to the choice of envelope solutions as energetically efficient, and integrated with envelope passive energy strategies. Specifically, the insulation is achieved with 100 mm hemp fibre placed within the studs, two panels of CELENIT for a total thickness of 75 mm on the exterior of the load bearing walls, an air gap of 35 mm and a cement-based panel, used as finishing.

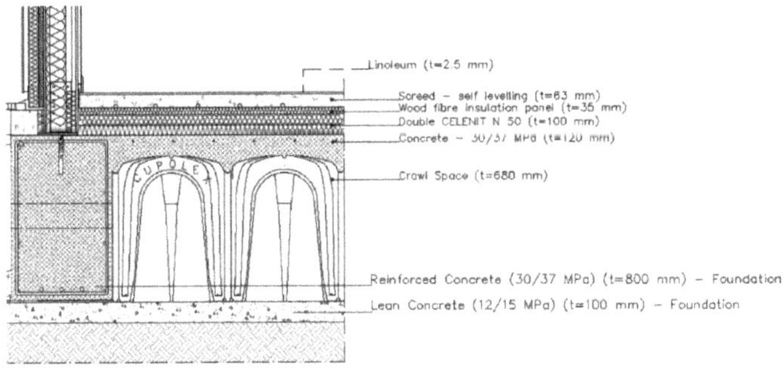

Figure 2. Foundation and ground floor sections.

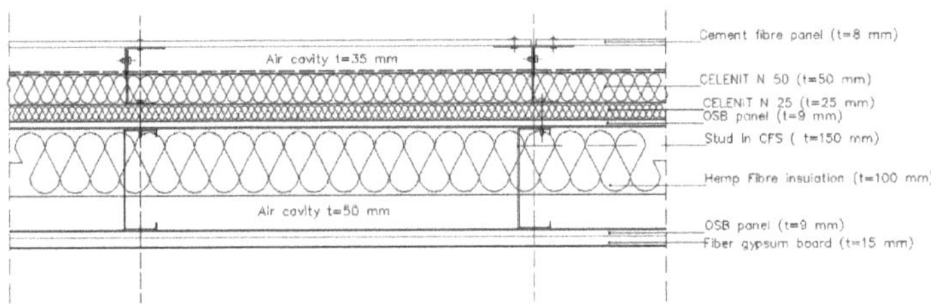

Figure 3. External wall section of the British Foundation school.

Internally, 15 mm gypsum-based boards are included to achieve the required fire performance. The internal walls are described in Figure 4, and are divided in load-bearing and not-load-bearing. The load-bearing are also composed of 150 mm wide studs, braced on both sides by 9 mm OSB type 3 panels, finished with two gypsum-based panels on both sides. The building has been structurally designed according to a "sheathing braced methodology" [12] that counts on the collaboration between studs and sheathing panels. The not-load-bearing walls [13] comprise 100 mm width studs, completed on both sides with two gypsum-based panels.

The roof is realized by 300 × 50 × 20 mm CFS joists, with thickness ranging between 1.50 and 3.00 mm, spaced at 600 mm. CFS tracks connected the joists at each end of the floor, and were made of 303 × 50 × 1.50 mm or 306 × 50 × 3.00 mm U sections. On the top flanges of floor joists, OSB panels are screwed and solid blocks, realized by 250 × 50 × 20 × 1.50 mm C-sections, are placed between the end joists to provide edge support to the sheathing. The floor, overall, acts as a diaphragm with joists as primary member and the OSB sheathing panels acting as bracing system.

The roof is insulated externally with 2 CELENIT boards of 25 mm thickness each, and one hemp fibre 100 mm thick, and completed with a soundproof corrugated metal sheet, placed on an adjustable load-bearing structure to obtain the required slope for the rain collection. Figure 5 includes all details.

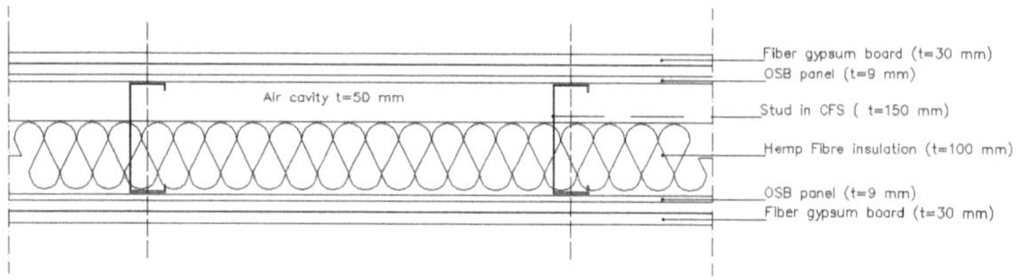

(a)

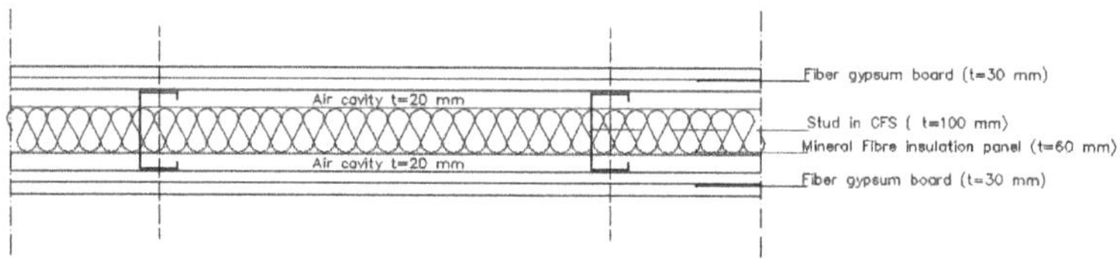

(b)

Figure 4. Internal wall sections: (**a**) load-bearing wall; (**b**) non-load-bearing wall.

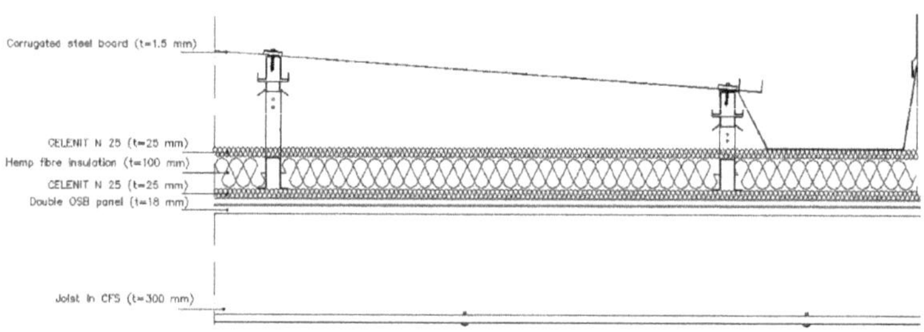

Figure 5. Roof section.

2.2. Cross Laminated Timber Technology

Cross laminated timber systems (CLT) are composed of CLT panels (walls and floors) having load-bearing function. CLT is a prefabricated system suitable for the construction of load-bearing walls as well as for the construction of attics and sloping roofs. It is ideal for anti-seismic constructions thanks to its excellent mechanical properties. The CLT panels are usually delivered to the construction site ready for assembly. CLT systems are an alternative dry-construction system to CFS, which are recently becoming widespread thanks to the eco-sustainability of the production process, the achievable high energy efficiency of the building system, and good weight-to-performance ratio and sound insulation properties. The CLT system is often considered for green building designs because the main raw materials are the wood and glues that do not emit solvents or formaldehyde. An example of application of CLT is the BNZEB, which is a nearly zero energy detached house built in Benevento, a city in Southern Italy. The building is located in the same climatic zone [14] as the previously presented CFS case study. The structure is designed for a seismic zone 1, characterized by very high seismic risk [9] and with a coefficient of structure equal to 1.5. The BNZEB is described in detail in Ascione et al. [15,16]. The vertical load-bearing structure is made of 9.5 cm thick CLT panels with five crossed layers. Each layer is composed of side-by-side solid wood boards, class C24, which have a thickness of 1.9 cm. The overlapping layers are crossed and rotated with respect to each other by 90°. The tree essence of which the layers are made is the spruce. The vertical load bearing panels represent the totality of the perimeter walls and internal partitions of the building. The panels are generally continuous and are interrupted only at the intersections with orthogonal walls. The continuity of the various panels was established by seeking the right compromise between production, construction and structural needs. The walls have the task of absorbing both the vertical and horizontal loads (accidental loads, earthquake and wind). In particular, the vertical panels have a membrane-like structural behaviour and are connected to the foundation by metal Hold-Downs anchors arranged on both panel faces at the beginning and end of each wall panel, as well as at the beginning and end of each doorway or window. The panels do not rest directly on the ribs of the concrete slab below the building, but between the latter and the panels themselves; 9.5 × 12 cm levelling strips are interposed, connected to the panels by means of anchoring screws. Wall-to-wall crossover connections are made using crossed anchor screws. The roof deck is also made with five-layer CLT panels whose thickness is equal to 12 cm. Each layer is composed of solid wood boards, class C24, which have the following thicknesses of 1.9–2.15–3.9–2.15–1.9 cm. The roof panels have a plate behaviour and are connected to the vertical wall panels with metal angles and screws. Strips of soundproofing material are interposed between the roof and wall panels. The CLT is coupled to one or more layers of wood wool thermal insulation, with different density and thermal conductivity, respectively, in the case of walls and roof, as shown in Figure 8b and Figure 9b. The thermal transmittance of the walls is 0.17 W/m^2 K while the thermal transmittance of the roof is 0.23 W/m^2 K.

2.3. *Traditional Construction*

The majority of traditional constructions in Southern Italy are reinforced masonry systems. Therefore, in this study, a traditional construction composed of perforated clay bricks (350 mm thick), filled with reinforced concrete, lined with 70 mm of mineral wool, and completed with render is considered. This configuration, with a total thickness of 550 mm, has a thermal transmittance of 0.22 W/m^2 K. The considered roof is a traditional flat roof made of concrete blocks, as retrieved by the independent Technical Report UNI/TR 11552, which is part of the update of UNI/TS 11300, developed as set of new regulations for the energy reduction in buildings by the European Commission [17–20]. For the aim of the comparison to the retrieved roof stratification, 100 mm of expanded polystyrene (EPS) with thermal conductivity (λ) equal to 0.035 W/mK is added to allow the roof to reach a U value equal to 0.28 W/m^{2}K [21].

3. Life Cycle Analysis (LCA)

3.1. *Goal and Scope Definition*

The goals of the LCAs developed in this study are: 1. to investigate the environmental impacts of the CFS system in a cradle-to-gate approach, to understand what are the components which are mostly responsible for the overall EC; 2. To compare the EC of a CFS system with that of a CLT system and a traditional reinforced masonry system.

3.2. *LCA Methodology*

The life cycle analysis (LCA) is carried out from "cradle-to-gate" and includes both structural and non-structural components of the investigated construction systems. The LCA boundaries include only the production stage (A1–A2–A3), as indicated in Figure 6.

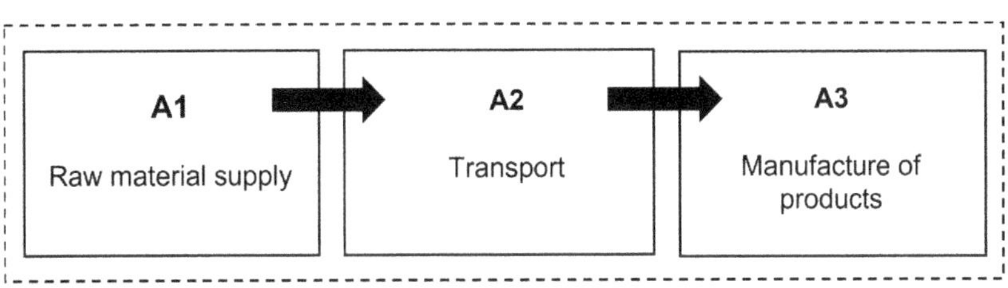

Figure 6. Cradle-to-gate system boundaries. According to EN 15978.

Modules A1, A2 and A3 are indicated as a single aggregated module A1–A3, which comprises the impacts of all materials and products' production and supply, as well as waste processing during the production stage. More specifically, the module "raw material supply" (A1) considers the emissions caused by raw materials when taken from nature, transported to industrial plants and processed. The "transport impacts" module (A2) involves the emissions due to the transport of raw materials from providers to the manufacturing plant and then to the construction site. The "manufacturing" module (A3) includes the impacts of production and fuels used by machines, plus transformation of the waste generated in the manufacturing production until end-of-waste state. A1–A3 module includes all the steps from the cradle-to-gate of the building components adopted for the case study construction. The total GHG emissions deriving from phases A1 to A3 represent the embodied emissions of the building.

The life cycle analysis is developed with the use of One Click LCA automated life cycle assessment software [22], according to the requirement of the EN 15978 standard [23], which is in line with the ISO 14040/44 standard [24]. Specifically, this study adopts the Center of Environmental Science of Leiden University (CML) methodology, to evaluate the environmental impacts in accordance with the EN 15978 and EN 15804 standards [23]. The CML methodology is integrated in One Click LCA and allows the development of environmental impact assessments of products and processes [25]. In this study, the life cycle impact analysis considers six standard impact categories, as reported in Table 1.

Table 1. Adopted impact categories.

Impact Category	Abbreviation	Unit
Global warming potential	GWP	$kgCO_{2\text{-}eq}$
Acidification potential	AP	$kgSO_{2\text{-}eq}$
Eutrophication potential	EP	$kgPO_{4\text{-}eq}$
Ozone depletion potential	ODP	$kgCFC_{11\text{-}eq}$
Formation of ozone of lower atmosphere	POCP	$kgC_2H_{4\text{-}eq}$
Total use of primary energy	TUPE	MJ

According to the LCA methodology, the bill of materials has been calculated for each study, and then for each material the corresponding cradle-to-gate impacts have been evaluated on the basis of the corresponding EPD, retrieved in the One Click databases. When the precise material was not available, then the environmental profiles of materials, which are similar in terms of thermophysical properties, country of production and production process were considered (Figure 7).

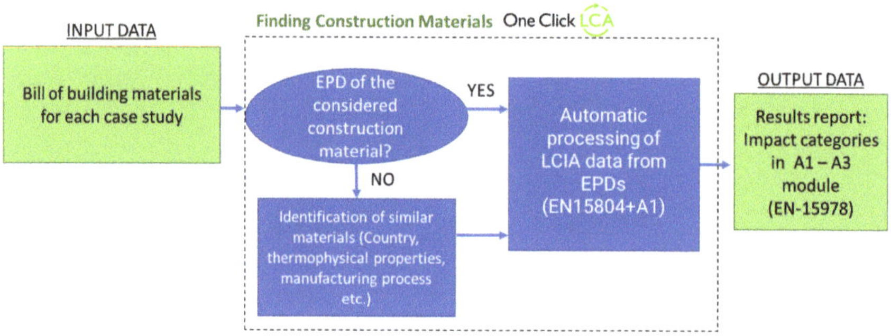

Figure 7. Workflow scheme.

Among all the impact categories, the global warming potential (GWP), which is defined as a relative measure of how much heat a greenhouse gas traps in the atmosphere, is calculated in carbon dioxide equivalents, meaning that the greenhouse potential of emission is given in relation to $CO_{2\text{-eq}}$. In addition, among the other EN standard impact categories, the total use of primary energy is considered, which represents the amount of both non-renewable and renewable primary energy (MJ), excluding those used as raw materials. For the sake of simplicity, in this study the Total Use of Primary Energy will be indicated by the acronym TUPE.

3.3. Functional Units for the LCA Comparisons

The functional units for the comparison of the three investigated systems (CFS, CLT and traditional construction) are 1 m^2 of wall without glazed elements and 1 m^2 of roof. All panels include both load-bearing and non-load-bearing elements.

Figures 8 and 9 show the sections of the three analysed construction typologies (i.e., CFS, CLT and traditional construction) for both walls and roofs, while Table 2 reports their thermal transmittance values and the whole thickness.

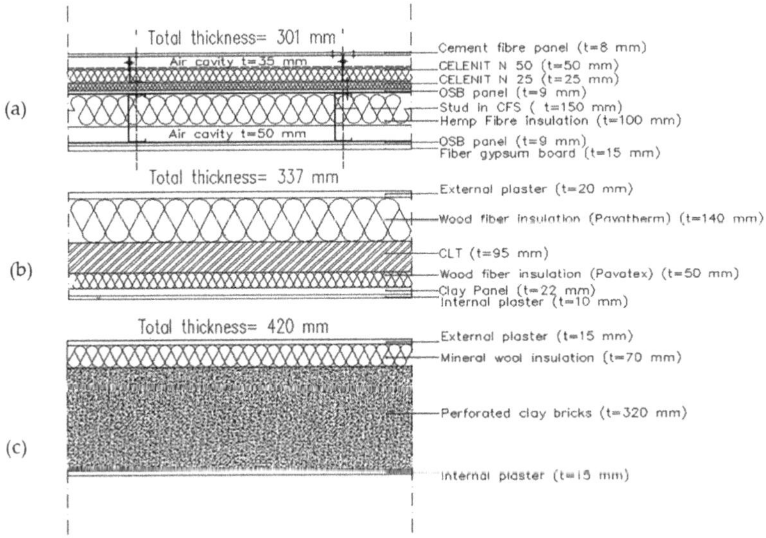

Figure 8. Wall configurations: (**a**) CFS; (**b**) CLT; (**c**) traditional.

Table 2. Total thickness and thermal transmittance values.

Wall	Total Thickness [mm]	U [W/m² K]
CFS	301	0.220
CLT	337	0.172
Traditional	420	0.217
Roof	**Total Thickness [mm]**	**U [W/m² K]**
CFS	169.5	0.290
CLT	498	0.229
Traditional	470	0.283

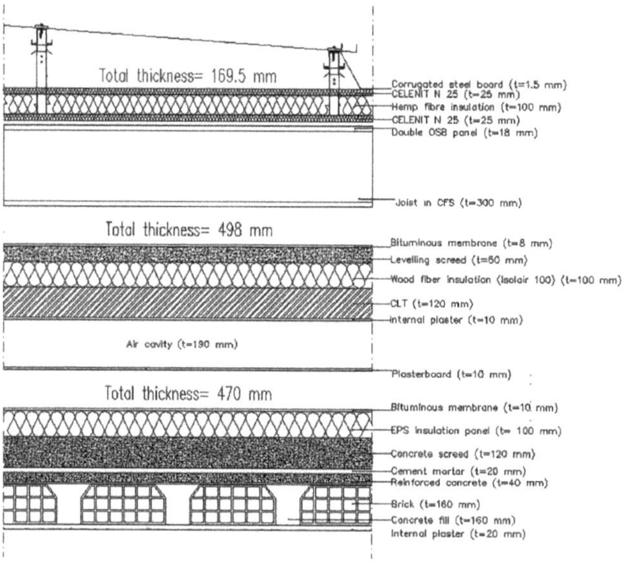

Figure 9. Roof configurations: (**a**) CFS; (**b**) CLT; (**c**) traditional.

3.4. Life Cycle Inventory (LCI for the A1–A3 Phase) for the CFS Case Study

In order to analyse the environmental impact of the CFS case study, the following Tables 3–7 show the quantities of materials used for the walls (both load-bearing and non-load-bearing walls) and roof of the British Force School, making the distinction between materials for structural and non-structural components.

Table 3. Bill of materials: foundation.

Construction Material	Quantity	Struct./Non-Struct.
Concrete (12/15 MPa)	199 m^3	Non-Struct.
Concrete (30/37 MPa)	424 m^3	Struct.
Reinforcement	18.3 ton	Struct.

The environmental data source and the relative upstream database for each investigated material is reported in Appendix A. Through this information, the life cycle inventory phase for the CFS system was completely characterized.

Table 4. Bill of materials: ground floor.

Construction Material	Quantity	Struct./Non-Struct.
Crawl space	1994 m^2	Struct.
Concrete (30/37 MPa)	240 m^3	Struct.
Wood wool panels—CELENIT N 50	199.4 m^3	Non-Struct.
Linoleum	1994 m^2	Non-Struct.
Wood fibre Insulation	80 m^3	Non-Struct.
Screed self levelling	126 kg	Non-Struct.

Table 5. Bill of materials: load-bearing walls.

Construction Material	Quantity	Struct./Non-Struct.
Cold Formed steel (CFS)	44.6 ton	Struct.
OSB panels	52.6 m^3	Struct.
Wood wool panels—CELENIT N 25	28 m^3	Non-Struct.
Wood wool panels—CELENIT N 50	56 m^3	Non-Struct.
Fibre-cement panels	8.9 m^3	Non-Struct.
Gypsum fibreboard—Knauf Vidifire	106 m^3	Non-Struct.
Hemp Fibre insulation	300 m^3	Non-Struct.

Table 6. Bill of materials: non load bearing walls.

Construction Material	Quantity	Struct./Non-Struct.
Cold Formed steel (CFS)	6.4 ton	Non-Struct
Mineral fibre insulation	46 m^3	Non-Struct.
Gypsum fibreboard—Knauf Vidifire	46 m^3	Non-Struct.

Table 7. Bill of materials: roof.

Construction Material		Quantity	Struct./Non-Struct.
Cold Formed steel (CFS)	ton	50.9	Struct.
OSB panels	m^3	45	Struct.
Wood wool panels—CELENIT N 25	m^3	126	Non-Struct.
Hemp Fibre insulation	m^3	252	Non-Struct.
Corrugated galvanized steel	m^2	2524	Non-Struct.

3.5. Life Cycle Inventory (LCI for the A1–A3 Phase) for the Comparison between CFS, CLT and Traditional System

For the second aim of this paper, which is the comparison between 1 m^2 of wall and roof of the CFS, CLT and traditional system, presented in Section 2, the bill of materials have been also calculated as reported in Tables 8–10. For each material, the environmental data source and the thermal conductivity are also reported.

Some criticalities can be underlined for the selection of the EDPs. The values available in the One Click LCA tool are mostly related to materials of recent construction with manufacturing processes that may have undergone important changes over the years, compared to the original construction period of the chosen traditional configurations (see Section 4.2). Building materials of the same type and with similar density and the thermal conductivity among those available are considered. For example, the perforated concrete blocks considered for the wall and the hollow roof bricks are produced by CERANOR S.A. [26] (Spain), with environmental product declaration (EPD 008-007). When the data were not available from the manufacturer, the professional database GABI/Thinkstep [27] was used.

Table 8. Bill of material, thermal conductivity and data source: CFS configurations for 1 m^2 of wall and roof.

	Construction Material	Mass per Unit [kg/m^2]	λ [W/m K]	Environment Data Source (EPD Number)
Wall	Fibre-cement panels	16	1.310	EPD Grossformatige Faserzementplatten pigmentiert beschichtet Carat/Reflex/Avera/Zenor/Aura/Integral/Plan Swisspearl Group AG (EPD-SWP-20180031-IAD1-DE)
	OSB panels (wall)	5.4	0.156	Oekobau.dat 2017-I, EPD SWISS KRONO OSB Panels SWISS KRONO Tec AG (EPD-KRO-20150067-IBD2-EN)
	Wood wool panels—CELENIT N 25 (wall)	11.5	0.070	EPD ACOUSTIC AND THERMAL PANELS CELENIT ABE, AE, AB, A, NB, N (S-P-00477)
	Wood wool panels—CELENIT N 50	18	0.070	EPD ACOUSTIC AND THERMAL PANELS CELENIT ABE, AE, AB, A, NB, N (S-P-00477)
	CFS	5.9	48	EPD Cold-Formed Steel Products (4789752901.101.1)
	Gypsum fibreboard—Knauf Vidifire	19.9	0.350	EPD Gypsum fibreboards Knauf Bulgaria (EPD-KNB-20130006-IAC1-EN)
	Hemp Fibre insulation (wall)	4	0.040	EPD EKOLUTION® HEMP FIBRE INSULATION (S-P-01961)
Roof	OSB panels (roof)	10.8	0.156	Oekobau.dat 2017-I, EPD SWISS KRONO OSB Panels SWISS KRONO Tec AG (EPD-KRO-20150067-IBD2-EN)
	Wood wool panels—CELENIT N 25 (roof)	23	0.070	EPD ACOUSTIC AND THERMAL PANELS CELENIT ABE, AE, AB, A, NB, N (S-P-00477)
	Hemp Fibre insulation (roof)	4	0.040	EPD EKOLUTION® HEMP FIBRE INSULATION (S-P-01961)
	CFS	13.8	48	EPD Cold-Formed Steel Products (4789752901.101.1)
	Corrugated galvanized steel	11.8	-	EPD for AlumiGard, MagnaFlow and ZinaCore pre-painted roofing and cladding for use in Australia (S-P-01540)

Table 9. Bill of material, thermal conductivity and data source: CLT configurations for 1 m^2 of wall and roof.

	Construction Material	Mass per Unit [kg/m^2]	λ [W/m K]	Environment Data Source (EPD Number)
Wall	Internal gypsum plaster (wall)	12	0.910	Oekobau.dat 2017-I, EPD GIPSPUTZ Bundesverband der Gipsindustrie e.V. (EPD-BVG-20140073-IAG1-DE)
	External gypsum plaster	20	0.910	Oekobau.dat 2017-I, EPD GIPSPUTZ Bundesverband der Gipsindustrie e.V. (EPD-BVG-20140073-IAG1-DE)
	Clay panel	35.2	0.353	MDEGD_FDES (INIES_DISO20161116_164615, 5730)
	Wood fibre insulation panel—Pavatex	2.8	0.038	EPD PAVAFLEX flexible woodfibre insulation material Pavatex SA (EPD-PAV-20150043-IBA4-EN)
	Wood fibre insulation panel—Pavatherm	18.2	0.04	EPD Woodfibre insulation materials produced in the wet process 135–200 kg/m3 PAVATEX SA (EPD-PAV-2013254-CBG2-EN)
	Cross Laminated timber (wall)	47.5	0.130	EPD del pannello in legno X-LAM Cross Laminated Timber (S-P-01408)
Roof	Cross Laminated timber (roof)	60	0.130	EPD del pannello in legno X-LAM Cross Laminated Timber (S-P-01408)
	Wood fibre insulation panel—Isolair 100	14.5	0.041	FDES (INIES_IPAN20191004_110244, 12569)
	Plasterboard	4.8	0.072	EPD PPC Italia Gyproc Wallboard 13 mm (S-P-00938)
	Internal gypsum plaster (roof)	12	0.910	Oekobau.dat 2017-I, EPD GIPSPUTZ Bundesverband der Gipsindustrie e.V. (EPD-BVG-20140073-IAG1-DE)
	Levelling screed	22.8	0.129	EPD for Ultraplan, Ultraplan Eco, Ultraplan Maxi, Novoplan Maxi (S-P-00908)
	Bituminous membrane	6.4	0.130	Single layer mechanically fastened modified bitumen roof waterproofing system, Bitumen Waterproofing Association (2014) (NEPD00268E)

Table 10. Bill of material, thermal conductivity and data source: traditional configurations for 1 m^2 of wall and roof.

	Construction Material	Mass per Unit [kg/m²]	λ [W/m K]	Environment Data Source (EPD Number)
Wall	Internal plaster (wall)	15	0.390	EPD Gypsum plasters ALFA, BETA, GAMMA, ZETA, SPRINT, TEMPO (INIES_CEND20201217_102244, 26218)
	External plaster	27	0.890	Oekobau.dat 2020-II
	Perforated clay bricks	252.7	0.140	DAP Ladrillos y bloques cerámicos para revestir (EPD 008-007)
	Mineral wool	3.5	0.035	ACUSTILAINE MD 50 60 mm, Saint Gobain Cristaleria 2013 (ES054277-4)
	Concrete, normal-strength (wall)	171.4	-	One Click LCA
	Bricks	111	-	EPD Ladrillos y Bloques cerámicos para revestir. Pieza P según norma UNE-EN 771-1 (EPD 008-007)
	Steel for reinforcement (wall)	7.7	-	EPD Hot-rolled reinforcing steel for concrete in bars and coils (S-P-00255)
	Internal plaster (wall)	15	0.390	EPD Gypsum plasters ALFA, BETA, GAMMA, ZETA, SPRINT, TEMPO (ITB EPD No 083/2018)
Roof	Concrete, normal-strength (roof)	147	-	One Click LCA (-)
	Steel for reinforcement (roof)	14	-	EPD Hot-rolled reinforcing steel for concrete in bars and coils (S-P-00255)
	Screed	240	1.160	EPD NHL based screed floor (Sella, DOMUSVR, OPUS-C Masetto) (EPD-Miniera San Romedio Srl-89-EN)
	Cement mortar	40		Oekobau.dat 2022 (-)
	Bituminous membrane	12	0.170	Single layer mechanically fastened modified bitumen roof waterproofing system, Bitumen Waterproofing Association (2014) (NEPD00268E)
	EPS	2.3	0.035	Enviromental Product Declaration: ECO ESPANSO K100 (EPDITALY0029)
	Internal plaster (roof)	28	0.900	FDES (INIES_CEND20201217_102244, 26218)

4. Results

4.1. Life Cycle Impact Assessment (LCIA) of CFS School

According to the adopted approach, the EC of the building amounts to approximately 767 tons of CO_2e (Table 11). As shown in Figure 10, 28% of these GHG emissions are due to the materials and quantities used in the roof, 26.9% to the walls (load-bearing and not-load-bearing walls), 21% to the ground floor composition, and 24.1% to the materials used in foundations. Therefore, considering the specificity of construction, it is possible to say that, for the CFS case study, about 55% of the GWP is concentrated in the building materials above ground, and about 45% in the foundation and ground floor. Among the other impact categories analysed, the contribution of the materials included in the ground floor almost completely covers the impacts quantified in AP, EP, ODP and POCP with percentages equal to 98%, 96%, 97.8% and 99.7%, respectively. Regarding TUPE, there is an incidence of about 32% of the materials used both in the walls and in the roof, of 27% in the case of the ground floor and of about 8% for the materials used in foundations.

Table 11. Global LCIA Results.

	GWP	AP	EP	ODP	POCP	TUPE
	$tonCO_{2\text{-eq}}$	$kgSO_{2\text{-eq}}$	$kgPO_{4\text{-eq}}$	$kgCFC_{11\text{-eq}}$	$kgC_2H_{4\text{-eq}}$	MJ
TOTAL	767	88	97	1.9	700	10,054,000

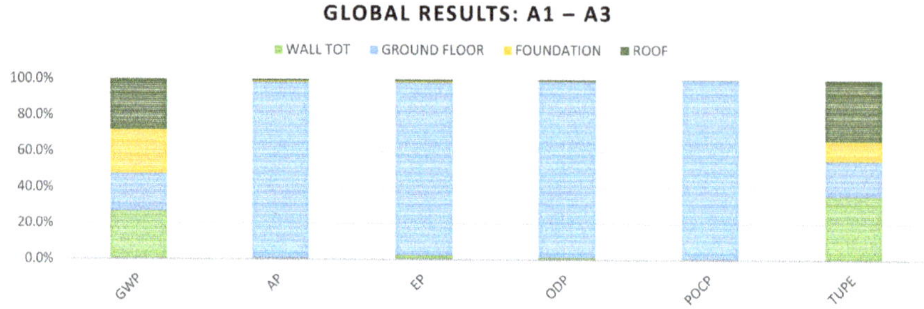

Figure 10. LCIA results: incidence percentages in each impact category.

Starting from the global results obtained in each impact category, the following paragraphs analyse in detail the percentage incidence of the individual materials within each building component with CFS framing system, distinguishing, where possible, between the impacts given by the structural and non-structural parts.

4.1.1. Walls

The impact assessment of the walls takes into account all materials included in both external and internal walls, i.e., structural and non-structural walls, with the exception of windows and doors. As stated in Section 2.1, the load-bearing walls includes the CFS profiles, as well as the OSB panels. The results obtained for the structural walls are shown in Figure 11 and Table 12. The structural components are responsible for 51.5% of the total GWP of the walls, of which 80% is due to CFS components (studs, tracks, flat straps, and blockings) and 20% to the OSB.

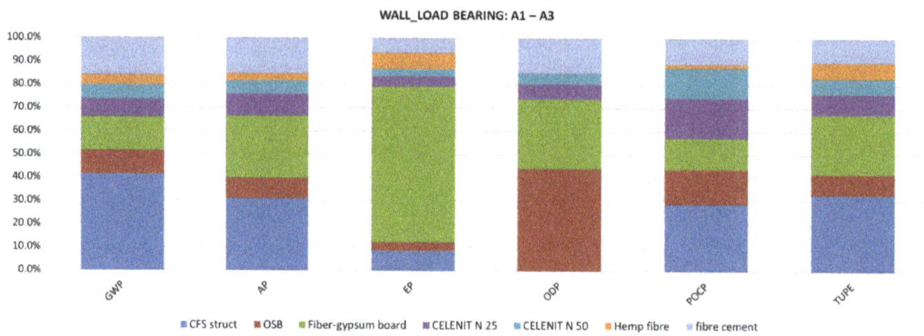

Figure 11. LCIA results: load-bearing walls.

Table 12. Results of Life Cycle impact assessment: load-bearing walls.

Material	**GWP**	**AP**	**EP**	**OPD**	**POCP**	**TUPE**
	ton$CO_{2\text{-eq}}$	**ton$SO_{2\text{-eq}}$**	**ton$PO_{4\text{-eq}}$**	**kg$CFC_{11\text{-eq}}$**	**ton$C_2H_{4\text{-eq}}$**	**MJ**
Structural	CFS struct	7.56×10^4	1.61×10^2	1.51×10	5.78×10^{-6}	1.38×10
	OSB	1.88×10^4	4.54×10	6.29	8.41×10^{-3}	7.16
Non-structural	Fibre-gypsum board	2.65×10^4	1.39×10^2	1.18×10^2	5.78×10^{-3}	6.49
	CELENIT N 25	1.42×10^4	4.85×10	7.56	1.24×10^{-3}	8.23
	CELENIT N 50	1.12×10^4	3.00×10	5.67	9.32×10^{-4}	6.17×10
	Hemp Fibre	7.76×10^3	1.64×10	1.22×10	3.48×10^{-6}	8.63×10^{-1}
	Fibre Cement	2.93×10^4	7.87×10	1.10×10	2.84×10^{-3}	5.20
Total		1.83×10^5	5.19×10^2	1.76×10^2	1.92×10^{-2}	4.79×10

The three thermal insulation panels cause about 18.1% of the embodied carbon of the walls. Despite the larger quantity, natural hemp fibre insulation is much more sustainable than the wood wool insulation panels. Furthermore, it is surprising that the CELENIT N 25 has a greater impact than the CELENIT N 50. Indeed, an emission factor equal to 509 $kgCO_2e/m^3$ emerges from the EPD of CELENIT N insulation panel for a thickness of 25 mm. Among the remaining materials of the load bearing walls, the fibre-gypsum board and the fibre cement panel are responsible for 14.5% and the 16% of EC, respectively. As for the other considered impact categories, the structural CFS is the most impacting building material in terms of AP (31%), followed by the fibre-gypsum board (26.8%). In addition, the fibre-gypsum board involves 67% of the EP of the walls. Regarding ODP, the impact of CFS is negligible, while the most impacting materials are the OSB and fibre-gypsum board, with percentages of 43.8% and 30.1%, respectively. Moreover, 30.1% of the photochemical ozone creation potential of the structural walls is caused by the insulating panels in wood wool (CELENIT N 50 and CELENIT N 25). Structural materials (structural CFS and OSB) account for about 43.7% (28.7% and 15%, respectively) in the POCP indicator, while the fibre-gypsum board represents about 14% of this. The materials that involve the greatest total use of primary energy are the structural CFS and the fibre-gypsum boards, respectively, with percentages of 32.9% and 25.7%. In this case, the percentage of incidence of non-structural materials is approximately 58.3%. Finally, with the exception of the GWP, the non-structural building materials present in the load-bearing walls of the BSF have overall larger impacts than the structural materials, with an average percentage of 61.3% considering all examined impact categories. This is mainly due to the high impact of the fibre gypsum boards and the CELENIT 25 and CELENIT N 50 insulating panels, as well as to the greater amount of materials with a non-structural function (see Tables 5 and 12).

In the same way, Figure 12 and Table 13 show LCA results for non-load-bearing walls in which there are no materials with structural function.

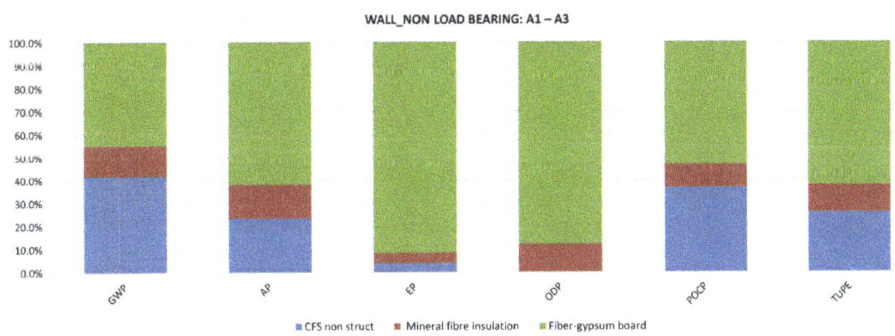

Figure 12. LCIA results: non-load-bearing walls.

Table 13. Results of Life Cycle impact assessment: non-load-bearing walls.

Material	GWP	AP	EP	OPD	POCP	TUPE
	$tonCO_{2\text{-eq}}$	$tonSO_{2\text{-eq}}$	$tonPO_{4\text{-eq}}$	$kgCFC_{11\text{-eq}}$	$tonC_2H_{4\text{-eq}}$	MJ
CFS non-struct.	1.09×10^4	2.33×10	2.17	8.31×10^{-7}	1.98	1.47×10^5
Mineral Fibre insulation	3.46×10^3	1.44×10	2.53	3.47×10^{-4}	5.31×10^{-1}	6.71×10^4
Fibre-gypsum board	1.17×10^4	6.12×10	5.18×10	2.50×10^{-3}	2.85	3.50×10^5
Total	2.60×10^4	9.89×10	5.65×10	2.85×10^{-3}	5.36	5.64×10^5

In general, it is observed that, in all analysed impact categories, the least sustainable material is the fibre-gypsum board, followed by the non-structural CFS. In terms of GWP, the fibre-gypsum board is responsible for 44.8% and the non-structural CFS for 41.9%. For EP and ODP indicators, the fibre-gypsum board is responsible for 91.7% and 87.8%, respectively. Furthermore, 62.1% of the total primary energy use is due to the fibre-gypsum board and 26% to non-structural CFS.

The mineral fibre insulation panel is the most sustainable material and affects an average of 11.1% of the indicators assessed. In general, the magnitude of the analysed impacts is clearly lower for non-structural walls than for structural ones. As can be seen from Tables 12 and 13 and from what is reported in Section 3.2, this is certainly due to the smaller number of materials involved and to the smaller quantities. For example, the total GWP of the non-load-bearing walls is equal to 14.2% of the total GWP calculated for load-bearing walls.

4.1.2. Roof

The roof is composed of CFS profiles (joists, floor tracks, blockings and flat straps) with the OSB panels acting as the structural components, and is completed by CELENIT N 50 insulation panels, hemp fibre insulation panels and corrugated galvanized steel sheets acting as non-structural materials. As can be seen in Figure 13, the materials that most affect the embodied carbon are the CFS (40.2%), CELENIT N 25 (29.9%) and corrugated galvanized metal sheets (19.3%); the respective impact is reported in Table 14. Structural materials cover 47.8% of the GWP. As in the case of the walls, the hemp fibre insulation has lower impact than CELENIT N, involving just 3% of the GWP. Despite the thickness of 100 mm, the hemp fibre insulation panel seems to be an extremely sustainable material (emission factor of 21.8 $kgCO_2e/m^3$) when compared to the impact caused by the double CELENIT N (25 mm) insulation panels used in the roof. Regarding AP, the percentages are equal to 45.1%, 26.8% and 21.8% for the corrugated galvanized metal sheets, CELENIT N 25 and CFS, respectively. Structural materials cover 29.9% of AP. The most impacting material in terms of eutrophication is CELENIT N 25 (37.7%), followed by the galvanized steel sheet (24.4%) and CFS (19.9%). This is the impact category in which the hemp fibre insulation panel has the greatest impact (11.8%).

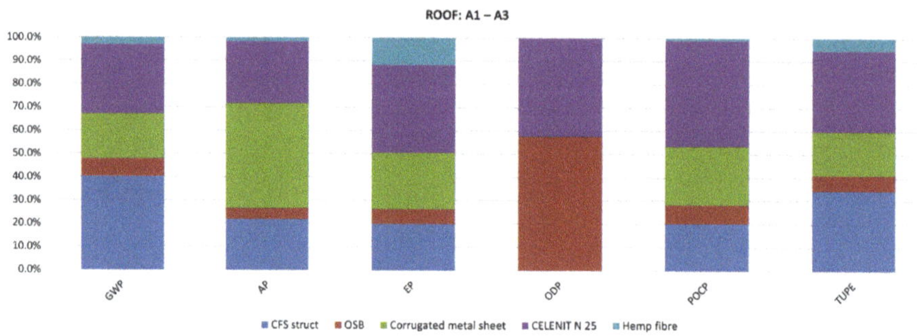

Figure 13. LCIA results: Roof.

Table 14. Results of Life Cycle impact assessment: Roof.

Material	GWP	AP	EP	OPD	POCP	TUPE
	$tonCO_{2\text{-eq}}$	$tonSO_{2\text{-eq}}$	$tonPO_{4\text{-eq}}$	$kgCFC_{11\text{-eq}}$	$tonC_2H_{4\text{-eq}}$	MJ
Structural	CFS structural	8.63×10^4	1.84×10^2	1.72×10	6.60×10^{-6}	1.57×10
	OSB	1.62×10^4	3.93×10	5.43	7.30×10^{-3}	6.19
Non- Structural	Corrugated metal sheet	4.14×10^4	3.81×10^2	2.11×10	6.10×10^{-8}	1.96×10
	CELENIT N 25	6.43×10^4	2.27×10^2	3.27×10	5.36×10^{-3}	3.56×10
	Hemp fibre insulation	6.52×10^3	1.38×10	1.03×10	3.00×10^{-6}	7.20×10^{-1}
Total		2.15×10^5	8.45×10^2	8.68×10	1.27×10^{-2}	7.78×10

Structural materials account for 23% of the eutrophication potential of the roof. As regards ODP, the impacts are due exclusively to OSB (57.6%) and CELENIT N 25 (42.3%), while in terms of the formation of ozone of the lower atmosphere, CELENIT N 25 accounts for 45.7%, followed by the galvanized steel sheet (25.2%). The structural materials, CFS and OSB, account for 20.2% and 8%, respectively, in POCP. Finally, in terms of TUPE, non-structural materials impact about 58.9%, with incidence percentages of 34.7%, 18.7% and 5.4% of CELENIT N 25, galvanized steel sheet and hemp fibre insulation, respectively. The CFS and the OSB account for 34.2% and 6.9%, respectively.

4.1.3. Most Contributing Materials and Components

Table 15 shows the most contributing materials in terms of global warming according to the approach used in this paper (cradle-to-gate) and on the basis of what has been assessed in detail for the CFS walls and roof.

Table 15. Overall and percentage impact of each British Defense School construction material.

Resource	Cradle to Gate Impacts (A1–A3) (Tons of CO_{2eq})	Cradle to Gate (A1–A3) Percentage Incidence
CFS	173	40.0%
Wood Wool panels	90	20.7%
Corrugated steel	41	9.6%
Fibre-gypsum board	38	8.8%
OSB	35	8.1%
fibre cement	29	6.8%
Hemp fibre	14	3.3%
Mineral fibre insulation	12	2.7%

The materials causing the greatest embodied carbon are the CFS and the wood wool panels, with 173 tons of CO_2e (40%) and 90 tons of CO_2e (20.7%), respectively.

In particular, the environmental profile of CFS manufactured by ClarkDietrich and declared in a corresponding EPD was chosen for the present study. In the environmental declaration, with reference to a declared unit of one metric tons of CFS Product (1000 kg), the LCIA results (CML v4.2) reported in EPD show a GWP equal to 1770 kgCO_2e in A1–A3 module. The production of raw material inputs (A1) accounts for most of the EC of CFS (93%) which is mainly attributed to the steel coil that is cold formed to realize the structural components. Energy and utility consumption (in the form of electricity, natural gas and water) are the most significant contributions to the impact of the manufacturing stage (A3).

Regarding wood wool thermal insulation panels, the investigated building adopts two typologies of wood wool panels, i.e., CELENIT N25 and CELENIT N50, which have a thickness of 25 mm and 50 mm, respectively.

As indicated in the EPDs available in One Click LCA, the CELENIT panels are made of spruce pine wood (47% in mass) coming from sustainable management forests and mineral binders (52% in mass), mainly composed of Portland cement and marble sawdust. The wood wool is exposed to a mineralization treatment, which reduces the natural deterioration, while keeping the mechanical properties, and increases the fire resistance. Portland cement is adopted as coating to provide high resistance to water and frost as well as to increase the wood wool resistance to bending and compression. In detail, the percentage composition of the raw materials included in 1 kg of panel is reported in Table 16.

Table 16. Wood wool—percentage composition (1 kg of panel).

Material	% in Weight
Portland cement	37%
Spruce wood and water (wood: 80%; water: 20%)	47.30%
Marble sawdust	15%
Calcium diformate	0.3%
Calcium chloride	0.2%
Alkylate	0.2%

Source: S-P-00477 (International EPD System).

The data elaboration has been performed with GABI software v. 7.3.0.40 (Sphera–GaBi solutions, Chicago, IL, USA). Primary data refer to 2015 and have been collected at CELENIT's plant located in Onara di Tombolo (ITA), whereas selected generic data have been retrieved from Ecoinvent 3.1, GaBi and ELCD databases. From the LCIA results (CML v4.6) the emission factors are about 509 $kgCO_2e/m^3$ and 200 $kgCO_2e/m^3$, respectively, for CELENIT N 25 and CELENIT N 50. As stated in the EPD, the biogenic carbon storage is calculated separately from the GWP results of building materials A1–A3 and not subtracted from the calculation.

These emission factors are interesting since, despite the reduced thickness, 1 m^3 of CELENIT N 25 is considerably more impactful than 1 m^3 of CELENIT N 50. Although not explicitly stated in the EPD, as Figure 14 shows, the reasons for this unexpected result are to be sought mainly in the processing phases of the insulating panels as the thickness decreases. Furthermore, it can be seen that the EC, due to the manufacturing (A3), is, in general, predominant and covers almost 93% in both cases.

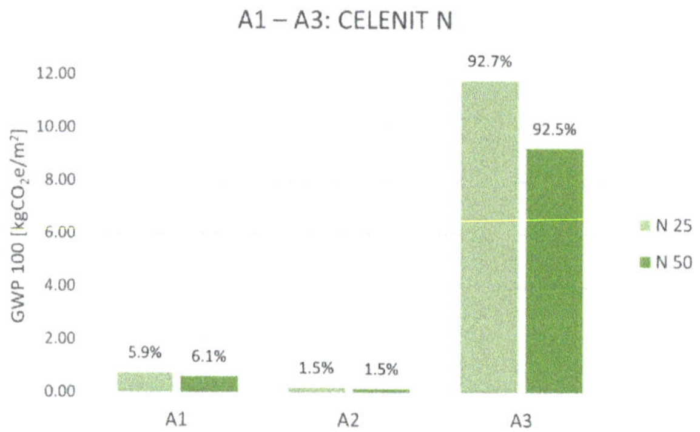

Figure 14. LCIA results for CELENIT N (25 mm and 50 mm).

4.2. *Results for the Comparison of CFS, CLT and Traditional*

Table 17 and Figure 15 show the life cycle assessment results for the square meters of the three investigated external wall configurations, while Table 18 and Figure 16 show those for the roof configurations. For external walls, in terms of GWP the CFS has the highest value with 123 $kgCO_2e/m^2$, although the result is very close to the GWP calculated for the traditional wall, which equates to 108 $kgCO_2e/m^2$. Among the building materials included in the traditional configuration, the main contribution is given by the perforated clay bricks which are responsible for 65% of the GWP. Other important contributions are given by concrete (20%) and steel reinforcement (24%) used as brick reinforcement. However, from the obtained results it can be seen that their impact is due to the high quantities per square metre of wall rather than the embodied emissions deriving from stages A1 to A3.

Table 17. LCA results: external walls.

		1 m^2 CFS	1 m^2 CLT	1 m^2 Traditional
GWP	$kgCO_{2\text{-eq}}$	1.23×10^2	4.88×10	1.08×10^2
AP	$kgSO_{2\text{-eq}}$	3.26×10^{-1}	1.45×10^{-1}	1.15×10^{-1}
EP	$kgPO_{4\text{-eq}}$	6.47×10^{-2}	1.05×10^{-1}	1.89×10^{-2}
OPD	$kgCFC_{11\text{-eq}}$	1.02×10^{-5}	3.38×10^{-6}	1.06×10^{-6}
POCP	kg Ethenee	3.44×10^{-2}	2.83×10^{-2}	1.64×10^{-2}
TUPE	MJ	1.83×10^3	8.98×10^2	1.44×10^3
GWP	$kgCO_{2\text{-eq}}$	1.23×10^2	4.88×10	1.08×10^2

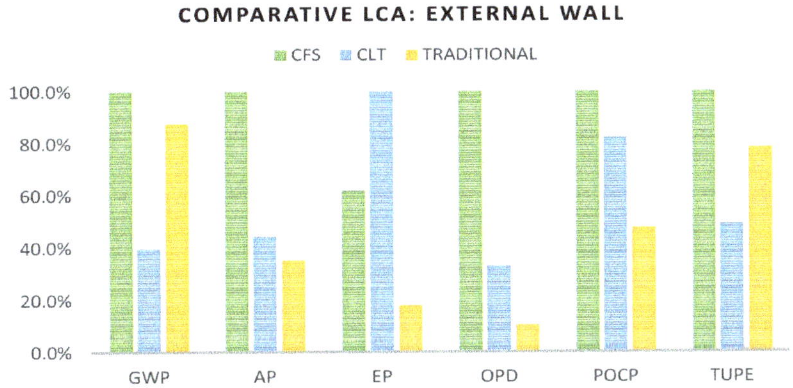

Figure 15. Results of Life Cycle impact assessment: External Wall.

Table 18. LCA results: roof.

		1 m^2 CFS	1 m^2 CLT	1 m^2 Traditional
GWP	kg $CO_{2\text{-eq}}$	8.51×10	6.42×10	9.76×10
AP	kg $SO_{2\text{-eq}}$	3.35×10^{-1}	8.96×10^{-2}	2.41×10^{-1}
EP	kg $PO_{4\text{-eq}}$	3.44×10^{-2}	2.29×10^{-1}	6.18×10^{-2}
OPD	kg $CFC_{11\text{-eq}}$	5.02×10^{-6}	5.12×10^{-6}	4.23×10^{-5}
POCP	kg $C2H_{4\text{-eq}}$	3.08×10^{-2}	4.41×10^{-2}	2.72×10^{-2}
TUPE	MJ	1.34×10^{3}	1.03×10^{3}	1.34×10^{3}
GWP	kg $CO_{2\text{-eq}}$	8.51×10	6.42×10	9.76×10

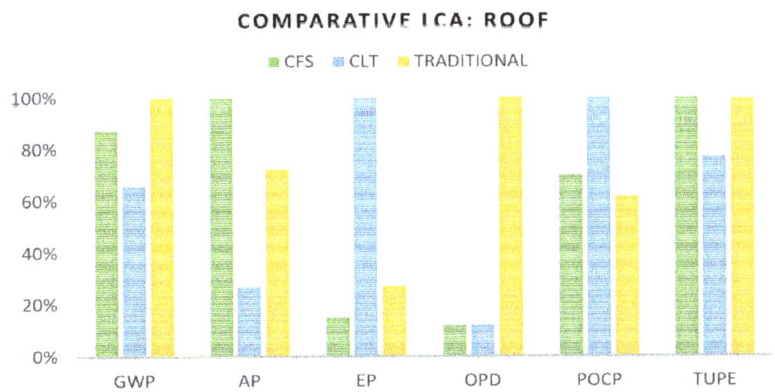

Figure 16. Results of Life Cycle impact assessment: Roof.

The internal and external plaster are responsible, respectively, for 1.2% and 5.6% of the estimated GWP for the wall. As reported in Table 10, the quantity of external plaster per square metre of wall is almost double the quantity of internal plaster. Furthermore, on the basis of the adopted environmental profiles, the EC coefficient of the internal plaster is equal to 0.0949 $kgCO_2e/kg$ while that of the external plaster is 0.23 $kgCO_2e/kg$. Finally, the ECC value of the mineral wool insulation panel of the wall is the highest (1 $kgCO_2e/kg$), and this is due to the very low density of this material (50 kg/m^3). The thickness of 70 mm corresponds to 7.5 kg of thermal insulation per m^2 of wall. Consequently, the insulation panel is responsible for 3.4% of the total GWP calculated for the traditional configuration.

In the case of the CLT technology, the EC emissions are equal to 49 $kgCO_2e$. However, it must be underlined that the external wall of the British School (CFS) also complies with the thermal transmittance limit value currently envisaged by Italian legislation for the construction of a nearly zero-energy building in the climatic zone to which the case study belongs. However, in terms of environmental sustainability, there is an important gap with respect to the current target for the new building.

For the roof, the results obtained in terms of $kgCO_2e/m^2$ show greater comparability between the three analysed cases. In this case, the value obtained with traditional solutions is the highest and equal to 98 $kgCO_2e/m^2$, i.e., +15.3% compared to CFS and +53% compared to CLT. The configuration foreseen for the BNZEB (CLT) is the best one, both in terms of sustainability and energy efficiency. The configuration envisaged for the CFS roof achieves a level of thermal insulation comparable to that obtained in the case of the CLT, with a total thickness reduced by more than 50% but with a carbon footprint per m^2 increased by 32.8%.

For the same configurations, Figure 15 shows the comparisons in terms of LCA considering all the impact categories. From the comparison it can be observed that the CFS wall has the higher impact in terms of AP, OPD, POCP and TUPE. In these impact categories, the traditional external wall, with a thermal transmittance equal to the CFS and a total thickness of 420 mm, has on average an impact equal to 42.9% of that of CFS. Instead, the external wall with laminated wood technology has an impact on average equal to 52.1% of CFS in the same indicators. The CLT wall is the least sustainable in terms of eutrophication potential (EP). For this indicator, the traditional wall shows an impact 82% lower than the wood wall. The eutrophication potential of the CFS wall is equal to 61.8% compared with the wood wall.

For the roof, from the obtained results, it is clear that the wood technology is the most sustainable in the GWP, AP and TUPE categories, while it is the most impactful in terms of EP and POCP categories. In addition to the GWP, the traditional solution is the most impactful in the of ODP indicators. The CFS and X-LAM configurations show lower ODP (of around 88%) than the traditional roof. In terms of TUPE, the impacts of the CFS roof and the traditional roof show negligible differences while the impact of the wood roof is lower by about 33%. Finally, as regards the AP, the least sustainable configuration is the CFS.

5. Conclusions

Sustainable buildings should have low or zero carbon emissions, considering both construction and operation. Choices of material can have a huge impact on this. As with many aspects of sustainability, comparing EC figures can be tricky. This is mainly true for innovative technologies for which there is a limited number of scientific studies exploring their environmental assessments. Therefore, this paper focuses on two innovative technologies: cold formed steel (CFS) systems and cross laminated timber (CLT). Specifically, it analyses in detail a full real case study in CFS, which is the first CFS school built in Italy in a medium/high seismic risk area, which has been designed according to a conservative approach (elastic design) due to inexistent seismic design codes specifically applicable to the investigated structural system in the EU and Italy. The study develops a detailed cradle-to-gate life cycle analysis, and shows that the CFS opaque building components amount to about 55% of the total GWP, with approximately 421 tons of CO_2e. Around 28% of these emissions are due to the materials and quantities used in the roof, and 26.9% to the walls. The impact assessment of the walls indicates that the structural components are responsible for 51.5% of the total GWP of the walls, of which 80% is due to CFS components (studs, tracks, flat straps, and blockings) and 20% to the OSB. The three thermal insulation panels cause about 18.1% of the embodied carbon of the walls, with the natural hemp fibre insulation more sustainable than the wood wool insulation panels. As regards the roof, the materials that most affect the embodied carbon are the CFS (40.2%), wood wool insulation (29.9%) and corrugated galvanized metal sheets (19.3%). Structural materials cover 47.8% of the GWP. As in the case of the walls, the hemp fibre insulation has lower impact than insulation, involving just 3% of the GWP. Therefore, the study clearly shows that, overall, the CFS components are primarily responsible for the EC of the full buildings. Hence, any energy efficient design of CFS constructions should pay very careful attention to optimizing the structural system, as any percentage of reduction of the structural components will have a significant impact on the overall EC. To this end, the development of structural codes that would allow the ductile seismic design of CFS structures in EU could also support and incentivize an optimized design of CFS structure, and then CFS systems with lower EC.

Moreover, when CFS is compared to CLT and traditional reinforced masonry structure, this finding appears even clearer. This study indeed shows that, when 1 m^2 of wall and roof for each construction system are considered, then the results indicate that in terms of global warming potential, the CFS is the one having higher GWP impacts (123 $kgCO_2e/m^2$) when the wall composition is considered, while the traditional system is the one having higher impact for the roof (97.6 $kgCO_2e/m^2$). X-LAM walls, that also allow better energy performance in terms of thermal transmittance, show about 60% less GWP impact then CFS and about 55% compared to the traditional system. In case of the roof's materials, the GWP decreases around 25% and 34%, respectively, compared with CFS and the traditional system. The eutrophication potential is the impact category for which the X-LAM gives the worst results for both wall and roof; at the same time, in only the case of the wall's materials, the formation of ozone of lower atmosphere is increased by around 43% and 62%, respectively, compared to CFS and the traditional system.

The decision to compare the three construction systems, considering the embodied carbon per 1 m^2 of wall and roof, allows the broadening of the applicability of the observed results. Although future sensitivity analysis could improve the reliability of the obtained results, this study demonstrates that CLT is, at the present, the system with lower EC, and that for CFS to fully deliver environmentally sustainable systems, then optimization of the structural system should be prioritized, to reduce the quantity of the material and the consequent EC.

Furthermore, for a wider diffusion of CLT and/or CFS buildings, it would also be important in the future to evaluate the end-of-life impacts, as they can be critical to reducing the overall life-cycle impacts.

The new knowledge generated by the in-depth analyses, discussed in this paper, can make a significant contribution to the early-stage decision-making design process for selecting structural and non-structural building materials with lower EC. Finally, this paper highlights the importance of integrating structural engineering solutions with life cycle assessment. This synergy, especially if implemented in the design phase, can help improve the sustainability performance and reduce the carbon footprint of buildings.

Author Contributions: Conceptualization, O.I.; methodology, O.I. and A.G.; software, A.G.; validation, O.I.; formal analysis, O.I., A.G.; resources, O.I., R.F.D.M.; writing—original draft preparation, O.I. and A.G.; writing—review and editing, O.I., R.F.D.M.; visualization, A.G. and O.I.; supervision, O.I. and R.F.D.M. All authors have read and agreed to the published version of the manuscript.

Funding: This research was funded by ESRC, grant number ES/V015923/1.

Data Availability Statement: Not applicable.

Conflicts of Interest: The authors declare no conflict of interest. The funders had no role in the design of the study; in the collection, analyses, or interpretation of data; in the writing of the manuscript; or in the decision to publish the results.

Appendix A

Table A1. CFS school—Life Cycle Inventory: source of data.

Construction Material	Environment Data Source	Upstream Database
Crawl space	EPD AIRCRAB H35	ecoinvent
Concrete (12/15 MPa)	One Click LCA	ecoinvent
Concrete (30/37 MPa)	One Click LCA	ecoinvent
Reinforcement	EPD Hot-rolled reinforcing steel for concrete in bars and coils	ecoinvent
Screed self-levelling	EPD weber.floor 110Fine/120Reno/130Core/140Nova	ecoinvent
Cold Formed steel (CFS)	EPD Cold-Formed Steel Products	GaBi
Fibre-cement panels	EPD Grossformatige Faserzementplatten pigmentiert beschichtet Carat/Reflex/Avera/Zenor/Aura/Integral/Plan Swisspearl Group AG	GaBi
OSB panels	EPD-KRO-20150067-IBD2-EN	GaBi
Wood wool panels—CELENIT N 25	EPD ACOUSTIC AND THERMAL PANELS CELENIT ABE, AE, AB, A, NB, N	ecoinvent
Cold Formed steel (CFS)	EPD Cold-Formed Steel Products	GaBi
Gypsum fibreboard—Knauf Vidifire	EPD Gypsum fibreboards Knauf	GaBi
Hemp Fibre insulation	EPD EKOLUTION® HEMP FIBRE INSULATION	GaBi
Mineral fibre insulation	EPD Glass Mineral Wool Insulation with ECOSE Technology (0.031–0.033 W/mK), Knauf Insulation 2015	ecoinvent
Linoleum	FDES	ecoinvent
Wood fibre Insulation	Oekobau.dat 2017-I, EPD Wood fibre insulation materials STEICO	GaBi
Corrugated galvanized steel	EPD for AlumiGard, MagnaFlow and ZinaCore pre-painted roofing and cladding	GaBi

References

1. Kylili, A.; Fokaides, P.A. Policy trends for the sustainability assessment of construction materials: A review. *Sustain. Cities Soc.* **2017**, *35*, 280–288. [CrossRef]
2. Marsh, E.; Orr, J.; Ibell, T. Quantification of uncertainty in product stage embodied carbon calculations for buildings. *Energy Build.* **2021**, *251*, 111340. [CrossRef]
3. Pakdel, A.; Ayatollahi, H.; Sattary, S. Embodied energy and CO_2 emissions of life cycle assessment (LCA) in the traditional and contemporary Iranian construction systems. *J. Build. Eng.* **2021**, *39*, 102310. [CrossRef]
4. Goggins, J.; Moran, P.; Armstrong, A.; Hajdukiewicz, M. Lifecycle environmental and economic performance of nearly zero energy buildings (NZEB) in Ireland. *Energy Build.* **2016**, *116*, 622–637. [CrossRef]
5. Kristjansdottir, T.F.; Stina Good, C.; Rose Inman, M.; Dahl Schlanbusch, R.; Andresen, I. Embodied greenhouse gas emissions from PV systems in Norwegian residential Zero Emission Pilot Buildings. *Sol. Energy* **2016**, *133*, 155–171. [CrossRef]
6. Peng, Z.; Lu, W.; Webster, C.J. Quantifying the embodied carbon saving potential of recycling construction and demolition waste in the Greater Bay Area, China: Status quo and future scenarios. *Sci. Total Environ.* **2021**, *792*, 148427. [CrossRef] [PubMed]
7. Iuorio, O.; Wallace, A.; Simpson, K. Prefabs in the North of England: Technological, Environmental and Social Innovations. *Sustainability* **2019**, *11*, 3884. [CrossRef]
8. Iuorio, O.; Fiorino, L.; Landolfo, R. Testing CFS Structures: The New School BFS Naples. *Thin-Walled Struct.* **2014**, *84*, 275–288. [CrossRef]
9. OPCM, n. *Criteri Generali per L'individuazione delle Zone Sismiche e per la Formazione e L'aggiornamento Degli Elenchi delle Stesse Zone*; Gazzetta Ufficiale: Rome, Italy, 2006; p. 3519. (In Italian)
10. Iuorio, O.; Napolano, L.; Fiorino, L.; Landolfo, R. The environmental impacts of an innovative modular lightweight steel system: The ELISSA case. *J. Clean. Prod.* **2019**, *238*, 117905. [CrossRef]
11. Simpson Strong-Tie Company. Available online: https://www.strongtie.com/ (accessed on 10 September 2021).
12. Iuorio, O. Cold-formed steel housing. *Pollack Period.* **2007**, *2–3*, 97–108. [CrossRef]
13. Landolfo, R.; Pali, T.; Bucciero, B.; Terracciano, M.T.; Shakeel, S.; Macillo, V.; Iuorio, O.; Fiorino, L. Seismic response assessment of architectural non-structural LWS drywall components through experimental tests. *J. Constr. Steel Res.* **2019**, *162*, 105575.

14. President of the Republic. *Regulation Containing Prescriptions for the Design, Installation, Operation and Maintenance of Heating Systems in Buildings in Order to Limit Energy Consumption, Implementing Art. 4, Par. 4 of Low 9 January 1991, n. 10*; Italian Legislative Decree n. 412/1993; Gazzetta Ufficiale: Rome, Italy, 1993; p. 1993.
15. Ascione, F.; Bianco, N.; De Masi, R.F.; de Rossi, F.; Vanoli, G.P. Concept, design and energy performance of a net zero energy building in Mediterranean climate. *Procedia Eng.* **2016**, *169*, 26–37. [CrossRef]
16. Ascione, F.; Borrelli, M.; De Masi, R.F.; de Rossi, F.; Vanoli, G.P. A framework for NZEB design in Mediterranean climate: Design, building and set-up monitoring of a lab-small villa. *Sol. Energy* **2019**, *184*, 11–29. [CrossRef]
17. UNI Ente Nazionale Italiano di Unificazione. *UNI/TS 11300-1 Part 1: Evaluation of Energy Need for Space Heating and Cooling*; UNI: Rome, Italy, 2008.
18. UNI Ente Nazionale Italiano di Unificazione. *UNI/TS 11300-Part 2: Evaluation of Primary Energy Need and of System Efficiencies for Space Heating and Domestic Hot Water Production*; UNI: Rome, Italy, 2008.
19. UNI Ente Nazionale Italiano di Unificazione. *UNI/TS 11300-Part 3: Evaluation of Primary Energy and System Efficiencies for Space Cooling*; UNI: Rome, Italy, 2010.
20. UNI Ente Nazionale Italiano di Unificazione. *UNI/TS 11300-Part 4: Renewable Energy and Other Generation Systems for Space Heating and Domestic Hot Water Production*; UNI: Rome, Italy, 2014.
21. Grazieschi, G.; Asdrubali, F.; Thomas, G. Embodied energy and carbon of building insulating materials: A critical review. *Clean. Environ. Syst.* **2021**, *2*, 100032. [CrossRef]
22. One Click LCA Software. Available online: https://www.oneclicklca.com (accessed on 5 June 2021).
23. *EN 15804*; Sustainability of Construction Works. Environmental Product Declarations. Core Rules for the Product Category of Construction Products. European Committee for Standardization: Brussels, Belgium, 2012.
24. *ISO 14040*; Environmental Management-Life Cycle Assessment—Principles and Framework. International Organization for Standardization: Geneva, Switzerland, 2006.
25. CML—Department of Industrial Ecology. CML-IA Characterisation Factors. 2016. Available online: https://www.universiteitleiden.nl/en/research/research-output/science/cml-iacharacterisation-factors (accessed on 30 May 2021).
26. CERANOR, S.A. (Spain). Available online: https://www.ceranor.es/ (accessed on 19 September 2021).
27. Sphera. GaBi LCA Database Documentation. Retrieved from Sphera Solutions. 2020. Available online: https://gabi.sphera.com/support/gabi/gabi-6-lci-documentation/ (accessed on 20 June 2021).

energies

Article

Evaluation of Acoustic Comfort and Sound Energy Transmission in a Yacht

Simon Kanka [1], Luca Fredianelli [1,2,*], Francesco Artuso [1], Francesco Fidecaro [1] and Gaetano Licitra [3,*]

[1] Physics Department, University of Pisa, Largo Bruno Pontecorvo 3, 56127 Pisa, Italy
[2] Institute of Chemical and Physical Processes of National Research Council, Via G. Moruzzi 1, 56124 Pisa, Italy
[3] Environmental Protection Agency of Tuscany Region, Via Vittorio Veneto 27, 56127 Pisa, Italy
* Correspondence: luca.fredianelli@ipcf.cnr.it (L.F.); g.licitra@arpat.toscana.it (G.L.)

Abstract: After being neglected for a long time, in the last years, ships have been recognized and studied as sound emitters. The sound energy they generate impacts the outside, but it can also affect the indoor quality of life if the environments are not properly designed. In fact, acoustic comfort plays a pivotal role, particularly in recreational crafts. In the present work, room acoustics and acoustic camera measurements were performed, inside a 50 m length overall yacht, chosen as a case study in order to evaluate the acoustic comfort. The Italian classification procedure UNI 11367:2010 for buildings was applied, and results have been compared to other international comfort classes. However, all of these are based on prescription for standard buildings, and the present work highlights that they do not account for the effective ship's acoustic issues: sound energy transfer from impacts over ceilings and sound energy leakage. While attention of shipbuilders in acoustic comfort is shown in the measured good reverberation times, the acoustic camera revealed sound energy leakages corresponding to hidden escape ways that have been poorly insulated. This compromises the standardized sound difference between contiguous compartments and also the thermal insulation, as leakage involves air passages. The present work attempts to evolve the classification procedure by also including, for the first time, the reverberation time, but future studies focused on finding correct standardized impact level noise for ship cases are needed. In fact, their values were very high and not comparable with those measured in actual buildings and for which reference values have been designed.

Keywords: sound energy; acoustic comfort; sound energy transmission; room acoustics; acoustic camera; noise; ships; sound energy leakage; reverberation time; acoustic performance indicator

1. Introduction

Recently, acoustic research on sound energy emissions produced by maritime transportation, both on the inside and on the outside of ships, has growth. In most of the articles, the ships are described as an energy-intense building or, in particular cases as a separate neighborhood. Ships' airborne sound energy emissions have been studied in recent projects of the INTERREG MARITTIMO Italy-France 2014–2020 program, focused on noise mitigation for people exposed to port noise. Different sound emission mechanisms have been studied in [1–6]. It has been shown that environmental issues arise when ships irradiate sound energy outside during their sailing, mooring, and loading/unloading operations in ports.

Comfort and quality of life are issues related to the interior of merchant ships and fishing vessels, or recreational vessels. These issues are mostly generated by the sound energy generated during engines and ventilations operations, which then propagate through both air and structure ways. While the highest levels are produced by the propulsion systems, other sound emitters like the hydraulic system and electric system can be significant contributors to the internal acoustic levels inside the ships. The majority of studies dedicated to internal acoustic in ships have been dedicated to crew's health and long-time exposition to high noise levels hazards [7–12]. However, the last years saw passenger ship acoustics receiving a larger consideration [13–17] and Goujard et al. [18] showed how acoustic comfort is the most critical factor perceived by passengers. As a result, ship-owners and their guests have demanded more attention not only to the functionality of the internal spaces but also to a better quality of life on the ships both in the private and recreational sector.

Given their size and different sectors usage, large ships are usually compared to buildings [19,20]. Room acoustics has been largely studied in the past [21] if compared to ship noise and, therefore, it can be considered as the reference for evaluating acoustic comfort on ships. As the acoustic comfort can be mainly affected by structure-borne noise and by airborne noise [22], different performance indicators are needed. These indicators belong to two main categories: insulation from noise coming from the outside or from the adjacent rooms, and the acoustic field inside the room itself. The inside comfort environment investigated is considered sufficient if specific indicators are higher or lower thresholds. Different ship classification organizations or societies started providing different class notations to vessels to label different ships in terms of the quality of life. However, in a recent study, Rocca et al. [23] showed the difference between the various acoustic classification schemes and the importance of different acoustic indexes. Different classification procedures have different indicators' thresholds or are applicable only to specific kinds of ships and their lifetime. The common reference elements for these classification procedures are specific acoustic indexes of a standard reference wall and floor (ISO 717-1:2020 [24] and ISO 717-2:2020 [25]). Unfortunately, these elements belong to solid buildings, and not ships.

Therefore, a proper universal approach to classify acoustic comfort inside ships is needed before defining a common methodology to improve the life quality onboard vessels and to localize the principal sound energy emitters. This cannot avoid following classification approaches for the acoustic comfort on ships based on buildings procedure methods. The overall issue does not only affect the ships' acoustic comfort but also their thermal energy consumption and insulation. In fact, a recent preliminary study reported on how safe passages inside recreational boat areas are affected by sound energy leakage phenomena and worsen the room acoustics of living areas [26]. Kang et al. [27] quantified this leakage in almost 5 dB(A) of reduction in insulation performances of walls. As this leakage involves air passages, it is expected that these do not only compromise the optimal cabin's acoustic quality but would also deteriorate the overall energy class sustainability by increasing thermal energy dissipation.

The present paper reports an overall determination of ships' comfort acoustic classification on the bases of the existing Italian's prescription of building classification. As a case study, the methodology is applied to measurements performed inside a 50 m length overall (LOA) yacht. The results are then compared with some of the most important comfort classes procedures. Acoustic camera measurements are also shown as support in highlighting the presence of sound energy leakage between partitions.

The analysis investigates all possible difficulties of a ship's acoustic classification, as well as the criticalities arising in room acoustic indicators in the different environments.

2. Background

2.1. Acoustic Classification of Buildings

Each country has its proper regulation for the acoustic classification of buildings, but they are generally based on the same performance indicators [28]. However, no uniformity in the threshold values among different countries can be found [23]. In Italy, UNI 11367:2010 [29] defines the acoustic classification of buildings referring to single rooms and, whenever possible, to the whole housing unit. The standard can be applied to all kind of buildings, both new and refurbished, except to industrial ones. The indicators considered in the present work are: reverberation time, standardized level difference, and impact sound pressure level, as described in the following chapter.

The prescription followed in the present work is the UNI 11367:2010 [29] one, which produces classification values for internal spaces. These are more restricted and, therefore, more in line with the idea that comfort has stricter requests than only health safety.

The classification can be applied to singular rooms and then extended to the entire study object (e.g., apartment, ship). For each room, the overall weighted apparent sound reduction index (R'_w) and the weighted normalized impact sound pressure level ($L'_{n,w}$) are computed with respect to different partition elements. The overall values are obtained with Equations (1) and (2), in which the index i represents the measured partition's acoustic performance indicator and n is the total number of partitions measured. The number of elements studied must be sufficient to describe the overall acoustic properties.

$$\overline{R'_w} = -10log\left(\sum_{i=1}^{n} 10^{-\frac{R'_{w,i}}{10}}\right) \quad (1)$$

$$\overline{L'_n} = 10log\left(\sum_{i=1}^{n} 10^{\frac{L'_{n,i}}{10}}\right) \quad (2)$$

The acoustic classification of the studied single room is derived with Equation (3), in which Z_i is the class weight and p is the number of acoustic characteristics considered. Then, the overall classification of the studied element can be taken as an arithmetic mean of the singular acoustic classification of the rooms. The acoustic classification classes acceptable for the results go from I to IV (from best to worst), plus another class named Non-Classifiable (NC). In the present work, the NC class is considered and taken as the worst case possible.

$$Z_{UI} = \frac{\sum_i^p z_i}{p} \quad (3)$$

2.2. Acoustic Comfort Classes

Naval agencies of different countries have given an acoustic comfort that does take into account the different life purposes of the ship's compartments. In this paper, the American Bureau of Shipping (ABS) [30], Germanischer Lloyd [31], and the Italian Naval Register (Rina) [32] classes are considered and compared with the results obtained with the Italian buildings classification. Each of the aforementioned classes differ on their ship type application, e.g., recreational boats or cargo vessel, and the different acoustic performance indicator used [24].

2.2.1. ABS Classification

The classification is intended for vessels falling under oil or chemical tankers, bulk or combination carriers, container carriers, multi-purpose cargo vessels, or crew areas on passenger's vessels. Vessels such as offshore support vessels, tugboats, towboats, dredgers, research vessels, drill ships, anchor handling vessels, or any other vessel providing service to offshore oil and gas exploration and production are not included in the scope of the ABS Crew habitability guide. Additionally, ABS considers the overall equivalent level on board moving vessels. In this Guide, comfort means the ability of the crew to use the space with minimal interference or annoyance from noise. The classes presented are three: Habitability (HAB), Habitability plus (HAB+), and Habitability plus plus (HAB++). The HAB class measurement method is based on the International Maritime Organization (IMO) and the measurement is considered close to non-compliance with the HAB notation if its value is in a range of 3 dB within the criteria. Acoustic measurements need to follow ISO 10140-1:2021 [33] and ISO 717-1:2020 [24] for passive acoustic indicators. The airborne sound insulation properties for bulkheads and decks need to satisfy the minimal criteria for R'_W. No more information about the classification procedure via acoustic performance indicators is reported.

In the present work, the HAB class is equivalent to the IV class in the UNI 11367.

2.2.2. Germanischer Lloyd Classification

This classification, which in 2013 joined the Derk Norske Veritas, has five comfort classes that goes from acceptable to excellent. The values are distinguished between passenger's and crew's spaces. Like in the procedure used in the present paper, the Germanischer Lloyd company follows the ISO 717-1:2020 [24] and ISO 717-2:2020 [25] procedure for the evaluation of the weighted acoustic indexes. The final class name obtained with the weighted sound reduction index is called acoustic privacy and depends on the cabin's scope. The overall quality class is the arithmetic sum of all the "Harmony Classes" (hc).

2.2.3. RINA

RINA formulated two different kinds of acoustic requirements for passenger and crew comfort on yachts, depending on the length between perpendicular (Lpp) of the vessel, namely, if it is greater or lower than 40 m. Airborne noise and structural vibrations levels are evaluated for yacht at berth and in navigation. For a yacht whose Lpp is greater than 40 m, which is interesting for the case study of the present work, the sound indicators considered are:

- the sound reduction index R_w evaluated according to EN ISO 717-1:2020 [24] and EN ISO 10140-2:2021 [33];
- the apparent sound reduction index R'_w evaluated according to ISO 717-1:2020 [24] and ISO 16283-1:2014 [34];
- the normalized impact sound pressure level $L'_{n,w}$ evaluated according to ISO 717-2:2020 [25] and ISO 16283-2:2020 [35].

The results of the measurements are divided into two class, A and B. The assigned comfort class is A if the measured R'_w is equal or greater than the reference values at the A score, and if the measured $L'_{n,w}$ is lower than or equal to the reference value at the A score. The same occurs for B score. Depending on the assignments of these labels to the different environments in the vessel, the yacht under analysis can be labeled with different notations, namely:

- Comfort Large Yachts Passengers (COMF LY PAX) A or B, if, for the guest accommodation, the acoustic indexes belong to the class A or to the class B;
- Comfort Large Yachts Crew (COMF LY CREW) A or B, if, for the crew accommodation, the acoustic indexes belong to the class A or to the class B;
- Comfort Large Yachts Passenger and Crew (COMF LY PACR) AA, AB, BA, or BB. AA, if both passengers and crew accommodations belong to class A; AB, if passengers' spaces belong to class A while crew spaces belong to class B; BA vice versa; while the label BB is used if both passengers and crew accommodation belong to the class B.

3. Materials and Methods

In order to give an acoustic classification of the ship with the proposed procedures, the ship's acoustic performance indicators should be measured. The present chapter describes the experimental setup and measurement procedure performed by following the most recent prescriptions for building's acoustic indexes and using class I instruments. Before describing in Section 3.3 the measurement procedure and the number of measurements per room, the ship's layout and the measurement conditions of the rooms are described in Section 3.1, while the rooms condition during the measurement are Section 3.2.

All measurements have been performed with a class I sound level meter, calibrated with a class I calibrator, a tapping machine, and an omnidirectional noise source.

3.1. Layout of the Yacht

The yacht structure develops vertically in four levels: the Lower Deck (LD), the Main Deck (MD), the Upper Deck (UD), and the Sun Deck (SD). The LD and the MD are the most concerning compartments of the ship for acoustic comfort issues, as they are those where owner, guests, and crew sleep or mostly spend their indoor time. The Vip and Guest Cabins are all in the LD and have partitions in common; Vip Cabins have a bulkhead with the Engine Room, the other Vip Cabin, and the Guest Cabin on the same side of the ship, port, or starboard (Stbd). Some of these partitions have "escape routes", which will be called "sound energy leakage" routes in this paper due to their properties. Figure 1 shows part of the LD; Figure 2 visualizes a piece of the layout of the MD.

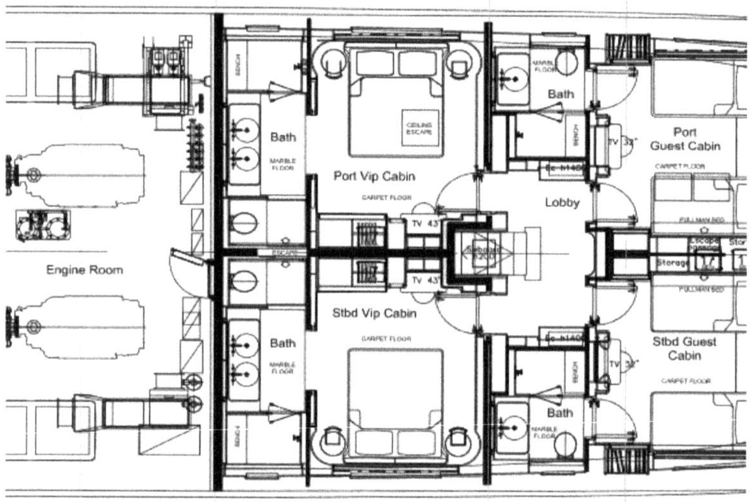

Figure 1. LD layout under analysis.

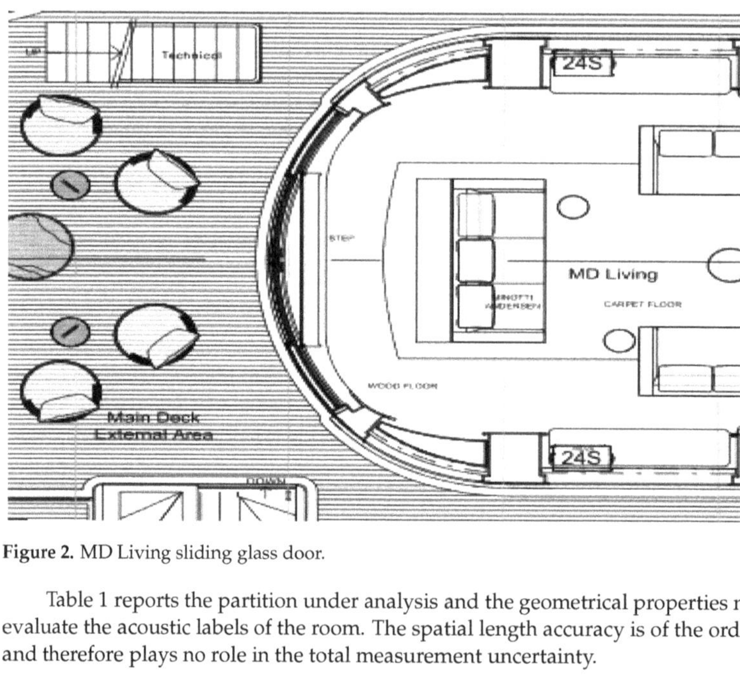

Figure 2. MD Living sliding glass door.

Table 1 reports the partition under analysis and the geometrical properties needed to evaluate the acoustic labels of the room. The spatial length accuracy is of the order of mm and therefore plays no role in the total measurement uncertainty.

Table 1. Values of the volume and area of the separating elements under analysis.

Room	$V\ (m^3)$	First near Room	Surface (m^2)
Master Cabin + Dressing Room	79.41	Galley	8.69
		Vip Cabin	9.47
Vip Cabin + Bath	30.68	Guest Cabin	6.78
		Engine Room	9.44
Guest Cabin + Bath	27.01	Guest Cabin	7.00
		Crew Mess	7.05
MD Living	127.44	Engine Room	26.35
MD Living	127.44	MD external area	16.24

3.2. *Measurement Condition of the Rooms*

In this subsection, the different measurement conditions of the rooms are described. During the measurements, the rooms were partially furnished, and a plastic layer covered the floors.

3.2.1. Master Cabin

The Master Cabin and the Dressing Room located near the Galley are considered as a whole. This arises from considering the Galley as a relevant sound energy emitter for the acoustic comfort of the Master Cabin. Therefore, the only door left open in the measurements was the one connecting the Dressing Room with the Master Cabin. Both doors connecting the Bath and the rest of the ship were closed.

3.2.2. Main Deck Living room

The Middle Deck Living room has the biggest problem arising from the presence of the upper deck rooms: the UD external zone and the Skylounge. The preferential noise transmissions are structural and therefore the impact is characterized by the impact level noise. These measurements are of most interest given the different floors that are present in both the source zones. The UD external zone has a wood floor, while the Skylounge has a wood floor with a carpet. The other two measured indexes in this paper are the standardized level difference measured for the vertical transmission, source positioned in the engine room, which is directly under the room, and the glass door, which divides the MD external zone from the MD Living.

3.2.3. Port Vip Cabin

The entire room taken for the evaluation of the acoustic performance indicators are the Port Vip Cabin with its bath. The shower is not considered in the entire room and the door is closed. In fact, the shower does not have common partition with the other compartments of the ship outside the engine room and the more realistic case is with the closed door. In this case, the airborne noise mitigation plays a bigger role in the overall acoustic comfort level. Nevertheless, impact level noise is measured with the tapping machine placed on the pavement of the MD Living and the Main Pantry.

3.2.4. Stbd Vip Cabin

The compartment considered as one in the analysis is the union of the Stbd Vip cabin and the bath with the shower door closed. The position of the cabin in the layout of the ship is the same as the Port Vip Cabins' and therefore the only acoustic index measured is the $D_{nT,w}$ with the Engine Room. In fact, the fundamental difference between the Vip cabins is the bulkhead between the Engine room and the Port Vip Cabin, which has an escape route that can be opened and, therefore, could worsen the overall sound mitigation.

3.2.5. Port Guest Cabin

No preferential sound energy transmission way is found in this room. Probably, the Crew Mess zone is the one that will have a larger sound emission and, therefore, the Bulkhead needs a bigger sound mitigation.

3.3. *Acoustic Performance Indicators*

According to what is described in Section 2, the acoustic performance indicators measured in this work are the reverberation time, the standardized sound reduction index, and the impact sound pressure level.

3.3.1. Reverberation Time

The reverberation time (T) of a compartment is the time needed for a sound pressure level to decay 60 dB after the source is stopped. An optimal room's reverberation time needs to consider the room's dimensions and its final use. As an example, the optimal T's limit (T_{ott}) Equation for speech comprehension is given in Equation (4), in which the only structural variable is the volume (V).

$$T_{ott} = 0.32log(V) + 0.03 \tag{4}$$

The reverberation time measurements of compartments follow the ISO 3382-2:2008 [36] prescriptions. If two compartments have the same configuration and furniture, then the operator can choose one reverberation time measurement as representative of both rooms. Given the small room dimension and the fast decay of the noise saturation level, measurements of the T_{20} have been carried in all cases. This has been done by positioning the omnidirectional noise source generator in one or more places of the room and measuring the sound level pressure in different places. Importance has been given to not following a regular scheme of the sound level meter disposition. Account of the measurement's uncertainty generated by the influence of the operator holding the sound level meter has been considered in the estimation of the uncertainty. The uncertainty is given by Equation (5), in which B is the bandwidth, N is the total number of measurements, and n is the number of noise measurements for noise source location. All measurements have been taken in third-octave band frequencies. Table 2 shows the number of the source position and the respective number of sound level meter positions.

$$\sigma_{T_{20}} = 0.88 \cdot T_{20}\sqrt{\left(1+\frac{1.9}{n}\right)/(NBT_{20})} \tag{5}$$

Table 2. Number of measurements of the reverberation time.

Cabin	#Pos. Noise Source	#Pos. Sound Level Meter
Master Cabin and Dressing	2	5
MD Living	3	5
Port/Stbd Vip Cabin	1	9
Port/Stbd Guest Cabin	1	6

3.3.2. Standardized Level Difference

The standardized level difference (D_{nT}) is equal to the reduction of sound level pressure characteristic of a separating element. This does account for the airborne sound transmission and does consider the structure-borne noise emission from the interaction with the sound wave. When measuring the characteristics of a bulkhead that cannot be isolated, the different sound transmission ways need to be considered. The overall effect is represented by the apparent sound reduction index (R'), which is used to give a standard value to the partitioning element characteristics. D_{nT} values are computed from R' and corrected considering the reverberation time T_0. As reported in Equation (6) room's receiving volume (V), the common surface dimension of the element between the emitting and receiving room are also considered in the calculation.

$$D_{nT} = R' + 10log\left(\frac{ST_0}{0.16V}\right) \quad (6)$$

The acoustic standardized sound difference best represents the passenger's annoyance. According to the measurement procedure described by ISO 16283-1:2014 [34], the measured value is standardized by the receiving room's reverberation time. Airborne horizontal and vertical propagation ways have been measured. Table 3 reports the number of measurements taken of the source, the receiving, and the background sound pressure level (SPL) measuring positions for different rooms combinations.

Table 3. Number of SPL measurements for D_{nT}.

Emission Room	Receiving Room	#Pos. Emission	#Pos. Receiving	#Pos. Background
Galley	Master Cabin and Dressing	5	5	5
Engine Room	Port Vip Cabin	6	5	5
	Stbd Vip Cabin	-	5	5
	MD Living	-	10	5
Stbd Vip Cabin	Port Vip Cabin	5	5	5
Port Guest Cabin	Port Vip Cabin	4	5	5
	Stbd Guest Cabin	-	5	5
Crew Mess	Stbd Guest Cabin	5	5	5
MD external area	MD Living	5	5	5

The final value assigned to the separation element is the weighted value obtained with the ISO 717-1:2020 [24] prescription and reference wall values.

3.3.3. Impact Sound Pressure Level

The impact sound pressure level is representative for the structure-borne sound pressure and considers all the transmission ways. It can be measured for a principal vertical transmission way, two rooms on different levels, or horizontal way, two rooms on the same level. The equivalent normalized impact sound pressure level (L'_n) does give a general characterization of the structure under analysis. Equation (7) gives its value as a function of the room's receiving volume V, the reference T_0, the reference equivalent absorption surface A_0, and the standardized impact sound pressure level L_{nT}.

$$L'_n = L_{nT} + 10log\left(0.16\frac{V}{T_0 A_0}\right) \quad (7)$$

Only the vertical transmissions have been measured in the present work. All measurements have been made by following the prescriptions given by ISO 16283-2:2020 [35]. As the tapping machine has a white noise spectrum, the resulting impact sound pressure level needs to be corrected with the coefficient C_I. For this reason, in the following, the paper for L'_n is meant the sum of $L'_n + C_I$. Table 4 shows the number of SPL for each noise source position for the different zones' combinations.

The representing standardized impact level is obtained by a weighting procedure described in the ISO 717-2:2020 [25].

Table 4. Number of source positions and SPL measurements for each source position.

Impact Floor	Receiving Room	Source Positions	SPL
UD Fwd external area (wood)	Master Cabin and Dressing	2	4
UD external area (wood)	MD Living	1	3
Skylounge (carpet)	MD Living	1	2
MD Living (wood floor)	Port Vip Cabin	1	5
Main Pantry (vynil floor)	Port Vip Cabin	1	5
Main Foyer (wood floor)	Stbd Guest Cabin	1	5

4. Results

In this section, the acoustic performance indicators measured and the classification procedure results are reported. Their values act as bases for an overall characterization of the rooms.

4.1. Acoustic Performance Indicators

The results are reported in the following Table 5 and Figure 3, ordered according to the room. All acoustic performance indicators are reported with the unilateral uncertainty, with $k = 1.95$ corresponding to a confidence level (CL) equal to 95%.

Table 5. Acoustic performance indicators for the different rooms.

Receiving Room	Source Position	$D_{nT,w}$(dB)	R'_w(dB)	$L'_{n,w}$(dB)
Master cabin	Galley	55.0 ± 0.7	50.3 ± 0.7	\
	UD external area	\	\	72.0 ± 1.0
MD living room	Skyloung	\	\	60 ± 1.2
	UD external area	\	\	71.0 ± 1.2
	Engine Room	56.2 ± 0.7	54.03 ± 0.7	\
Port Vip cabin	Engine Room	61.5 ± 0.7	61.3 ± 0.7	\
	Stbd Vip Cabin	43.7 ± 0.7	43.5 ± 0.7	\
	Port Guest Cabin	49.7 ± 0.7	48.1 ± 0.7	\
	MD Living (wood floor)	\	\	67.9 ± 1.0
	MD Living (vinyl floor)	\	\	70.5 ± 1.0
Stbd Guest cabin	Port Guest Cabin	38.9 ± 0.7	38.0 ± 0.7	\
	Crew Mess	53.8 ± 0.7	52.9 ± 0.7	\
	Main Foyer (covered marble floor)	\	\	70.1 ± 1.0

The Stbd Vip cabin's acoustic performance indicator $D_{nT,w}$ for the partition with the Engine Room is equal to 56.7 ± 0.7 dB. All the other acoustic performance indicators are taken as equal to the Stbd Vip cabin, given the same disposition of the cabin in the ship.

4.2. Comfort Acoustic Classification

In this section, the yacht's cabin acoustic classifications are computed with respect to the different classification threshold values. The ship's acoustic classification arises from an arithmetic mean of the different classification results obtained.

Tables 6 and 7 report the different classification values obtained by following UNI 11367:2010 [29]. The values $D_{nT,v}$ and $L'_{n,v}$ are the validation values and are equal to the rounded weighted value with the unilateral uncertainty up to the nearest integer number to the worst case. This means that $D_{nT,v}$. are rounded up to the lowest value in the 95% CL and $L'_{n,v}$ to the highest value. The MD living standardized difference level ($D_{ls,2m,nT,v}$) of the glass door in common with the MD external area is also reported and does not account to the mean standardized level difference of the room.

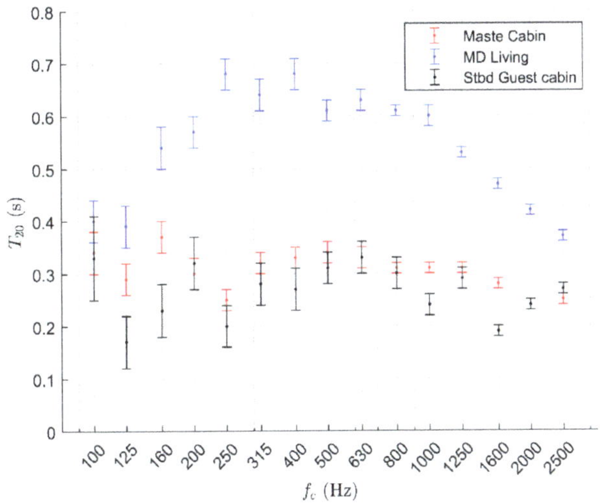

Figure 3. Reverberation time of the different rooms.

Table 6. Ship's compartment acoustic evaluation indicators.

Receiving Room	Source Room	Indicator (dB)	Acoustic Class
Port Vip Cabin	Engine Room	$D_{nT,v} = 61$	I (≥ 56 (dB))
	Stbd Vip Cabin	$D_{nT,v} = 43$	NC (< 45 (dB))
	Port Guest Cabin	$D_{nT,v} = 49$	IV (≥ 45 (dB))
	MD Living	$L'_{n,v} = 70$	NC (> 68 (dB))
	Main Pantry	$L'_{n,v} = 72$	NC (> 68 (dB))
Stbd Vip Cabin	Engine Room	$D_{nT,v} = 56$	I (≥ 56 (dB))
Stbd Guest Cabin	Port Guest Cabin	$D_{nT,v} = 38$	NC (< 45 (dB))
	Crew Mess	$D_{nT,v} = 53$	III (≥ 50 (dB))
	Main Foyer	$L'_{nT,v} = 71$	NC (> 68 (dB))
MD Living	Engine Room	$D_{nT,v} = 56$	I (≥ 56 (dB))
	MD external Area	$D_{Is,2m,nT,v} = 26$	NC (< 32 (dB))
	UD external area	$L'_{n,v} = 72$	NC (> 68 (dB))
	Skylounge	$L'_{n,v} = 61$	II (≤ 58 (dB))
Master Cabin	Galley	$D_{nT,v} = 54$	II (≥ 53 (dB))
	UD external area	$L'_{n,v} = 73$	NC (> 68 (dB))

Table 7. Acoustic classification of the ship's compartments.

Compartment	Mean Evaluation Index and Classification (dB)	
Master Cabin	L'_n	73 NC
	D_{nT}	54 II
MD Living	L'_n	70 NC
	D_{nT}	56 I
	$D_{IS,2m,nT}$	26 NC
Port Vip Cabin/Stbd Vip Cabin	L'_n	71 NC
	D_{nT}	48 IV
Stbd Vip Cabin	L'_n	71 NC
	D_{nT}	47 IV
Stbd Guest Cabin	L'_n	71 NC
	D_{nT}	42 NC
Port Guest Cabin	L'_n	72 NC
	D_{nT}	41 NC

5. Discussions

In a usual classification procedure, the final acoustic performance indicator is taken as a weighted value in function of a reference curve. UNI 11367:2010 [29] takes the ISO 717-2:2020 [25] heavyweight floors reference curve for the normalized impact sound pressure level and the unique ISO 717-1:2020 [24] curve of reference values for airborne sound. As an example, Figure 4 reports the one-third octave spectrum of the measured D_{nT} between the Port Vip Cabin and the Engine Room, with the reference curve reported in order to mark the difference with it in terms of spectral components.

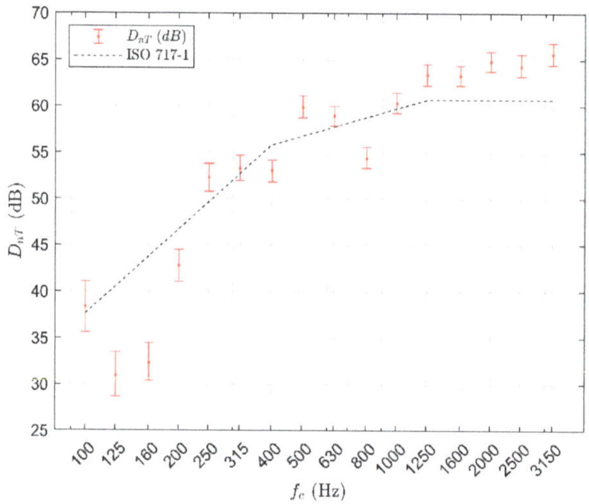

Figure 4. Standardized level difference of the Stbd Vip Cabin—Engine Room partition and reference values according to ISO 717-1:2020 [24].

All the different spectral components of the measured D_{nT} for all the partitions and the ISO 717-1:2020 [24] reference curve are reported in Figure 5.

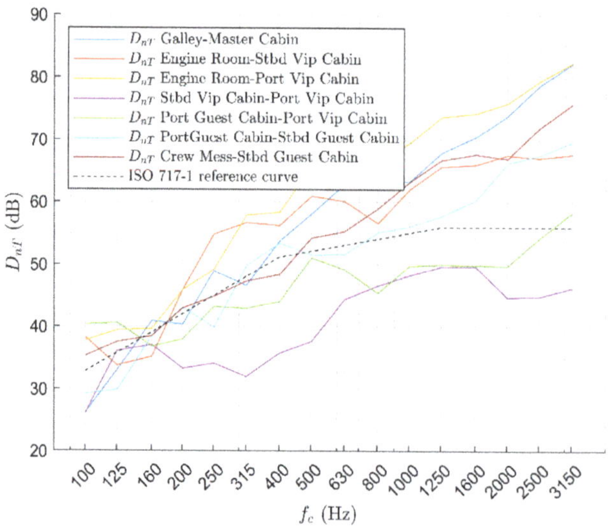

Figure 5. Spectrum behavior of the different D_{nT} of the analyzed partitions and reference values according to ISO 717-1:2020 [24].

Although most of the measured acoustic performance indicators exceed the UNI 11367:2010 [29] threshold values, they effectively represent the specific and overall acoustic performance of the different rooms under analysis. In fact, an overall ship's classification class NC is probably a non-realistic final class notation if considering that the classification is set for buildings and not for ships.

The reported result highlights that the standardized impact level noise results are the main cause of noise inside the investigated rooms, and they will worsen the overall acoustic comfort inside the vessel. However, while the present study has been only conducted on a singular vessel, it is expected that all others would show similar behavior. Thus, the results should warn ships' shipbuilders to seriously take this indicator into consideration for future designs. A correct acoustic classification procedure would help pinpoint the ship's zones needing more care, but the reference standardized impact level noise values used for buildings are actually too different with respect to what is needed for ships' comfort acoustic classification. For this reason, the authors encourage future studies aimed at defining proper threshold levels. In order to support this thesis, Tables 8 and 9 report the class assigned to the different rooms according to the Germanischer Lloyd classification procedure, which classifies the acoustic comfort in five different classes as in the Italian one. As the Germanischer Lloyd threshold values are dependent on the room use, Pax cabins to public spaces are the most similar ones to the threshold used in UNI 11367:2010 [29]. Following a one-to-one correspondence between the two regulations, the paper's NC class corresponds to the IV Germanischer Lloyd class, and so on.

Table 8. Standardized level difference with the Germanischer Lloyd classification.

	Passenger's Space	
Receiving Space	**Emission Space**	**hc**
Port Vip Cabin	Engine Room Stbd Vip Cabin Port Guest Cabin	II E E
Stbd Vip Cabin	Engine Room	IV
Stbd Guest Cabin	Port Guest Cabin	I
	Crew's Space	
MD Living	Engine Room MD External Area	III /
Master Cabin	Galley	II
Stbd Guest Cabin	Crew Mess	III

Table 9. Impact level classification with the Germanischer Lloyd classification.

	Crew's Space	
Receiving Space	Emission Space	hc
Port Vip Cabin	MD Living Main Pantry	IV IV
	Passenger's Space	
UD external area	Master Cabin	/
Main Foyer	Stbd Guest Cabin	IV

The results show that in most cases, both classification procedures do assign the worse class to the different rooms, although they have different values for compartments utilization.

Only the Geramnischer Lloyd classification is used to confront the different results because of the similarity of the classification procedure between the UNI 11367:2010 and Germanischer Lloyd classification, while the other two classification procedures are different.

In fact, the ABS classification and the RINA classification do not classify single partitions in different classes or consider different vessels and operations. Figure 6 highlights the different occurrences by class for the standardized level difference for the Germanishcer and UNI 11367:2010 classification procedure.

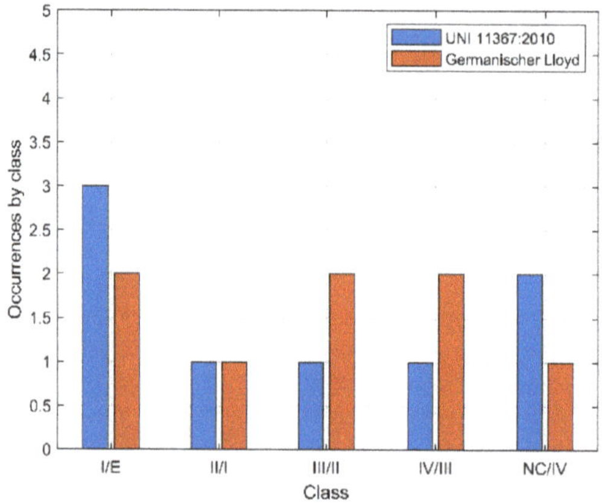

Figure 6. Occurrences by class for the standardized level difference.

The difference between the translated reference curve and the measure third-octave band spectrum of D_{nT}, reported in Figure 5, is a possible sentinel of the presence of sound energy leakage, responsible for the worsening of the overall standardized sound difference index of up to almost 5 dB according to Kang et al. [27]. However, in [27], the measurements performed to verify such behavior were taken under controlled situation, thus only accounting for the direct propagation ways of air-borne noise. According to the previously presented work [26], sound energy leakage of different kinds can be found with the acoustic camera. Figure 7 reports the localized sound energy leakage in the Stbd Vip Cabin—Engine Room partition. This, as well as all the others, correspond to the escape ways and the acoustic camera measurements helped on revealing them, which are voluntarily hidden for recreational reasons. However, the combined acoustic camera and room acoustic analysis shows that their design has not been carried out to take into account the attenuation of the sound energy, and they are responsible for the scarce apparent sound reduction index. Moreover, poor sound energy insulation surely corresponds to poor thermal insulation, thus, also in this case, more attention must be paid by the designers.

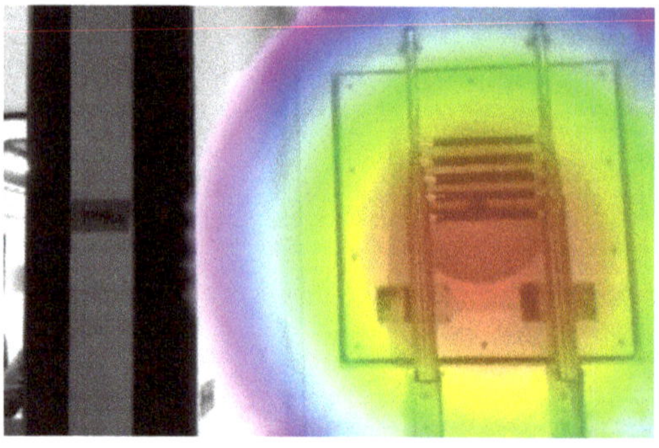

Figure 7. Sound energy leakage of the Stbd Vip Cabin—Engine Room partition. Frequency: 250–500 Hz.

6. Conclusions

An attempt to determine ships' comfort acoustic classification has been done in the present studies on the bases of the existing Italian's legislative framework on building classification. Room acoustics and acoustic camera measurements have been performed, as a case study, inside a 50 m LOA yacht, also in order to evaluate sound energy transmission.

The results of this work have shown that the acoustic classification of ships based on the existing prescription for the normal buildings does not account for the effective ship's acoustic issues: sound energy transfer from impacts over ceilings and sound energy leakage. While the other parameters are well performed, showing attention of shipbuilders to some aspects, all the evaluated standardized impact levels do result as non-classifiable (NC) or equivalent. This confirm that impact sounds are a concrete issue for acoustic comfort inside recreational boats and, at the same time, that the actual classification prescription has been projected to give an acoustic classification to ground buildings, with values that should be considered only as lower boundary values. In fact, they are also not referred to acoustic comfort. The present work proposes also to use the reverberation time as an acoustic performance indicator itself, and not only as a parameter to standardize the different acoustic performance indicators. Until now, it has found a small place in the acoustic classification of rooms that do not require a specific utilization. Nevertheless, it plays a fundamental role in the acoustic comfort of any kind of room. In small rooms, a high reverberation time could generate high sound level pressures that can drastically deteriorate the overall comfort in a room.

The standardized sound difference between contiguous compartments did not show any critical behavior in the classification, with their values ranging from class I to NC. Nevertheless, with the aid of an acoustic camera, sound energy leakages were revealed, coinciding with hidden escape ways that were poorly insulated, from both sounds and thermal ways. A specific localization of sound-leakage routes could not only improve the acoustic properties of the separating elements but with further studies, it could show energy transmission loss ways, which do worsen the energy consumption of a ship. The work is expected to raise awareness in shipbuilders to place attention on the design of all aspects that can influence acoustic comfort of interior spaces. Finally, for proper classification, future studies should focus on finding correct standardized impact level noise for ship cases, whose values are not comparable with those measured in actual buildings and for which reference values have been designed.

Author Contributions: Conceptualization, L.F.; Data curation, F.A.; Formal analysis, S.K. and F.A.; Funding acquisition, F.F. and G.L.; Investigation, S.K. and L.F.; Methodology, L.F.; Project administration, F.F.; Resources, F.F.; Supervision, L.F., F.F. and G.L.; Validation, L.F.; Visualization, G.L.; Writing—original draft, S.K., L.F. and F.A.; Writing—review and editing, L.F., F.A., F.F. and G.L. All authors have read and agreed to the published version of the manuscript.

Funding: This research was funded by CNR-INM within the project CTN01_00176_163601 "Tecnologia e Ricerca Industriale per la Mobilità Marina" (TRIM) funded by the Italian Ministry of Education, University and Research.

Data Availability Statement: Not applicable.

Conflicts of Interest: The authors declare no conflict of interest.

References

1. Borelli, D.; Gaggero, T.; Rizzuto, E.; Schenone, C. Holistic control of ship noise emissions. *Noise Mapp.* **2016**, *3*, 107–119. [CrossRef]
2. Nastasi, M.; Fredianelli, L.; Bernardini, M.; Teti, L.; Fidecaro, F.; Licitra, G. Parameters affecting noise emitted by ships moving in port areas. *Sustainability* **2020**, *12*, 8742. [CrossRef]
3. Fredianelli, L.; Bolognese, M.; Fidecaro, F.; Licitra, G. Classification of noise sources for port area noise mapping. *Environments* **2021**, *8*, 12. [CrossRef]
4. Fredianelli, L.; Nastasi, M.; Bernardini, M.; Fidecaro, F.; Licitra, G. Pass-by characterization of noise emitted by different categories of seagoing ships in ports. *Sustainability* **2020**, *12*, 1740. [CrossRef]

5. Fredianelli, L.; Gaggero, T.; Bolognese, M.; Borelli, D.; Fidecaro, F.; Schenone, C.; Licitra, G. Source characterization guidelines for noise mapping of port areas. *Heliyon* **2022**, *8*, e09021. [CrossRef] [PubMed]
6. Borelli, D. Maritime airborne noise: Ships and harbours. *Int. J. Acoust. Vib.* **2019**, *24*, 631–632. [CrossRef]
7. Rapisarda, V.; Valentino, M.; Bolognini, S.; Fenga, C. Il rischio rumore a bordo dei pescherecci: Alcune considerazioni sulla prevenzione e protezione degli esposti. *G. Ital. Med. Lav. Ed Ergon.* **2004**, *26*, 191–196. (In Italian)
8. Vukić, L.; Fredianelli, L.; Plazibat, V. Seaferers' perception and attitudes towards noise emission on board ships. *Int. J. Environ. Res. Public Health* **2021**, *18*, 6671. [CrossRef] [PubMed]
9. Mansi, F.; Cannone ES, S.; Caputi, A.; De Maria, L.; Lella, L.; Cavone, D.; Vimercati, L. Occupational exposure on board fishing vessels: Risk Assessments of biomechanical overload, noise and vibrations among worker on fishing vessels in southern Italy. *Environments* **2019**, *6*, 127. [CrossRef]
10. Turan, O.; Helvacioglu, I.H.; Insel, M.; Khalid, H.; Kurt, R.E. Crew noise exposure on board ships and comparative study of applicable standards. *Ships Offshore Struct.* **2011**, *6*, 323–338. [CrossRef]
11. Zytoon, M.A. Occupational noise exposure of fishermen aboard small and medium-scale fishing vessels. *Int. J. Ind. Ergon.* **2013**, *43*, 487–494. [CrossRef]
12. Kurt, R.E.; Khalid, H.; Turan, O.; Houben, M.; Bos, J.; Helvacioglu, I.H. Towards human-oriented norms: Considering the effects of noise exposure on board ships. *Ocean. Eng.* **2016**, *120*, 101–107. [CrossRef]
13. Borelli, D.; Gaggero, T.; Rizzuto, E.; Schenone, C. Onboard ship noise: Acoustic comfort in cabins. *Appl. Acoust.* **2021**, *177*, 107912. [CrossRef]
14. Badino, A.; Borelli, D.; Gaggero, T.; Rizzuto, E.; Schenone, C. Criteria for noise annoyance evaluation on board ships. In *Sustainable Maritime Transportation and Exploitation of Sea Resources*; CRC Press/Balkema: Leiden, The Netherlands, 2011.
15. Borelli, D.; Schenone, C.; Di Paolo, M. Application of a simplified "source-path-receiver" model for hvac noise to the preliminary design of a ship: A case study. In Proceedings of the 21st International Congress on Sound and Vibration, Beijing, China, 13–17 July 2014.
16. Vladimir, N.; Lončar, I.; Ančić, I.; Senjanović, I. Prediction of noise performance of Ro-Ro passenger ship by the hybrid statistical energy Analysi. *Pomor. Zb.* **2018**, *2*, 29–45. [CrossRef]
17. Shin, Y. Study on reduction of excessive noise and vibration of aft part of high speed Ro-Ro passenger ship. *J. Ocean. Eng. Technol.* **2019**, *33*, 196–202. [CrossRef]
18. Goujard, B.; Sakout, A.; Valeau, V. Acoustic comfort on board ships: An evaluation based on a questionnaire. *Appl. Acoust.* **2005**, *66*, 1063–1073. [CrossRef]
19. Rasmussen, H.L.; Jensen, P.A. Tools and methods to establish a feed-forward loop from operation to design of large ships and buildings. In Proceedings of the European Facility Management Conference 2018, Sofia, Bulgaria, 5–8 June 2018; pp. 75–84.
20. Casareale, C.; Bernardini, G.; Bartolucci, A.; Marincioni, F.; D'Orazio, M. Cruise ships like buildings: Wayfinding solutions to improve emergency evacuation. In *Building Simulation*; Tsinghua University Press: Beijing, China, 2017; Volume 10, pp. 989–1003. [CrossRef]
21. Vardaxis, N.G.; Bard, D.; Persson Waye, K. Review of acoustic comfort evaluation in dwellings—part I: Associations of acoustic field data to subjective responses from building surveys. *Build. Acoust.* **2018**, *25*, 151–170. [CrossRef]
22. Azar, E.; O'Brien, W.; Carlucci, S.; Hong, T.; Sonta, A.; Kim, J.; Zhou, J. Simulation-aided occupant-centric building design: A critical review of tools, methods, and applications. *Energy Build.* **2020**, *224*, 110292. [CrossRef]
23. Rocca, M.; Di Puccio, F.; Forte, P.; Leccese, F. Acoustic comfort requirements and classifications: Buildings vs. yachts. *Ocean. Eng.* **2022**, *255*, 111374. [CrossRef]
24. *ISO 717-1:2020*; Acoustics-Rating of Sound Insulation in Buildings and of Building Elements—Part 1. Airborne Sound Insulation: Geneva, Switzerland, 2020.
25. *ISO 717-2:2020*; Acoustics-Rating of Sound Insulation in Buildings and of Building Elements—Part 2. Impact Sound Insulation: Geneva, Switzerland, 2020.
26. Licitra, G.; Fredianelli, L.; Kanka, S.; Artuso, F.; Fidecaro, F. Acoustic comfort in yachts: Measurements with acoustic camera. In Proceedings of the 28th International Congress on Sound and Vibration, Singapore, 24–28 July 2022.
27. Kang, H.J.; Kim, J.S.; Kim, H.S.; Kim, S.R. Influence of sound leaks on in-situ sound insulation performance. *Noise Control. Eng. J.* **2001**, *49*, 113–119. [CrossRef]
28. Rasmussen, B. Acoustic classification of buildings in Europe–Main characteristics of national schemes for housing, schools, hospitals and office buildings. In Proceedings of the EuroNoise 2015, Maastricht, The Netherlands, 31 May–3 June 2015.
29. *UNI 11367:2010*; Building Acoustics-Acoustic Classification of Building Units-Evaluation Procedure and in Situ Measurements. UNI: Milan, Italy, 2010.
30. HAB, A.C. *Guide for Crew Habitability on Ships*; American Bureau of Shipping: Houston, TX, USA, 2016.
31. *Rules for Classification and Construction Ship Technology, 1 Seagoing Ships, 16 Harmony Class-Rules on Rating Noise and Vibration for Comfort Cruise Ships (v<= 25 kn)l*; Germanischer LLoyd AG: Hamburg, Germany, 2003.
32. Registro Italiano Navale (RINA). *Rules for the Classification of Pleasure Yacht-Part E: Additional Class Notations*; RINA: Genova, Itly, 2022.
33. *ISO 10140-2:2021*; Acoustics-Laboratory Measurement of Sound Insulation of Building Elements-Part 2: Measurement of Airborne Sound Insulation. International Organization for Standardization: Geneva, Switzerland, 2021.

34. *ISO 16283-1:2014*; Acoustics-Field Measurement of Sound Insulation in Buildings and of Building Elements—Part 1: Airborne Sound Insulation. International Organization for Standardization: Geneva, Switzerland, 2014.
35. *ISO 16283-2:2020*; Acoustics-Field Measurement of Sound Insulation in Buildings and of Building Elements—Part 2: Impact Sound Insulation. International Organization for Standardization: Geneva, Switzerland, 2020.
36. *ISO 3382-1:2008*; Acoustics-Measurement of Room Acoustic Parameters—Part 2: Reverberation Time in Ordinary Rooms. International Organization for Standardization: Geneva, Switzerland, 2008.

energies

MDPI

Article

Proposal of a Simplified Tool for Early Acoustics Design Stage of Classrooms in Compliance with Speech Intelligibility Thresholds

Paolo Croce [1,*], Francesco Leccese [1], Giacomo Salvadori [1] and Umberto Berardi [2]

[1] Department of Energy, Systems, Territory and Construction Engineering, University of Pisa, 56122 Pisa, Italy
[2] Department of Architectural Science, Toronto Metropolitan University, Toronto, ON M5B 2K3, Canada
* Correspondence: paolo.croce@phd.unipi.it

Abstract: The speech intelligibility properties of classrooms greatly influence the learning process of students. Proper acoustics can promote the inclusion of foreign students and children with learning or hearing impairments. While awareness of the topic is increasing, there is still no parameter that can describe all aspects of speech transmission inside a room. This complicates the design of classrooms and requires designers to have extensive knowledge of theory and experience. In the scientific and technical literature, there is a lack of predictive tools, easy to use by designers, which can guide the choices in the early design stages in order to move towards technical solutions able to ensure adequate levels of speech intelligibility. For this reason, in this paper, the most relevant speech intelligibility parameters found in the literature were collected and discussed. Among these, the Clarity index and Speech Transmission Index were singled out as the most effective ones, whose prediction can be made with relatively simple methods. They were then analyzed through their prediction formulas, and a tool was proposed to allow an easy estimation of the minimum total equivalent sound absorption area needed in a classroom. This tool greatly simplifies the early acoustics design stage, allowing the intelligibility of speech within a classroom to be increased without requiring much theoretical effort on the part of the designers.

Keywords: room acoustics; classroom acoustics; students learning; speech intelligibility; clarity index; prediction diagram

1. Introduction

Classroom acoustics is one of the main goals in classroom design to ensure a fruitful learning environment. Neglecting sound in design can easily lead teachers to vocal diseases [1] due to high phonation times and high sound pressure levels of speech [2] and can worsen students' performance and create annoyance [3]. The increase in sound levels of speech under conditions of high background noise, as described in ISO 9921:2003 [4], is due to what is now known as the Lombard effect [5], an unconscious vocal change that leads to an increase in pitch and lengthening of word duration [6]. At the same time, noise in classrooms has a dual origin: it can come from the external environment or students' activity [7] and consequently from an untreated decay of sound in the environment. For this reason, Reverberation Time (*RT*) has often been considered one of the indispensable parameters in proper acoustic design [8,9], although not directly correlated with student performance [10,11].

Parameters of greater relevance are those that consider the contribution of early reflections related to the listening experience [12], such as the Clarity index C_{50} (ISO 3382-1:2009 [13]), or parameters that include both room acoustics and speech-to-noise-ratios measurements, such as the Speech Transmission Index *STI* (IEC 60268-16:2020 [14]) or the Useful-to-detrimental sound ratio U_{50} [15]. Describing these parameters in international [13,16] and national acoustic standards [17,18], it has become necessary to unambiguously establish which of these is needed as the main parameter of the acoustic quality of learning environments, even considering the expensive instrumentation required for proper in situ measurement. The lack of adequate instrumentation and the presence of several parameters complicate the design of these environments for those who do not specialize in the acoustics of learning environments.

To minimize the number of measurement points, Astolfi et al. [19] proposed to characterize classrooms with the Clarity index in a central location and a spatial average of Reverberation Time (T20), while a global indicator of classroom acoustics was proposed by Chetoni et al. [20]. These parameters, however, can and should be able to be predicted at the early acoustics design stage so that materials and their arrangement can be consciously chosen to ensure speech intelligibility within the school environment. This can be achieved through acoustic simulation of the classroom [21,22] and through prediction formulas [23], of which those that allow the estimation of Reverberation Time [24], Clarity index [25,26], and Speech Transmission Index [27–29] were considered in this article, as conducted in previous research by the same authors [30]. However, the advantages associated with being able to predict the acoustic behavior of the environment are not obtainable without effort. Simulations require knowledge of specific software that are not always freeware, with high modeling times of the environment and calculation times highly dependent on the type of simulation and the required accuracy. Prediction formulas, on the other hand, are many, and their choice is not always obvious. They also, in some cases, require a nontrivial amount of hand calculations, often having to consider different variables. These difficulties prevent a simple preliminary assessment and do not directly address the main objective of acoustic design: the choice and arrangement of materials.

For these reasons, the authors found it useful to propose a simplified tool to be used in the early acoustics design stage. The tool would allow for the rapid estimation of the total equivalent sound absorption area (*A*) required to ensure compliance with the thresholds of speech intelligibility parameters necessary for the construction of fruitful learning environments. Firstly, the parameters of interest were selected based on their importance in describing the acoustics of classrooms. Secondly, the prediction formulas of the respective parameters were selected according to the criterion of effectiveness versus simplicity of use. Thirdly, the chosen prediction formulas were made a function of the same variables (volume *V* and total equivalent sound absorption area *A*) so that they could be easily calculated and compared. Finally, in the same prediction *A*–*V* diagram, the contour lines of the parameter thresholds suggested by standards and the technical literature for classrooms were drawn in order to divide the total sound absorption area (*A*) into different zones with different speech intelligibility ratings. The result was then compared with measurements taken in two medium-sized university classrooms at the University of Pisa ($300 < V < 800$ m^3) and with data from four small-sized classrooms of first-grade schools ($V < 300$ m^3) taken from Astolfi et al. [19].

2. Materials and Methods

In creating a simplified tool for the estimation of speech intelligibility parameters, the first part of the research focused on trying to determine which parameters best represent speech intelligibility and their relationship to actual student learning. Of those set out in the literature, they cannot all be taken indiscriminately, but the parameters must also be selected on the basis of the most recognized and widespread prediction models, adhering to the study's objective of creating a simplified tool that is a shortcut in the use of different prediction theories.

Taking for granted the need to consider *RT* as a predictive parameter for classroom design, it must be evaluated what else to consider. For the contribution of the first reflections, C_{50} was considered the best parameter, which is solely dependent on classroom characteristics and not affected by external noise, which is difficult to estimate. At the same time, *STI* and U_{50} can be considered complete predictors of speech intelligibility because they both include a measure of room acoustic quality and a measure of speech-to-noise ratio [31]. Since *STI* is easier to predict with formulas found in the literature and easier to measure in situ, it was chosen as the parameter of interest. These considerations, and thus the choice of selecting *RT*, *STI*, and C_{50} as characteristic parameters. The choices made agree with what was suggested by Astolfi et al. [19], who also suggested the possibility of considering the C_{50} in the center of the room instead of the average C_{50}, recommending threshold values for both.

Reverberation Time (*RT*) can be easily estimated from classical theory with the equation:

$$RT(A, V) = 0.161\frac{V}{A} \quad (1)$$

where V is the volume of the room (m^3) and A is the total equivalent sound absorption area (m^2). A was intentionally chosen as a variable so that one could indiscriminately choose whether to use Sabine's ($A = S\overline{\alpha}$) or Eyring's ($A = -S\ln(1-\overline{\alpha})$) formula. S is the total area of the room (m^2), and the equation is considered to neglect the air absorption contribution under the assumption of small and medium-sized room conditions.

The clarity index (C_{50}) can be estimated with Barron's revised theory [32], also in the Italian standard UNI 11532-1:2018 [33]:

$$C_{50}(RT,\ V,\ r) = 10\log\left(\frac{\frac{100}{r^2} + \left(\frac{31200RT}{V}\right)\left(1 - e^{\frac{-0.691}{RT}}\right)e^{\frac{-0.04r}{RT}}}{e^{\frac{-0.04r}{RT}}\left(\frac{31200RT}{V}\right)\left(e^{\frac{-0.691}{RT}}\right)}\right), \quad (2)$$

where r (m) is the source-receiver distance, this theory was first developed for concert halls and has been tested under reasonably diffuse room conditions [25]. In the case of no-uniform absorption, Barron's revised theory could be corrected by including the reverberation radius proposed by Arau-Puchades and Berardi [34] within the formula. Berardi expanded his studies to look at the sound energy decay in space with no uniform absorption and proposed several new models to take in account this circumstance in highly reverberant rooms like churches [35,36]. The use of Equation (2) involves two problems: the presence of three variables and the fact that, among them, *RT* and r are unknown at the early acoustics design stage. In accordance with the criterion of effectiveness versus simplicity, a change of variables is deemed necessary so that the choice of variables is consistent with what is made explicit in the choice of the *RT* prediction formula. In addition, V and A are the designer's starting and target information, respectively, placing the choice of materials as the first necessity and their layout at a later stage.

In accordance with the revised theory and substituting *RT* into Equation (2) using Equation (1), the following is obtained:

$$C_{50}(A,\ \mathrm{V},\ r) = 10\log\left(\frac{\frac{100}{r^2} + \frac{1600\pi}{A}e^{-\frac{rA}{4V}}\left(1 - e^{-\frac{4.29A}{V}}\right)}{\frac{1600\pi}{A}e^{-\frac{rA}{4V}}e^{-\frac{4.29A}{V}}}\right) \quad (3)$$

The only remaining problem is to be able to somehow correlate the volume V to the source-receiver distance r so that all formulas can be considered within the same graph. In this regard, the proposal of Astolfi et al. [19] was considered, according to which both the mean value C_{50M} and the value measured at the center of the classroom C_{50C} can be chosen.

Considering the C_{50C} helps to significantly reduce the range of source-receiver distances that can be considered within a classroom of fixed volume. For example, if the position of the teacher (source) within a regular-shaped classroom is fixed on a symmetry axis 1.5 m away from the wall, the distance to the center of the classroom (receiver) can easily be calculated as a function of area (Figure 1). Following this intuition, two different situations of regular-shaped classrooms were assumed: in the first case (Figure 1a), a square-shaped classroom; in the second case (Figure 1b), a rectangular-shaped classroom with sides in golden ratio (ca. 1.61), with the source-receiver distance parallel to the short side.

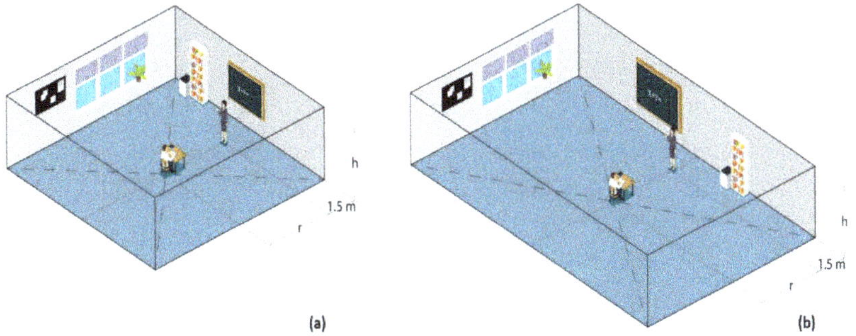

Figure 1. Axonometric diagram of the assumed arrangement of the location of the source and the receiver within a: (**a**) Square-shaped classroom; (**b**) Rectangular-shaped classroom.

Then, consider the following:

- a volume for classrooms ranging from 150 to 300 m^3;
- a net height varying from 3 to 4 m (3 is the minimum by law in Italy according to Ministerial Decree 18/12/1975 for newly built classrooms);

Plausible r ranges can be easily found (Figure 2).

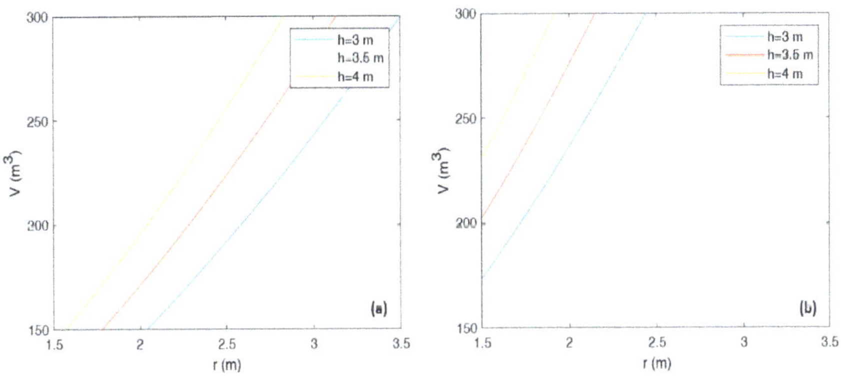

Figure 2. Graph of the correlation between volume (V) and source-receiver distance (r), for different heights of the classrooms, within a: (**a**) Square-shaped classroom; (**b**) Rectangular-shaped classroom.

With these considerations, it seems reasonable to consider a value of r = 2 m as the source-receiver distance. This assumption is also made in view of the fact that the aim of the study is to find plausible A-value zones that can meet the threshold requirements of speech intelligibility values. Otherwise, in cases where it is necessary to carry out detailed calculations, it will be necessary to use Equation (2) or Equation (3).

STI can be quickly estimated with several equations proposed in the literature using RT [28,29]. Among the various models, the logarithmic equation proposed by Leccese et al. [27], and obtained by regression analysis, is considered here as the reference equation. However, nothing prohibits the use of another model among those proposed. To easily compare the results with data found in the literature, it was chosen to correlate the STI with the mean RT values of the 0.5–1 kHz frequencies, resulting in the Formula:

$$STI\ (RT_{0.5-1\ \text{kHz}}) = 0.6158 - 0.2106\ln(RT_{0.5-1\ \text{kHz}}) \tag{4}$$

The same expedient should also be considered in the evaluation of A and C_{50}. Similarly to what was performed with C_{50}, STI is made a function of the variables V and A as well, substituting Equation (1) in Equation (4):

$$STI(A, V) = 0.6158 - 0.2106 \ln\left(0.161 \frac{V}{A}\right) \quad (5)$$

The found equations can thus be easily plotted as a function of the same variables and easily compared.

3. Results

The proposed simplified tool for estimating speech intelligibility parameters is based on Barron's revised theory [25] and the logarithmic equation by regression analysis proposed by Leccese et al. [27] on the assumption that r can be set as a plausible value and on considering conditions under which the acoustics of the classroom depend primarily on its physical characteristics. This was to highlight the fact that background noise from external noise sources was not considered. In fact, in order to ensure the reliability of the proposed tool, it is taken for granted that the lowering of background noise, predominantly from outside, requires the correct design of façade sound insulation, which is not the subject of this study, although it is of relevant importance.

3.1. Graph of STI and C_{50} as a Function of A and V

STI (calculated with Equation (5)) and C_{50} (calculated with Equation (3)) predictive graphs can be plotted as a function of A and V (Figure 3), considering for C_{50} (A, V, r) a fixed r = 2 m value. As can be easily seen from the graphs, the two parameters show similar trends, demonstrating some correlation. By also plotting the graph of *RT*, the strong negative correlation of *RT* values with C_{50} and speech intelligibility parameters [19,37] can also be confirmed.

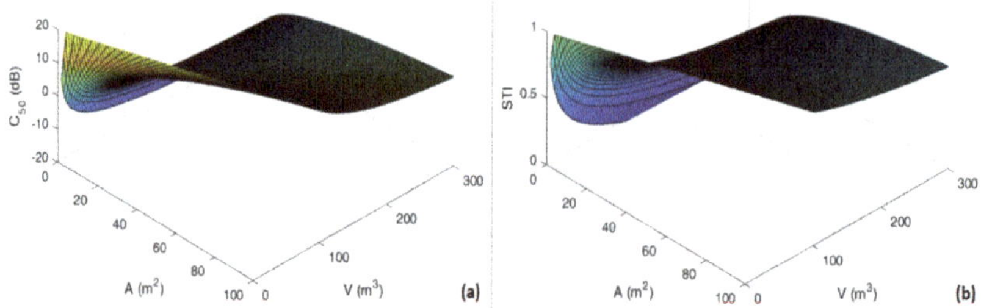

Figure 3. Graphs of two-variable functions of (**a**) Clarity index C_{50} (dB) for a fixed r = 2 m value; (**b**) Speech Transmission Index *STI*.

Given the trend of the functions, contour lines can be plotted for significant values, dividing the graph into two areas: one where the value will be less than the value of the contour line and one where it will be greater (Figure 4). Thus, given a certain volume V, a range of A values for which the speech intelligibility parameters are met can easily be found on the graph. Another thing that can be noticed from this type of graph is that by plotting different contour lines at the same C_{50}, and changing the source-receiver distance, the curves flatten out after about 4.5 m, making the difference in values after that distance more negligible.

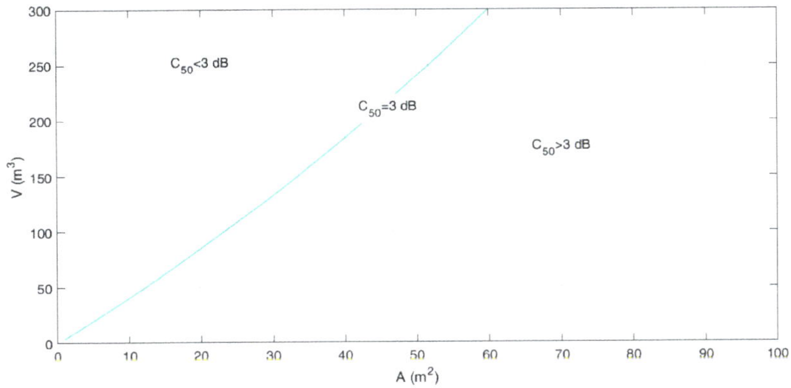

Figure 4. Prediction *A–V* Diagram: contour line for C_{50} = 3 dB, which divides the graph into two zones, one for C_{50} < 3 dB and the other for C_{50} > 3 dB.

3.2. Prediction A–V Diagram of Speech Intelligibility Parameters in Small-Medium Classrooms

Now the question remains as to what values the contour lines (Figure 4) should be drawn. Not considering *RT*, for which reference values have been repeatedly proposed in national standards [9,18], it is considered necessary to establish limit values for C_{50} and *STI*. For example, the Italian standard UNI 11367:2010 [38] recommends values for speech rooms of $C_{50} \geq 0$ dB and $STI \geq 0.6$; values are generally accepted as a minimum threshold.

Referring again to the proposal by Astolfi et al. [19], the C_{50} threshold (both as spatial average and as a single central value) for *moderate requirements* should be 3 dB, while the C_{50C} threshold (single value in the central position) for *severe requirements* should be 6 dB, thus ensuring a high component of first reflections on late sound, necessary for good speech intelligibility [39].

On the other hand, the *STI* thresholds for *good* (0.6–0.75) and *excellent* (>0.75) ratings can simply be taken in IEC 60268-16:2020 [14]. It should be mentioned here, however, that in the case of non-native listeners, the thresholds should be increased to 0.68 and 0.86 for the two ratings, according to the same standard.

Then, set the thresholds of speech intelligibility parameters at:

- C_{50} = 0, 3, and 6 dB;
- *STI* = 0.6, and 0.75;

The contour lines can be plotted. The prediction *A–V* diagram shown in Figure 5 thus divides the plane into areas with greater or lesser values than the selected thresholds. This makes it easy to determine which combinations of volume *V* and total equivalent sound absorption area *A* allow a greater chance of achieving desired levels of speech intelligibility. Considering, for example, a room volume of $V = 150$ m^3, a total equivalent sound absorption area greater than $A = 50$ m^3 is considered necessary to facilitate the condition that C_{50} > 6 dB occurs. Clearly, this consideration must be supported by the material layout provisions in the standards or by considerations based on experience. Figure 5 is, therefore, the prediction *A–V* diagram of the speech intelligibility parameters underlying the simplified tool and can be used in the early acoustics design stage to have a starting point on which to choose the materials with which to furnish a classroom in order to improve its speech intelligibility. It could also be used to improve the speech intelligibility of an existing classroom by finding its position within the prediction *A–V* diagram and moving on the *x*-axis to find the minimum difference in the *A* values required to reach the next threshold. To perform this in a simple way, the geometric and material survey of the room can be bypassed, and the *A* values of the room can be obtained directly using the inverse Sabine formula while always making sure that the room is in diffuse field conditions.

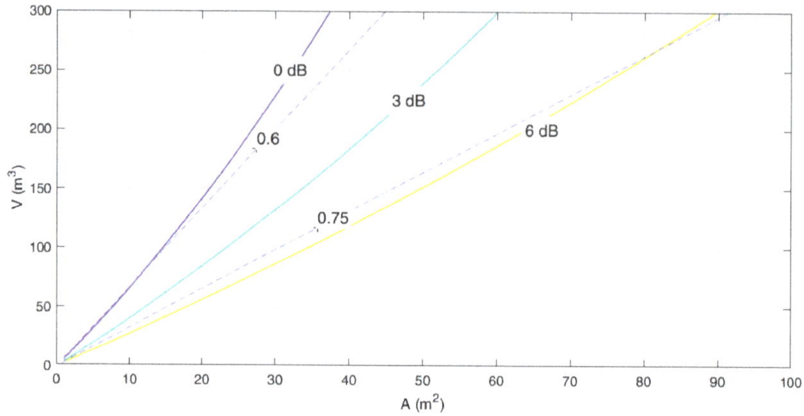

Figure 5. Prediction *V–A* diagram of speech intelligibility parameters in small-medium classrooms. C_{50} contour lines are continuous; *STI* contour lines are dashed.

3.3. Notes on the Applicability of the Model

For the proposed simplified tool here to be applicable, notes on the geometric and material characteristics of the classrooms must be considered. Some of these are due to the applicability of the theories on which the method is based, while others are due to the choices made within this paper on the definition of the problem.

The characteristic of the classrooms for which this method is reputed to be valid is summarized as follow.

1. The floor plan should be as regular as possible (square or rectangular).
2. Side surfaces should not be excessively absorbent.
3. Side walls should have elements that allow the avoidance of the occurrence of standing waves.
4. Ceilings should not be shaped in such a way as to create significant sound focus.
5. Ceilings should not be excessively diffuse.
6. Dimensions of the classrooms should be proportionate (comparable height, width, and depth).
7. Source-receiver distance (receiver placed in the center of the room) should be comparable to 2 m.
8. The environment should be properly isolated from external noise sources.

These notes mostly represent the well-known diffuse field conditions. Considering the aim of making the proposed tool as simple as possible, it was considered useful to be able to easily distinguish which environments are not suitable to be estimated with the models used. It is, in any case, considered that these notes exclude only a small percentage of the real classrooms and that the proposed simplified tool can be used extensively.

4. Discussion

In order to verify that the component models of the proposed simplified tool were plotted correctly, data obtained from measurements of classrooms were inserted into the prediction *A–V* diagram. The data chosen do not serve to verify the reliability of the prediction models of the C_{50} and *STI*, which have already been discussed and recognized, but only as counter-evidence of the ease and reliability of using the prediction *A–V* diagram to find the total equivalent sound absorption area of a classroom required to ensure speech intelligibility. As can be seen in Minelli et al. [40], it is not easy to find data from measurements of classrooms in the literature that present at the same time *V*, *A*, *RT*, C_{50}, and *STI* values. In this regard, to validate the proposed simplified tool and consequently verify the reliability of the prediction *A–V* diagram of the speech intelligibility parameters, the values of six different classrooms were considered (Table 1). In particular, measurements of two university classrooms were taken from an earlier work by the authors [30] (larger in size but relevant to the required characteristics) and four primary school classrooms from the work by Astolfi et al. [19].

Table 1. Classroom data taken from the authors' previous work and found in the literature.

ID	Ref.	V (m^3)	RT (s)	A_H (m^2)	A_C (m^2)	C_{50M} (dB)	C_{50C} (dB)	STI
C1	[30]	773	2.3 (1)	51.5	53.1	−3.83	—	0.45
C4	[30]	438	1.6 (1)	42.7	43.4	−2.09	—	0.52
A4	[19]	233	0.7 (*)	—	53.6	4.7	4.4	—
B2	[19]	201	0.5 (*)	—	64.7	7	8.1	—
F2	[19]	261	1.7 (*)	—	24.7	−0.1	−1.1	—
G3	[19]	138	0.8 (*)	—	27.8	4.4	4.7	—

Note: V (m^3) indicates the volume of the classroom. RT (s) the measured Reverberation Time (T30 or T20) for frequencies between 0.5 and 1 kHz; the values with apex (1) are T30 measured in unoccupied condition; the values with apex (*) are T20 measured in unoccupied condition. A_H (m^2) is the total equivalent sound absorption area assumed from the geometric survey and estimation of absorption coefficients by tables of known values. A_C (m^2) is the total equivalent sound absorption area calculated by inverse formula from the measured RT value. C_{50M} (dB) is the spatial average of measured Clarity index values for frequencies between 0.5 and 1 kHz. C_{50C} (dB) is the single measured Clarity index value at the room center position for frequencies between 0.5 and 1 kHz. STI is the measured Speech Transmission Index value. To use the data of the last four classrooms, the A values were found by inverse formula.

The proposed simplified tool assumes that the classroom can be approximated by the prediction formulas of classical theory and, most frequently, by Sabine's own theory. Consequently, it was seen that the prediction A–V diagram had only a little match with classrooms whose assumed values of A did not meet Sabine's prediction of RT, in accordance with what has already been presented in Section 3.3.

The classroom data entered within the prediction A–V diagram show more than acceptable agreement with the measured results, as can be seen in Figure 6. The only classroom discordant with the thresholds represented in the prediction A–V diagram is G3, the smallest, whose C_{50C} is 4.7 dB, although it is in the region between 0 and 3 dB. While the discordance is acceptable because it was underestimated, it is believed that this is probably due to the small size, which results in the lower accuracy of Barron's revised theory formulated for much larger environments [25]. In smaller classrooms, less reliance can be placed on the diffuse field hypothesis, and additional arrangements are required. However, using this prediction A–V diagram only at an early acoustics design stage to choose a value of A and verify the speech intelligibility afterward during construction, the proposed tool can still be considered reliable. Once the total equivalent sound absorption area A of the room has been established, materials can then be selected based on their random incidence absorption coefficients.

In any case, it should be remembered that the propagation of sound waves within the classroom to ensure good levels of speech intelligibility depends not only on the sound absorption area but also on the arrangement of materials, for which it is advisable to follow the suggested indications in the literature or the technical standards. It is precisely because of the dependence of speech intelligibility parameters on other factors that this proposed simplified tool can be used at the early acoustics design stage and provides a considerable starting point, but it cannot replace more accurate survey methods. The obtained values should be verified by in situ measurements during construction and the final testing phase.

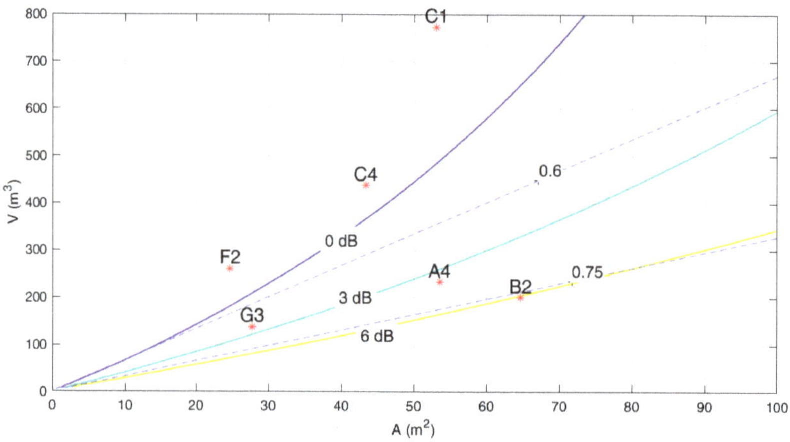

Figure 6. Prediction *A–V* diagram of speech intelligibility parameters in small-medium classrooms. C_{50} contour lines are continuous; *STI* contour lines are dashed. Classroom values taken from previous work by the authors (C1, C4) [30] and found in the literature (A4, B2, F2, G3) [19] were included.

5. Conclusions

In this paper, a simplified tool has been proposed to estimate the minimum total equivalent sound absorption area in the early acoustic design stage of a classroom. The estimate of the minimum sound absorption area, with the proposed tool, allows directing of the design solutions (from the early stages) towards those capable of ensuring compliance with the thresholds of the Clarity index and Speech Transmission Index values necessary to ensure an adequate level of speech intelligibility. The simplified tool is a prediction diagram composed of contour lines that divide the *A–V* plane into areas where speech intelligibility parameters are above or below the selected thresholds. At the early acoustics design stage, knowing the volume of the room, a starting value of the total equivalent sound absorption area can then be estimated from which materials can be chosen. Alternatively, it can be used in existing classrooms by checking how much the sound absorption area found with the inverse Sabine formula deviates from that in the diagram to achieve the desired threshold. This reliable, quick, and simple tool can be used at an early acoustics design stage, but it cannot replace more precise survey methods and does not provide an indication of the correct arrangement of materials within the environment. It also does not directly account for background noise and requires that the sound insulation of the envelope has already been treated separately.

In the future, the method can be developed by adding corrections to the prediction formulas based on considerations due to the non-uniformity of classroom sound absorption or the addition of the dependence on the signal-to-noise ratio. A version of the prediction *A–V* diagram under fully occupied room conditions could also be proposed.

Author Contributions: Conceptualization, P.C., G.S., F.L. and U.B.; methodology, P.C., G.S., F.L. and U.B.; investigation, P.C., G.S., F.L. and U.B.; data curation, P.C., G.S., F.L. and U.B.; writing—original draft preparation, P.C., G.S., F.L. and U.B.; writing—review and editing, P.C., G.S., F.L. and U.B.; visualization, P.C.; funding acquisition, G.S. and F.L. All authors have read and agreed to the published version of the manuscript.

Funding: This research was partially funded by the University of Pisa (Italy) within the financing activities for multidisciplinary research projects developed at the University of Pisa (ID research project PRA_2022_70, resolution of the board of directors 375/2022).

Data Availability Statement: Not applicable.

Conflicts of Interest: The authors declare no conflict of interest. The funders had no role in the design of the study; in the collection, analyses, or interpretation of data; in the writing of the manuscript; or in the decision to publish the results.

References

1. Mogas Recalde, J.; Palau, R.; Márquez, M. How Classroom Acoustics Influence Students and Teachers: A Systematic Literature Review. *J. Technol. Sci. Educ.* **2021**, *11*, 245. [CrossRef]
2. Durup, N.; Shield, B.; Dance, S.; Sullivan, R.; Gomez-Agustina, L. How Classroom Acoustics Affect the Vocal Load of Teachers. *Energy Procedia* **2015**, *78*, 3084–3089. [CrossRef]
3. Shield, B.M.; Dockrell, J.E. The Effects of Noise on Children at School: A Review. *Build. Acoust.* **2003**, *10*, 97–116. [CrossRef]
4. *EN ISO 9921*; Ergonomics—Assessment of Speech Communication. ISO International Organization for Standardization: Geneva, Switzerland, 2003.
5. Luo, J.; Hage, S.R.; Moss, C.F. The Lombard Effect: From Acoustics to Neural Mechanisms. *Trends Neurosci.* **2018**, *41*, 938–949. [CrossRef] [PubMed]
6. Brumm, H.; Zollinger, S.A. The Evolution of the Lombard Effect: 100 Years of Psychoacoustic Research. *Behaviour* **2011**, *148*, 1173–1198. [CrossRef]
7. Levandoski, G.; Zannin, P.H.T. Quality of Life and Acoustic Comfort in Educational Environments of Curitiba, Brazil. *J. Voice* **2020**, *36*, 436.e9–436.e16. [CrossRef] [PubMed]
8. Mealings, K. The Effect of Classroom Acoustic Conditions on Literacy Outcomes for Children in Primary School: A Review. *Build. Acoust.* **2022**, *29*, 135–156. [CrossRef]
9. Education & Skills Funding Agency. *BB93: Acoustic Design of Schools—Performance Standards*; Education & Skills Funding Agency: Manchester, UK, 2015.
10. Ronsse, L.M.; Wang, L.M. Relationships between Unoccupied Classroom Acoustical Conditions and Elementary Student Achievement Measured in Eastern Nebraska. *J. Acoust. Soc. Am.* **2013**, *133*, 1480–1495. [CrossRef]
11. Puglisi, G.E.; Prato, A.; Sacco, T.; Astolfi, A. Influence of Classroom Acoustics on the Reading Speed: A Case Study on Italian Second-Graders. *J. Acoust. Soc. Am.* **2018**, *144*, EL144–EL149. [CrossRef]
12. Campbell, C.; Nilsson, E.; Svensson, C. The Same Reverberation Time in Two Identical Rooms Does Not Necessarily Mean the Same Levels of Speech Clarity and Sound Levels When We Look at Impact of Different Ceiling and Wall Absorbers. *Energy Procedia* **2015**, *78*, 1635–1640. [CrossRef]
13. *EN ISO 3382-1*; Acoustics—Measurement of Room Acoustic Parameters—Part 1: Performance Spaces. ISO International Organization for Standardization: Geneva, Switzerland, 2009.
14. *EN 60268-16*; Sound System Equipment. Objective Rating of Speech Intelligibility by Speech Transmission Index. IEC International Electrotechnical Commission: Geneva, Switzerland, 2020.
15. Bradley, J.S.; Reich, R.D.; Norcross, S.G. On the Combined Effects of Signal-to-Noise Ratio and Room Acoustics on Speech Intelligibility. *J. Acoust. Soc. Am.* **1999**, *106*, 1820–1828. [CrossRef]
16. *EN ISO 3382-2*; Acoustics—Measurement of Room Acoustic Parameters—Part 2: Reverberation Time in Ordinary Rooms. ISO International Organization for Standardization: Geneva, Switzerland, 2008.
17. *UNI 11532-2*; Internal Acoustical Characteristics of Confined—Design Methods and Evaluation Techniques—Part 2: Educational Sector. UNI Ente Nazionale Italiano di Unificazione: Milan, Italy, 2020.
18. *DIN 18041*; Acoustic Quality in Rooms—Specifications and Instructions for the Room Acoustic Design. DIN Deutsches Institut für Normung: Berlin, Germany, 2016.
19. Astolfi, A.; Minelli, G.; Puglisi, G.E. A Basic Protocol for the Acoustic Characterization of Small and Medium-Sized Classrooms. *J. Acoust. Soc. Am.* **2022**, *152*, 1646–1659. [CrossRef] [PubMed]
20. Chetoni, M.; Ascari, E.; Bianco, F.; Fredianelli, L.; Licitra, G.; Cori, L. Global Noise Score Indicator for Classroom Evaluation of Acoustic Performances in LIFE GIOCONDA Project. *Noise Mapp.* **2016**, *3*, 157–171. [CrossRef]
21. Pulella, P.; Hladek, L.; Croce, P.; Seeber, B.U. Auralization of Acoustic Design in Primary School Classrooms. In Proceedings of the 2021 IEEE International Conference on Environment and Electrical Engineering and 2021 IEEE Industrial and Commercial Power Systems Europe (EEEIC/I&CPS Europe), Bari, Italy, 7–10 September 2021; pp. 1–6.
22. Zhang, D.; Tenpierik, M.; Bluyssen, P.M. Individual Control as a New Way to Improve Classroom Acoustics: A Simulation-Based Study. *Appl. Acoust.* **2021**, *179*, 108066. [CrossRef]
23. Hodgson, M.; Rempel, R.; Kennedy, S. Measurement and Prediction of Typical Speech and Background-Noise Levels in University Classrooms during Lectures. *J. Acoust. Soc. Am.* **1999**, *105*, 226–233. [CrossRef]
24. Stephenson, U.M. *Different Assumptions—Different Reverberation Formulae*; Institute of Noise Control Engineering: New York, NY, USA, 2012; p. 12.
25. Barron, M.; Lee, L. Energy Relations in Concert Auditoriums. I. *J. Acoust. Soc. Am.* **1988**, *84*, 618–628. [CrossRef]
26. Arau-Puchades, H. Sound Pressure Levels in Rooms: A Study of Steady State Intensity, Total Sound Level, Reverberation Distance, a New Discussion of Steady State Intensity and Other Experimental Formulae. *Build. Acoust.* **2012**, *19*, 205–220. [CrossRef]
27. Leccese, F.; Rocca, M.; Salvadori, G. Fast Estimation of Speech Transmission Index Using the Reverberation Time: Comparison between Predictive Equations for Educational Rooms of Different Sizes. *Appl. Acoust.* **2018**, *140*, 143–149. [CrossRef]

28. Nowoświat, A.; Olechowska, M. Fast Estimation of Speech Transmission Index Using the Reverberation Time. *Appl. Acoust.* **2016**, *102*, 55–61. [CrossRef]
29. Gómez Escobar, V.; Barrigón Morillas, J.M. Analysis of Intelligibility and Reverberation Time Recommendations in Educational Rooms. *Appl. Acoust.* **2015**, *96*, 1–10. [CrossRef]
30. Croce, P.; Leccese, F.; Zannin, P.H.T. Evaluation of Predictive Methods of Acoustic Comfort Parameters in University Classrooms. In Proceedings of the 2021 IEEE International Conference on Environment and Electrical Engineering and 2021 IEEE Industrial and Commercial Power Systems Europe (EEEIC/I&CPS Europe), Bari, Italy, 7–10 September 2021; pp. 1–6.
31. Cho, Y.-J. Comparison of Two Types of Combined Measures, STI and U50, for Predicting Speech Intelligibility in Classrooms. *Arch. Acoust.* **2017**, *42*, 527–532. [CrossRef]
32. Barron, M. Theory and Measurement of Early, Late and Total Sound Levels in Rooms. *J. Acoust. Soc. Am.* **2015**, *137*, 3087–3098. [CrossRef] [PubMed]
33. *UNI 11532-1*; Internal Acoustical Characteristics of Confined Spaces—Design Methods and Evaluation Techniques—Part 1: General Requirements. UNI Ente Nazionale Italiano di Unificazione: Milan, Italy, 2018.
34. Arau-Puchades, H.; Berardi, U. A Revised Sound Energy Theory Based on a New Formula for the Reverberation Radius in Rooms with Non-Diffuse Sound Field. *Arch. Acoust.* **2015**, *40*, 33–40. [CrossRef]
35. Berardi, U.; Cirillo, E.; Martellotta, F. A comparative analysis of acoustic energy models for churches. *J. Acoust. Soc. Am.* **2009**, *126*, 1838–1849. [CrossRef]
36. Berardi, U. Simulation of acoustical parameters in rectangular churches. *J. Build. Perform. Simul.* **2014**, *7*, 1–16. [CrossRef]
37. Bradley, J.S. Speech Intelligibility Studies in Classrooms. *J. Acoust. Soc. Am.* **1986**, *80*, 846–854. [CrossRef] [PubMed]
38. Building Acoustics—Acoustic Classification of Building Units—Evaluation Procedure and in Situ Measurements. *UNI 11367*; UNI Ente Nazionale Italiano di Unificazione: Milan, Italy, 2010.
39. Barron, M. *Auditorium Acoustics and Architectural Design*, 2nd ed.; Spon Press/Taylor & Francis: London, UK; New York, NY, USA, 2010; ISBN 978-0-419-24510-0.
40. Minelli, G.; Puglisi, G.E.; Astolfi, A. Acoustical Parameters for Learning in Classroom: A Review. *Build. Environ.* **2022**, *208*, 108582. [CrossRef]

energies

MDPI

Article

Exergoeconomic and Environmental Evaluation of a Ground Source Heat Pump System for Reducing the Fossil Fuel Dependence: A Case Study in Rome

Fabio Nardecchia *, Laura Pompei, Edoardo Egidi, Riccardo Faneschi and Giuseppe Piras

Department of Astronautical, Electrical and Energy Engineering (DIAEE), Sapienza University of Rome, 00184 Rome, Italy; laura.pompei@uniroma1.it (L.P.); egidi.1795329@studenti.uniroma1.it (E.E.); faneschi.1796336@studenti.uniroma1.it (R.F.); giuseppe.piras@uniroma1.it (G.P.)
* Correspondence: fabio.nardecchia@uniroma1.it

Abstract: By 2050, the European Commission aims to achieve a 90% reduction in greenhouse gas emissions within the construction sector due to new targets set for greenhouse gases by the Commission. One of the most effective solutions for mitigating the environmental impact of buildings is to integrate renewable-energy systems such as air-to-water heat pumps or geothermal heat pumps. Several works in the literature investigated the advantages of heat pumps, particularly ground source heat pumps, for specific contexts. Furthermore, the evaluation can encompass not only energy considerations but also exergetic aspects, and this paper makes a significant contribution to the latter. The study presented here applies exergy analysis to a geothermal heat pump system that is interconnected with photo-voltaic and solar collector systems in a building located in Rome. Feasibility and environmental assessments were also conducted. It is evident that the exergy efficiency obtained is much lower than the energy efficiency. The heat pump demonstrates higher exergy efficiency by producing high-temperature thermal power compared to a constant dead state at relatively low temperatures. Following the heat pump, the adiabatic mixer exhibits the second highest exergy efficiency, trailed by two heat exchangers. Compared to a conventional plant case, the money saved is EUR 16,772 per year, translating to a Payback Period of 23 years. Furthermore, the average annual reduction in fossil fuel emissions is estimated at 26.2 metric tons of CO_2.

Keywords: energy renovation of buildings; exergy analysis; ground source heat pump; solar collector; photovoltaic plant; energy transition

1. Introduction

Since 2010, the European Commission has established new targets for greenhouse gas emissions from the construction sector, aiming to reduce carbon dioxide equivalent emissions (CO_2eq) by 90% by 2050 [1]. According to the 2014/2015 European Work Program [2], it is projected that over 17% of the EU's energy-saving potential for 2050 will be dedicated to retrofitting buildings. Throughout the lifetime of a building, energy and raw materials are continuously utilized. As a results, the operation phase of the building requires significant in situ energy consumption, while the construction, operation, retrofit and demolition stages consume minor energy, alongside the production of building materials and technical equipment.

A critical strategy for enhancing building energy efficiency is recognizing that 60% of the energy consumption is attributed to space conditioning. This concern is particularly relevant in countries like Italy, where a majority of residential and commercial structures were erected prior to the implementation of stringent energy efficiency regulations. Numerous studies have emphasized the urgent need for comprehensive energy and environmental redesign of the entire urban district [3–6]. The integration of renewable energy systems, such as air-to-water heat pumps or geothermal heat pumps (GSHP), represents the most effective solution, especially to mitigate the environmental impacts of buildings [7–12].

Exergy analysis is a thermodynamic analysis methodology that assesses the quality or value of the useful work generated by a system or process. In essence, exergy analysis evaluates a system's efficiency not only in terms of energy but also in terms of energy quality [13,14]. When comparing air–water and geothermal heat pumps using exergy analysis, key factors to consider encompass efficiency, heat source temperatures, and environmental effects. The exergy efficiency of a heat pump indicates the ratio of useful work produced to the total work supplied to the system. Exergy efficiency is influenced by variables such as the heat source temperature and the temperature of the heating or cooling fluid. In the scientific community, the comparison of exergy efficiency has been extensively studied over time. As early as 1994, Groll and Braun [15] conducted a comparative exergy analysis between air–water and geothermal heat pumps, assessing exergy efficiency and comparing outcomes between the two systems. Ibrahim and Al-Sulaiman [16] performed an exergy analysis of a geothermal heat pump system, evaluating exergy efficiency and identifying key factors affecting system performance. Conversely, Sopian and Yatim [17] focused on the exergy efficiency of heat pump systems used for drying purposes, examining different refrigeration cycles and evaluating performance in terms of exergy. Exergy analysis and exergy efficiency evaluation also serve for comparisons with other heating technologies [18].

Both air–water heat pumps and GSHPs are technologies employed for heating and cooling buildings, leveraging the heat pump principle to transfer heat from a low-temperature source to a high-temperature sink using a refrigerant. Both types have been examined from an exergy standpoint. For air–water heat pumps, studies such as Al-Sulaiman and Dincer [19] have examined the exergy efficiency of air–water heat pumps and evaluated the environmental impact associated with the use of these technologies. Al-Sulaiman and Hamdan [20] focused on an exergy analysis of an air–water heat pump, assessing the exergy efficiency of the system and identifying the main sources of exergy losses.

Similarly, investigations have explored the exergy efficiency and performance of geothermal heat pumps [21,22], as well the exergy losses of various components connected to building heating and cooling systems [23,24]. Additionally, Wang and Zhao [25] and Villafuerte and Coronas [26] conducted an exergy analysis of geothermal heat pump systems for building heating and cooling, gauging system exergy efficiency and comparing it with other heating and cooling technologies.

GSHPs utilize the ground's constant temperature as a heat source, while air–water heat pumps use the outside air as a heat source. Since the ambient air temperature can vary significantly, this can affect the efficiency of air–water heat pumps [27].

Cui, Zhao and Fang [28] and Sheng, Zhao and Wu [29] compared the performance of air–water and geothermal heat pumps in heating-dominated operation mode by analyzing the temperatures of heat sources and evaluating the system efficiency. Jeong, Jeong and Kang [30] addressed the performance disparities between geothermal and air–water heat pumps, factoring in outdoor air temperature fluctuations and their influence on heat source temperatures and system efficiency. The ground temperature also affects the design of the ground heat exchanger in geothermal heat pumps and consequently the system efficiency [31].

Compared to air–water heat pumps, GSHPs are often deemed more environmentally sustainable. Because the ground's temperature remains relatively constant year-round, geothermal heat pumps necessitate less energy to reach desired temperatures [32]. Bales and Ramesh [33] assessed the environmental performance of geothermal heat pumps in heating-dominated climates in North America, examining the benefits in terms of greenhouse gas emissions' reduction and the impact on energy and water consumption. Similar studies analyze the environmental and economic implications of air–water heat pumps in Chinese residential buildings, juxtaposed with other heating and cooling technologies, assessing their environmental advantages [34–36].

Within this context, this study endeavors to evaluate the energy systems of twelve residences situated in Rome, Italy. Exergy analysis is employed to evaluate the energy quality of the GSHP, photovoltaic system (PV), and solar collector (SC) proposed. Simultaneously, economic and environmental assessments are conducted. In summary, the study aims to (1) perform an exergy evaluation of an energy system comprising GSHP, PV and SC applied to a Rome building, (2) outline feasibility and environmental analyses. The Materials and Methods section (Section 2.1) outlines the characteristics and energy load demands of the case study. Section 2.2 describes the proposed energy system, while Section 2.3 applies an exergy evaluation to the energy system and its components, alongside economic analysis. Section 3 presents primary outcomes of the energy system, including exergy results, while the chief economic and environmental savings are detailed in Section 3.3.

2. Materials and Methods

2.1. Case Study and Energy Loads Evaluation

The case study involves a building characterized by high energy performance (class A4 of the energy performance certificate) with a thermal performance index for winter conditioning of 25 kWh/m^2/year. This building is situated in Rome within climatic zone D, with a heating degree days (HDD) value of 1415. Comprising 12 flats, each apartment has a floor area of 90 m^2, and the operational schedule of the energy system adheres to regulations [37,38]. The heat generated by the GSHP serves two thermal purposes: it supplies the DHW thermal storage tank and fulfills the thermal demand for domestic heating during the winter operation. Throughout the summer period, the GSHP produces heat used for the DHW storage tank, and it subsequently utilizes the water mains as a cold source to facilitate summer cooling. The initial step involves calculating the total annual energy consumption for winter heating under the most severe conditions (Equation (1)), [37].

$$Q_w = \frac{ETh \cdot (T_{set\ point} - T_{min})}{HDD} \tag{1}$$

where $T_{set\ point}$ is the temperature to guarantee inside the zone [37] and T_{min} is the outside minimum temperature (0 °C), Figure 1. During the summer period, the technical plant must guarantee the indoor temperature equal to the summer set point temperature, assumed to be 26 °C (Equation (2)).

$$Q_s = \frac{ETh \cdot (T_{max} - T_{set\ point})}{HDD} \tag{2}$$

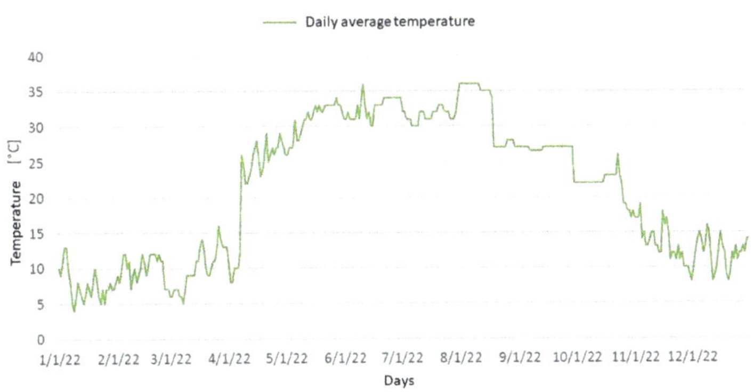

Figure 1. The annual temperature profile in Rome.

Moreover, the overall average efficiency of the heating system (downstream of the GSHP) is calculated as the product of the efficiency of the distribution system (95%), the efficiency of the heat emission system (with a radiator efficiency of 96%), and the efficiency of the regulation system (96%), resulting in a total of 87.5%. By dividing the Q_w and Q_s

values with the global average efficiency and considering the surface (90 m^2) and operating hour of the thermal plant, the thermal and electrical power can be obtained.

The UNI 9182 [38] standard necessitates the definition of the duration of the "peak periods", contingent upon the specific type of user under consideration. Proper sizing of the heating system requires the DHW consumption calculation to be performed within these peak periods. For this scenario, the duration can be assumed to be 2 h.

The estimation of domestic hot water consumption is derived by considering the installed sanitary appliances and their frequency of utilization [38]. The calculation of the maximum hot water usage is carried out according to Equation (3), in which the numerator represents the consumption of each device (based on daily usage), while the total consumption is denoted in the denominator.

$$Q = \sum_{i=1}^{devices} \frac{Devices\ consumption\ \left[\frac{l}{use}\right]_i \cdot n^\circ\ of\ usage\ \left[\frac{use}{day}\right]_i \cdot n^\circ\ of\ devices}{consumption\ \left[\frac{h}{HDD}\right]_i} \quad (3)$$

The outcome corresponds to the peak identified by the apex of the dotted red line (approximately 0.2 kg/s at 19:00) in Figure 2. Subsequently, the daily DHW load profile was derived based on the typical patterns of domestic usage, resulting in the dashed red line (Figure 2). By employing an iterative approach to the daily load profile, the replenishment curve profile was established. The volume of water within the tank proved adequate to fulfill user demands, as indicated by the solid blue line in Figure 2. The simultaneity factor, established in relation to the number of dwellings considered, is set at 0.48. The comprehensive DHW load diagram is described in Figure 2.

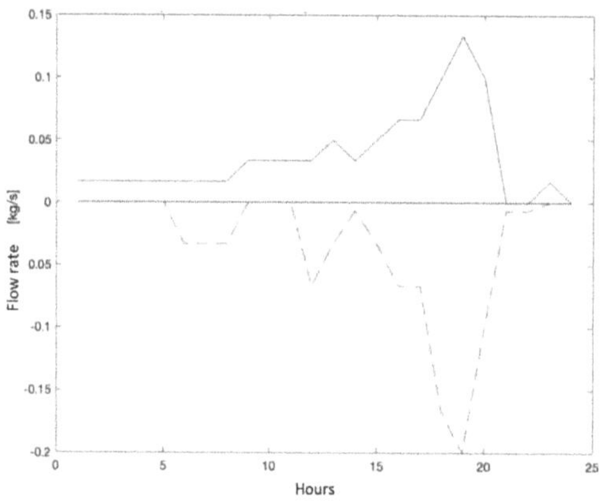

Figure 2. Profile of mass flow rates in thermal storage.

The daily DHW demands per user are approximately 260 L/day. The thermal storage tank necessitates a volume of 50 L. The daily filling pattern of the tank was computed and depicted in Figure 3 to validate the volumetric sizing. This was achieved by utilizing the daily load profiles (represented by the red dashed line in Figure 2) and replenishment profiles (depicted by the solid blue line in Figure 2).

To calculate the thermal power required for storage daily, it is necessary to define the operating conditions:

- Average daily temperature at the outlet of the accumulation tank equal to 318.15 K.
- Tank replenishment water temperature equal to 283.15 K in winter operating mode, and about 292 K in summer operating mode.
- Pressure of the make-up water entering the tank equal to 240 kPa.
- DHW pressure at tank outlet equal to 200 kPa.
- Overall heat-exchange coefficient of the tank equal to 2 W/K.

The energy balance equation referring to the heating (Equation (4)), takes into account the mass flows inlet/outlet and the thermal losses from the tank to the outside (Figure 4).

$$Q = Q_{HP} + Q_{solar} = m_{DHW} \cdot dh_{out} + m_{heating} \cdot dh_{in} + \frac{d}{dt}\left[\int e \cdot \rho \cdot dV\right] + Q_{loss} \tag{4}$$

where Q is the daily trend of the required thermal power, composed of two different components such as Q_{HP} (heat supplied by the GSHP) and Q_{solar} (heat supplied by the solar collector); m is the mass flow rate; dh is the thermal power and e is the specific energy contribution to the mass in the system; ρ is the density of water contained in the tank; dV is the variation in water's volume and Q_{loss} is the thermal losses from the tank to the outside.

The temperature of the contents within the tank during the heat pump's operation (from 5:00 to 20:00) is maintained at 318.15 K. The system has been engineered to operate continuously throughout the day to mitigate fluctuations in temperature. Consequently, the daily trajectory of the necessary thermal power can be derived, as depicted in Figure 5.

Both winter and thermal losses (Figures 4–7) were calculated according to the following parameters:

- Overall heat transfer coefficient of the tank is 2 W/K.
- Ambient temperature.
- Temperature of the water mass within the boiler.

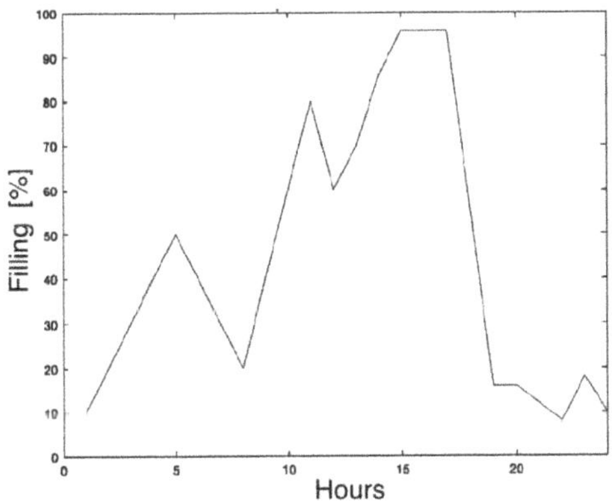

Figure 3. Daily store of thermal storage.

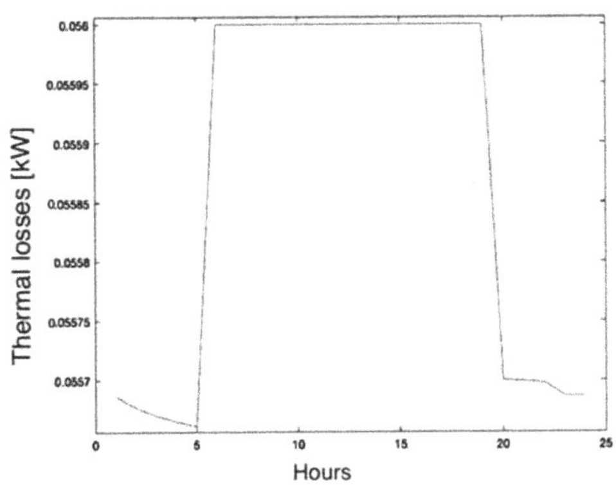

Figure 4. Losses towards the outside of the thermal storage.

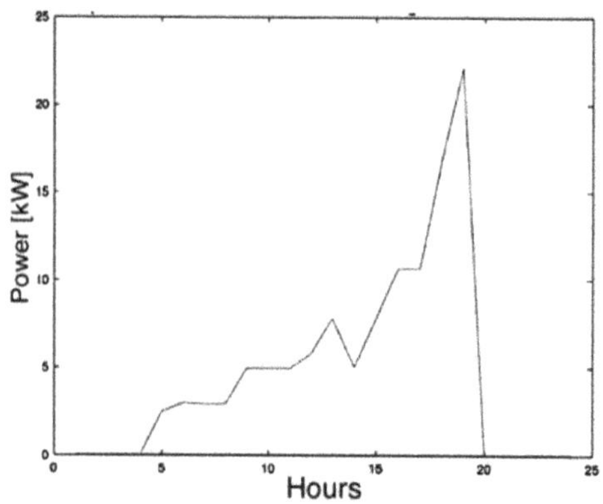

Figure 5. Thermal power is required for thermal storage in winter operating mode.

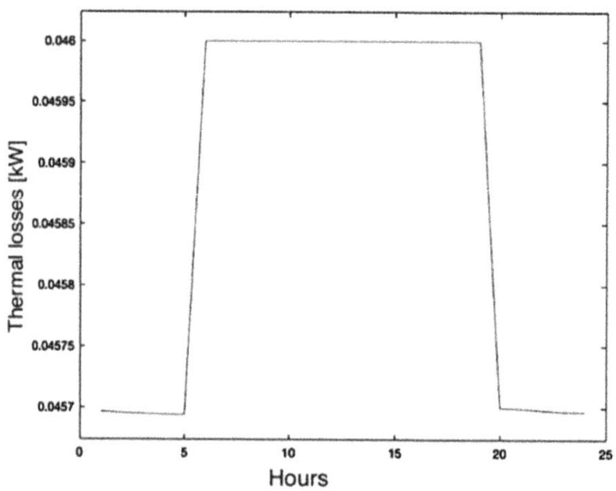

Figure 6. Losses towards the outside of the thermal storage during the summer period.

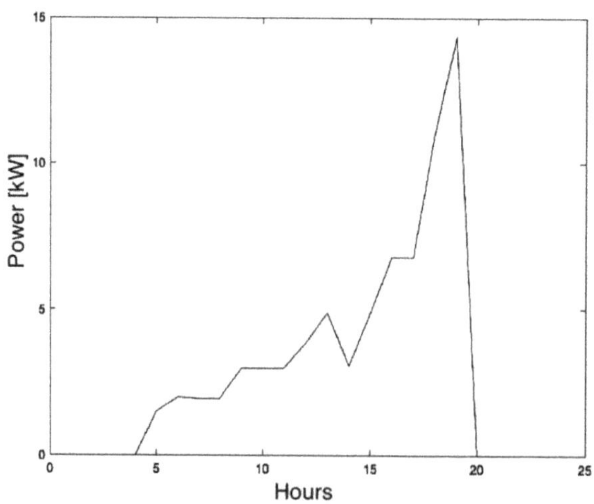

Figure 7. Thermal power is required for thermal storage in the summer period.

Since the aim of the project is the annual analysis of the average daily values proposed by the system, the average daily thermal load values of DHW are assumed to be constant during the year. These values have been calculated as 2/3 of the peak power (Figure 8).

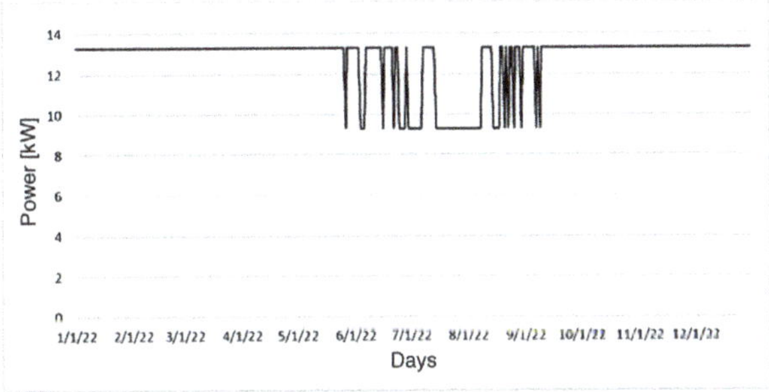

Figure 8. Average daily total thermal power required by the thermal storage.

2.2. Energy Plant Description

The low enthalpy geothermal heat pump sends hot water to the users at a set temperature of 321.15 K. The system is regulated by controlling the mass flow from the GSHP, which varies according to the average daily heat requirement of the user. The COP of the GSHP is 5.8. The entire energy system is shown in Figure 9.

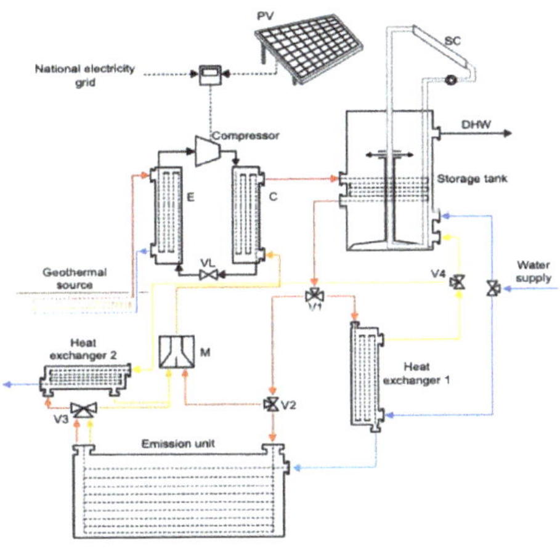

Figure 9. Technical layout plant.

The low enthalpy geothermal heat pump delivers hot water to users at a predetermined temperature of 321.15 K (Figure 9). The system is regulated by controlling variable mass flow in line with the user's average daily heat demand. The GSHP conveys the heat transfer fluid to the boiler, which is heated through a closed circuit, causing an imposed thermal drop of 4 K upon crossing the boiler. Simultaneously, the boiler receives thermal power from the solar thermal collector through a separate closed circuit to ensure fluid separation. The total thermal power supplied to the boiler facilitates the distribution of DHW to users via a specialized, unillustrated distribution system. Water replenishment within the storage occurs through two methods:

- During the winter season, heat exchanger 1 remains inactive, directing water directly from the "Water Supply" into the boiler.
- In the summer, the need for cooling mandates a reduction in the heat transfer fluid's temperature as it passes through the radiator. This cooling process transpires within

heat exchanger 1. The heat extracted from the fluid in heat exchanger 1 is repurposed for preheating water entering the boiler, aiding in reducing the overall thermal load. In this instance, replenishment occurs through the V4 valve.

In this case study, a vacuum solar collector is employed, possessing an absorption coefficient of $F_r = 0.94$. The collector consists of 14 double-walled glass vacuum tubes, each enclosing a copper tube. These copper pipes are interconnected in parallel and traversed by a heat transfer fluid with a specific nominal mass flow of 0.008 kg/s·m^2. The overall heat transfer coefficient towards the external environment is 1.15 W/m^{2}K. The average daily solar radiation in Rome throughout the year is shown in Figure 10.

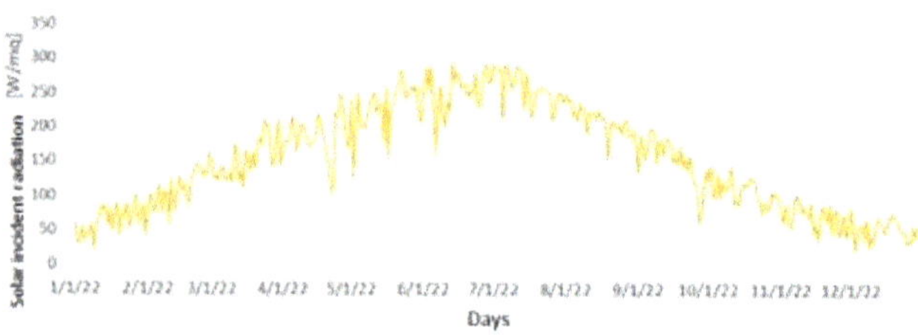

Figure 10. Average daily incident solar radiation.

Given a panel surface ($A_m = 2.7$ m^2 per module), the energy flows to the solar thermal collector can be obtained using Equations (5)–(7).

$$Q_{sun} = F_r \cdot A * Rad_{sun} = F_r \cdot n_{mod} \cdot Rad_{sun} \quad (5)$$

where F_r is the absorption coefficient of the vacuum solar collector; A is the collector surface; Rad_{sun} is the average solar radiation in Rome; n_{mod} is the number of collector modules. The maximum surface available is 9 m^2 and n_{mod} (number of the module) is three.

$$Q_{loss} = U \cdot A \cdot (T_{collector} - T_{outside}) \quad (6)$$

$$Q_{used} = Q_{sun} - Q_{loss} \quad (7)$$

where Q_{loss} is the thermal power lost; U is the transmittance of the panel and Q_{sun} is the thermal power obtained from solar radiation incident on the collectors.

By setting the inlet temperature of the collector equal to the average daily temperature of the heat storage (45 °C), the outlet temperature from the collector is obtained. The energy efficiencies of the solar thermal collector are calculated using Equation (8).

$$\eta_{energy,solar\ collector} = \frac{Q_{out}}{Q_{absorbed}} \quad (8)$$

After calculating the thermal power provided by the solar source and the thermal power needed for the DHW thermal storage, it became feasible to assess the thermal power entering the boiler from the GSHP (Equation (9)) and the mass flow exiting the heat pump (Equation (10)). An aggregated thermal loss was identified at the outlet of the boiler, encompassing the inefficiencies of the heating system's distribution. Consequently, this mass flow experiences a marginal temperature reduction (Equation (11)), as illustrated in Figure 11.

$$Q_{boiler,HP} = Q_{boiler} - Q_{u,solar} \quad (9)$$

$$m_{out,HP} = \frac{Q_{boiler,HP}}{cp \cdot dT} \quad (10)$$

$$T = T_{out,Boiler} - \frac{Q_{loss,destroyed}}{cp \cdot m_{out,HP}} \quad (11)$$

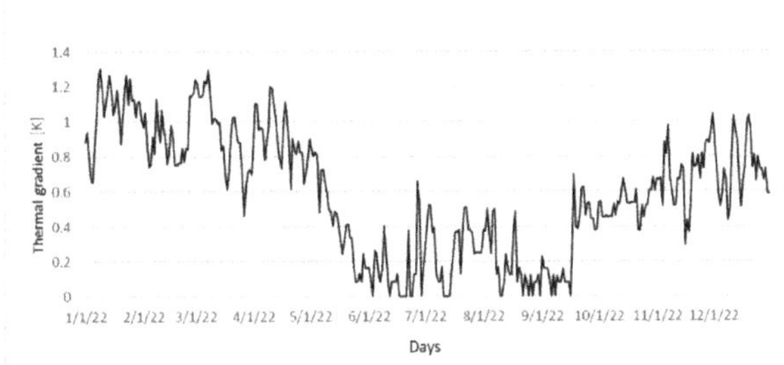

Figure 11. Concentrated thermal jump caused by inefficiencies of the distribution system.

To maximize the energy performance and enhance user self-consumption, the geothermal heat pump heating system is powered not solely by the national electricity grid but also by a photovoltaic panel system. In the case of monocrystalline silicon solar panels, the overall electrical system efficiency stands at 86%, while the panels have a TILT angle of 35° and an azimuth of 0°. Firstly, the average daily energy demand is established.

By considering the power demanded by the compressor and the daily operational hours of the heat pump (HP), it becomes possible to compute the average daily energy consumed. Then, to calculate the average energy produced daily, it is necessary to take into account the efficiency of the chosen module, which can be assumed equal to 18.5%. The average energy produced annually is then determined by multiplying the efficiency of the global electricity system, the efficiency of a single panel, and the average annual radiation on an inclined surface (1737.40 kWh/m^2 year). To calculate the installed kWp, i.e., the power extracted from the modules in standard conditions (1000 W/m^2 of irradiation, module temperature of 25 °C), it is enough to consider the efficiency of the system to pass from the STC (Standard Test Condition) to actual operating conditions.

2.3. Exergy and Economic Evaluations

The exergy analysis of the plant was carried out following Equations (12)–(14).

$$EX_{QU} = 0 \tag{12}$$

$$EX_Q = |Q| \cdot (1 - \frac{T_0}{T_{Hot}}) \tag{13}$$

$$EX_L = L \tag{14}$$

where T_{Hot} is set to the flow temperature of the GSHP. Through the exergy balance equations [14], it is possible to calculate the annual average exergy efficiency (Equation (15)).

$$\varepsilon_{GSHP} = 1 - \frac{EX_{destroyed}}{EX_{fuel}} \tag{15}$$

The exergies related to the heat flows entering the boiler are the inlet exergy from solar thermal collector, GSHP and the outlet exergy from the boiler. Then, it was possible to calculate the exergy destroyed and the efficiency (Equation (16)).

$$EX_{distroyed} = EX_{in,boiler} + EX_{out,boiler} \tag{16}$$

The exergy efficiency of solar collector and heat exchanger is calculated as Equations (17) and (18), respectively.

$$\varepsilon_{solar\ collector} = \frac{EX(Q_{out})}{EX(Q_{absorbed})} \tag{17}$$

$$\varepsilon_{heat\ exchang.} = \frac{\left(1 - \frac{T_0}{T_{f,ML}}\right)}{\left(1 - \frac{T_0}{T_{c,ML}}\right)} \tag{18}$$

where T_0 is dead state temperature (15 °C); $T_{h,ML}$ and $T_{c,ML}$ are the temperatures hot- and cold-related to the heat exchanger.

At the inlet to the mixer are the flow rates m_{mix}, relative to the fluid with higher temperature ($T_{in,elem}$), and m_{elem} output from the emission element at a lower temperature ($T_{in,elem}$, dT_{elem}). Table 1 reports the average values involve, where h is enthalpy, T is temperature, s is heating capacity (Table 1).

Table 1. Enthalpy, temperature and heat capacity of mixers, emission element.

$h_{element}$ [kJ/kg]	h_{mixer} [kJ/kg]	h_{out} [kJ/kg]	T_{elem} [K]	T_{mix} [k]	T_{out} [K]	s_{elem} [kJ/kg/K]	s_{mix} [kJ/kg/K]	s_{out} [kJ/kg/K]
179.5	154.5	160	316	310	312	0.6	0.53	0.545

The exergy destroyed is described in Equation (19); then, the exergy efficiency can be evaluated, wherein $EX_{m,elem.}$ refers to the exergy of the considered element inside the system, while $EX_{m,out}$ refers to the exergy leaving the system.

$$EX_{distroyed} = EX_{m,mix} + EX_{m,elem.} - EX_{m,out} \tag{19}$$

To perform an economical evaluation, the marginal costs of the heating and cooling system will be analyzed to establish a quantitative comparison of the average annual cost compared to a conventional gas-fired heating system, and electricity (domestic air conditioners) for cooling requests. The traditional gas boiler has an energy efficiency of 90%, a PCI of 9.59 kWh/m^3 and the gas natural price is set to EUR 1.24/m^3 (ARERA). On the other hand, the price of electricity is EUR 0.361/kWh. Table 2 collects the cost of installation of each component employed within the system.

Table 2. Installation cost of the components listed.

Component	Cost (EUR)
Horizontal probe	5000
HP and storage thank	12,000
Underfloor system	60,000
Heat exchanger and mixer	1350
Solar collector	12,000
Photovoltaic panels	3,000,000

3. Results

3.1. Results of the Solar Collector, HP and PV Plant

The ultimate winter and summer energy demands amount to 5.5 kWh/m^2 and 2.2 kWh/m^2, respectively. The thermal power profiles necessary for the system, along with losses accounting for inefficiencies in water distribution and heat transmission, are depicted in Figure 12. The average daily power absorbed by the solar collector is shown in Figure 13.

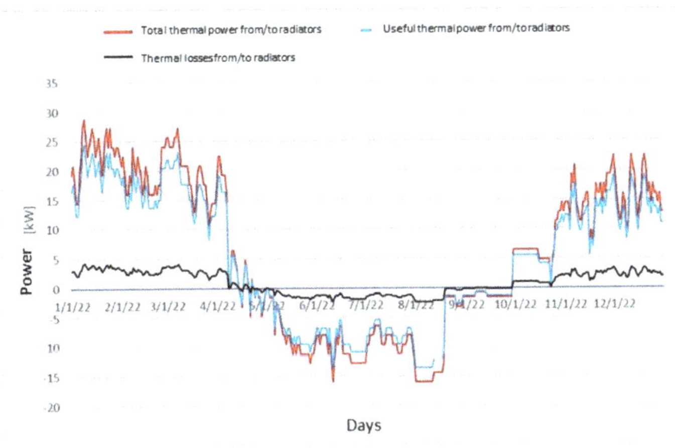

Figure 12. Power values to the emission element.

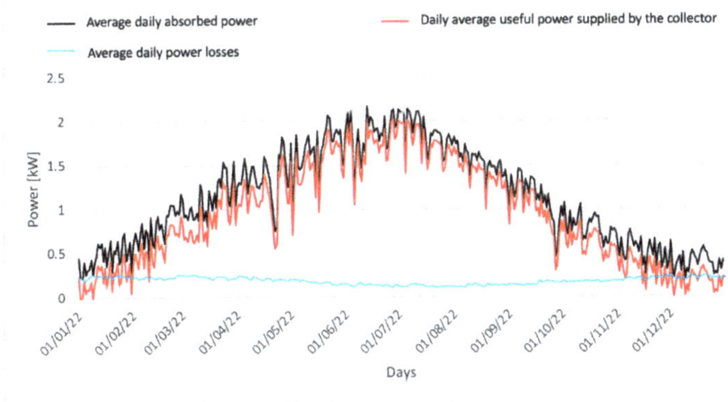

Figure 13. Energy flows to the solar thermal collector.

By setting the inlet temperature to the collector equal to the thermal accumulation's average daily temperature (45 °C), the resulting outlet temperature from the collector is determined (Figure 12). The annual average performance of the solar collector is 74.73%. The average power supplied to the storage tank and the mass flow deriving from the GHSP are reported in Figure 14. Figure 15 outlines the daily average energy fluxes of the GSHP.

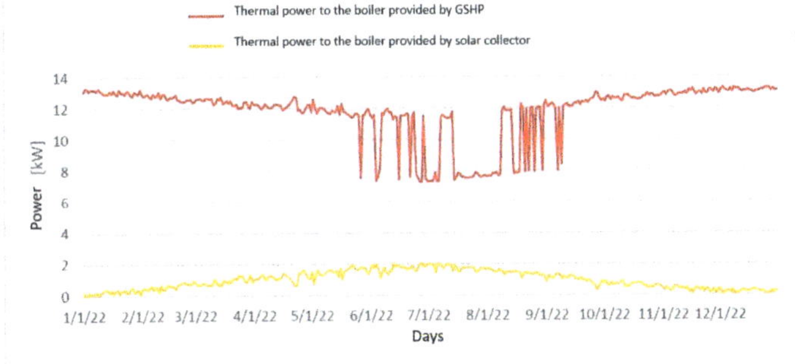

Figure 14. Average daily thermal power supplied to the thermal storage.

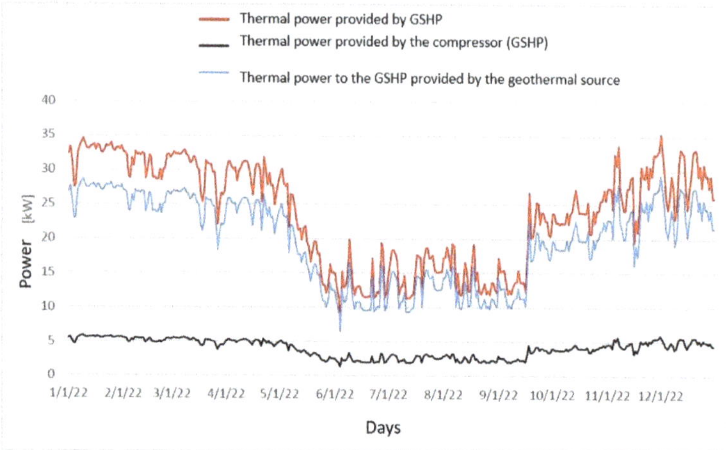

Figure 15. Powers exchanged by GSHP.

3.2. Results of Exergy Analysis

Firstly, the exergy efficiency of each component of the systems is collected in Table 3.

Table 3. Installation cost of the components listed.

Component	ε (%)
Photovoltaic system	8.00
HP and storage thank	16.80
GSHP	64.30

The exergy efficiency of the GSHP is lower when compared to its energy efficiency (97.1%). This discrepancy arises from the dissipated exergy. The increase in exergy destruction within the system, signifying its inefficiency, is intricately tied to three factors:

- The temperature of the reference dead state.
- The COP of the heat pump.
- The external temperature (which influences the thermal power demanded by the user).

In this case the temperature of the dead state is constant, as we assume that the heat capacity of the soil is infinite. It can be assumed EX_{destr} is a function of $T_{outside}$ and COP (Figure 16).

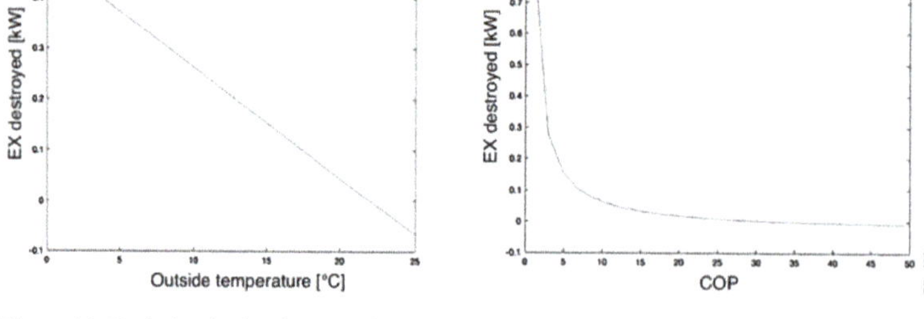

Figure 16. Variation in the destroyed exergy in function of (**a**) outside temperature and (**b**) COP.

As the COP increases to around 12, the value of exergy destruction experiences a significant reduction, followed by a more gradual decline. This behavior can be at-tributed to the coefficient of performance, which denotes the amount of mechanical power required to generate a specific quantity of thermal power. This phenomenon is characterized by the degradation of valuable energy, a degradation that diminishes as the COP of the machine increases.

Regarding the exergy destruction linked to ambient temperature, an increase in temperature corresponds to a linear decrease in exergy destruction. This phenomenon can be elucidated by the reduction in thermal energy, which, if generated from forms of mechanical energy, signifies a loss of thermodynamic efficiency. Furthermore, the breakdown of contributions from each system component to exergy destruction is presented in Figure 17.

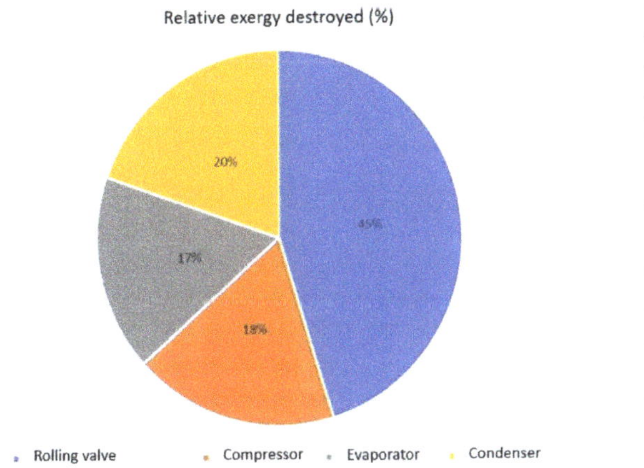

Figure 17. Relative exergy destroyed results of GSHP components.

The exergy destroyed and efficiency of both heat exchangers are reported in Table 4. The exergy efficiency of the mixer is 62%. The chart shown in Figure 18 compared the exergy efficiency of each component.

Table 4. Exergy destroyed and exergy efficiency of heat exchangers.

Element	Exergy Destroyed	ε (%)
Heat exchanger 1	0.65	17.9
Heat exchanger 2	0.43	21.7

The heat pump stands out with superior exergy efficiency, primarily due to its ability to generate thermal power at a higher temperature compared to the constant and relatively lower temperature of the dead state throughout the year, capitalizing on the heat capacity of the ground. The adiabatic mixer ranks as the second most exergetically efficient component, owing to the minimal logarithmic mean temperature difference between inlet fluids during year-round operation. In contrast, the two heat exchangers, which encounter higher thermal differentials, face drawbacks stemming from the irreversibilities inherent in heat exchange as the temperature variance between sources expands.

Photovoltaic panels, however, encounter notable constraints due to inherent material limitations; this results in considerably diminished energy efficiency, which subsequently reflects in their exergy efficiency.

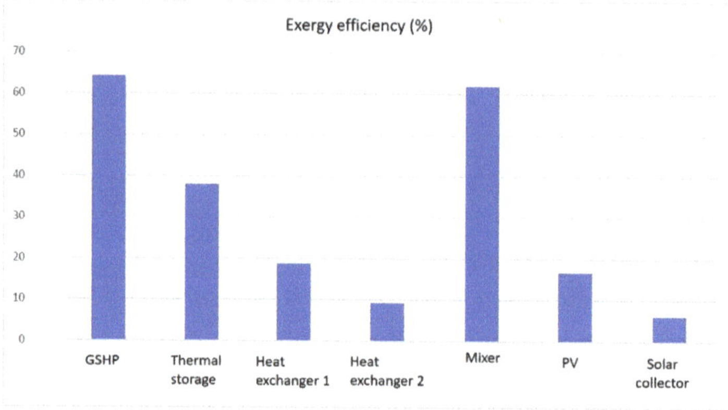

Figure 18. Exergy efficiency of each component of the energy plant.

3.3. Results of Economic Evaluation

The annual gas consumption of the boiler chosen is equal to 6483.60 m^3/year for a cost of EUR 8039.67/year. Considering an EER of 3, the electricity demand is 913 kWh/year, for a cost of EUR 329.14/year. Moving to the DHW, the annual cost is EUR 8404.3/year. The total annual savings is around EUR 16,772/year. Based on the prices collected in Table 3, the total installation cost (PV not included) is EUR 390,350. The savings compared to the case of the conventional plant is equal to EUR 16,772 a, this means that the PBP is equal to 23 years.

The use of no fossil fuels in the plant implies a negligible CO_2 emission into the atmosphere. If we assumed for domestic heating an average emission factor derived from the use of natural gas equal to 1.95 kg CO_2 m^3, the average emissions into the atmosphere avoided are 2.85 tCO_2/year. During the summer period, the avoided emission account for 0.396 tCO_2/year, assuming a factor of 0.433 kg CO_2/kWh for a total of 26.2 tCO_2/year saved.

4. Conclusions

The predominant source of energy consumption within buildings arises from space conditioning, constituting approximately 60% of their overall energy usage. To mitigate the environmental footprint of buildings, the integration of renewable-energy systems like air-to-water heat pumps or geothermal heat pumps (GSHP) during the design phase can yield significant reductions. This study focus on the analysis of a heat pump serving a twelve-unit building situated in Rome. The investigation encompasses an energy/exergy analysis, supplemented by an eco-nomic evaluation, and culminates in an assessment of the system's environmental im-pact. To optimize energy performance and elevate user self-consumption, the geo-thermal heat pump heating system draws power not only from the national electricity grid but also from a photovoltaic panel system. Furthermore, a vacuum solar collector is considered for hot water generation. The system operates across two modes: summer and winter. During winter, the GSHP provides heat for domestic hot water (DHW) and space heating, while in summer, it supports DHW and facilitates cooling through the water network.

The exergy analysis reveals a substantial discrepancy between exergetic and energy efficiency. This outcome stems from the notable exergy destruction observed in the system. Further analysis dissects the exergetic performance of individual components, highlighting those with high efficiency and those that negatively impact overall exergetic efficiency. Notably, the heat pump demonstrates heightened exergetic efficiency, benefiting from its capacity to generate high-temperature thermal power in contrast to the constant and relatively low-temperature dead state. The adiabatic mixer follows as the second most exergetically efficient component, trailed by two heat exchangers. The inherent material limitations of photovoltaic panels contribute to their relatively low energy efficiency, as evidenced in their exergetic efficiency.

Savings in comparison to a conventional plant configuration amount to EUR 16,772 per annum, equating to a Payback Period (PBP) of 23 years. Regrettably, substantial costs may deter widespread adoption, despite the notable thermal system efficiency. Nonetheless, the benefits extend beyond monetary savings, as the system's absence of fossil fuel usage translates to negligible (if any) CO_2 emissions. The system's annual average greenhouse gas emissions reduction tallies up to 26.2 metric tons of CO_2 per year. Future iterations of this study will explore extending the exergy analysis to an entire energy district, assessing the potential of GSHP and similar energy plants for the decarbonization process of urban settlements.

Author Contributions: Conceptualization, F.N., E.E. and R.F.; methodology, F.N., E.E. and R.F.; validation, F.N. and L.P.; data curation, F.N. and G.P.; writing—original draft preparation, F.N., L.P. and G.P.; writing—review and editing, L.P.; visualization, E.E. and F.N.; supervision, G.P. All authors have read and agreed to the published version of the manuscript.

Funding: This research received no external funding.

Data Availability Statement: No new data were created or analyzed in this study. Data sharing is not applicable to this article.

Conflicts of Interest: The authors declare no conflict of interest.

References

1. European Commission. *Roadmap to a Competitive Low Carbon Economy in 2050*; COM (2011) 112 Final. Commission Communication; European Commission: Brussels, Belgium, 2011.
2. European Commission. *Draft Horizon 2020 Work Programme 2014–2015 in the Area of Secure, Clean and Efficient Energy*; European Commission: Brussels, Belgium, 2015.
3. Mauri, L.; Vallati, A.; Ocłoń, P. Low impact energy saving strategies for individual heating systems in a modern residential building: A case study in Rome. *J. Clean. Prod.* **2019**, *214*, 791–802. [CrossRef]
4. Schneider, S.; Zelger, T.; Sengl, D.; Baptista, J. A Quantitative Positive Energy District Definition with Contextual Targets. *Buildings* **2023**, *13*, 1210. [CrossRef]
5. van der Leer, J.; Calvén, A.; Glad, W.; Femenías, P.; Sernhed, K. Energy systems in sustainability-profiled districts in Sweden: A literature review and a socio-technical ecology approach for future research. *Energy Res. Soc. Sci.* **2023**, *101*, 103118. [CrossRef]
6. Pompei, L.; Nardecchia, F.; Bisegna, F. A new concept of a thermal network for energy resilience in mountain communities powered by renewable sources. *Sustain. Energy Grids Netw.* **2023**, *33*, 100980. [CrossRef]
7. Kim, S.; Kim, M. Exergy analysis of a ground-source heat pump system with horizontal ground heat exchangers. *Energy Build.* **2017**, *152*, 109–116.
8. Basso, G.L.; Pastore, L.M.; de Santoli, L. Power-to-Methane to Integrate Renewable Generation in Urban Energy Districts. *Energies* **2022**, *15*, 9150. [CrossRef]
9. González-Prieto, D.; Fernández-Nava, Y.; Megido, L.; Prieto, M.M. Economic and environmental prioritisation of potential retrofitting interventions in electricity decarbonisation scenarios: Application to a heritage building used as offices. *J. Build. Eng.* **2023**, *72*, 106561. [CrossRef]
10. Elaouzy, Y.; El Fadar, A. Investigation of building-integrated photovoltaic, photovoltaic thermal, ground source heat pump and green roof systems. *Energy Convers. Manag.* **2023**, *283*, 116926. [CrossRef]
11. Pompei, L.; Rosa, F.; Nardecchia, F.; Piras, G. The Environmental and Energy Renovation of a District as a Step towards the Smart Community: A Case Study of Tehran. *Buildings* **2023**, *13*, 1402. [CrossRef]
12. Pastore, L.M.; Basso, G.L.; Ricciardi, G.; de Santoli, L. Smart energy systems for renewable energy communities: A comparative analysis of power-to-X strategies for improving energy self-consumption. *Energy* **2023**, *280*, 128205. [CrossRef]
13. Júnior, J.A.; Sanches, A.O.; Martins, F.R. Exergetic analysis of a geothermal heat pump system. *Energy Procedia* **2018**, *152*, 398–405.
14. Pompei, L.; Nardecchia, F.; Mattoni, B.; Gugliermetti, L.; Bisegna, F. Combining the exergy and energy analysis for the assessment of district heating powered by renewable sources. In Proceedings of the 2019 IEEE International Conference on Environment and Electrical Engineering and 2019 IEEE Industrial and Commercial Power Systems Europe (EEEIC/I&CPS Europe), Genova, Italy, 11–14 June 2019; pp. 1–5. [CrossRef]
15. Groll, E.A.; Braun, J.E. Exergy analysis of air-source and ground-source heat pump systems. *Int. J. Energy Res.* **1994**, *18*, 369–383.
16. Ibrahim, M.M.; Al-Sulaiman, F.A. Exergy analysis of a ground-source heat pump system. *Energy Convers. Manag.* **2000**, *41*, 45–63.
17. Sopian, K.; Yatim, B. Exergy analysis of heat pump systems for drying applications. *Appl. Therm. Eng.* **2002**, *22*, 583–597.

18. Jeong, H.I.; Jeong, M.J. Comparative exergy analysis of a geothermal heat pump and an air source heat pump for space heating. *Energy* **2013**, *61*, 398–405.
19. Al-Sulaiman, F.A.; Dincer, I. Thermodynamic and environmental analysis of ground-source heat pump systems. *Appl. Therm. Eng.* **2002**, *22*, 185–194.
20. Al-Sulaiman, F.A.; Hamdan, M.A. Performance analysis of an air source heat pump using exergy. *Energy Convers. Manag.* **2009**, *50*, 53–59.
21. Hrnjak, P.; Braun, J.E. Exergy analysis of a ground-source heat pump system. *Energy* **1997**, *22*, 167–174.
22. Madani, H.; Ahmadi, M.H. Exergy analysis of a ground-coupled heat pump system with earth tube heat exchanger. *Energy Convers. Manag.* **2012**, *54*, 94–101.
23. Kwak, H.Y.; Kim, Y. Exergy analysis of a ground source heat pump system for heating and cooling applications. *Energy Build.* **2010**, *42*, 396–402.
24. Wang, R.Z.; Zhao, Y. Exergy analysis of a ground-source heat pump system for building heating. *Energy Build.* **2006**, *38*, 828–835.
25. Villafuerte, N.L.C.; Coronas, A. Exergy analysis of geothermal heat pump systems for heating and cooling of buildings. *Energy Build.* **2014**, *72*, 30–40.
26. Kilkis, B. Geothermal heat pump efficiency. *Appl. Energy* **2014**, *133*, 250–256.
27. Cui, X.; Zhao, L.; Fang, Z. Performance comparison of air source and ground source heat pump systems in heating-dominated operating mode. *Energy Build.* **2014**, *69*, 65–72.
28. Sheng, W.; Zhao, C.Y.; Wu, J.Y. Performance comparison of air- and ground-source heat pumps with different heat sources. *Energy Build.* **2011**, *43*, 3509–3517.
29. Jeong, S.; Jeong, M.; Kang, H. Comparative analysis of ground source and air source heat pumps considering the effects of air temperature variations. *Appl. Therm. Eng.* **2013**, *52*, 361–366.
30. Zeng, J.; Fang, Z.; Xu, Y. The influence of ground heat exchanger design on the performance of ground source heat pump systems. *Energy Procedia* **2017**, *142*, 1545–1550.
31. Lund, J.W.; Boyd, T.L. Direct utilization of geothermal energy 2015 worldwide review. *Geothermics* **2016**, *60*, 66–93. [CrossRef]
32. Bales, C.; Ramesh, G. Environmental performance of ground-source heat pumps in North American heating-dominated climates. *Renew. Energy* **2016**, *85*, 627–637.
33. Yang, M.; Zhai, Z.; Liu, G.; Li, N.; Fang, Z. Life cycle environmental and economic assessment of air source heat pumps for residential buildings in China. *Energy* **2018**, *163*, 699–707.
34. Zhai, Z.; Fang, Z.; Liu, G.; Li, N. Comparative life cycle assessment of ground source and air source heat pumps for residential buildings in China. *Energy Build.* **2016**, *129*, 183–190.
35. Wang, Y.; Yang, H.; Zhao, X. Life cycle energy and environmental analysis of ground source heat pump system: A case study of a residential building in Tianjin, China. *Renew. Energy* **2017**, *112*, 98–107.
36. *UNI 10339*; The Exchange of Air in Aeraulic Systems. Italian Organization for Standardization: Milan, Italy, 1995.
37. *UNI/TS 11300*; Reference Technical Standards for the Estimation of the Energy Performance of Buildings. Italian Organization for Standardization: Milan, Italy, 2014.
38. *UNI 9182*; The Technical Criteria, and Parameters to be Considered for the Sizing of the Distribution Networks of Water Intended for Human Consumption. Italian Organization for Standardization: Milan, Italy, 2014.

energies

MDPI

Article

Modelling Porous Cementitious Media with/without Integrated Latent Heat Storage: Application Scenario

Mona Nazari Sam [1,*], Jens Schneider [2] and Holger V. Lutze [1,3,4]

1 Department of Civil and Environmental Engineering, Institute IWAR, Chair of Environmental Analytics and Pollutants, Technical University of Darmstadt, 64297 Darmstadt, Germany; h.lutze@iwar.tu-darmstadt.de
2 Institute of Structural Mechanics, and Design (ISM+D), Technical University of Darmstadt, 64297 Darmstadt, Germany; schneider@ismd.tu-darmstadt.de
3 IWW Water Centre, 45476 Mülheim an der Ruhr, Germany
4 Centre for Water and Environmental Research (ZWU), 45141 Essen, Germany
* Correspondence: nazarisammona@gmail.com

Abstract: This paper presents a methodological approach for the evaluation of the thermal behavior of cementitious porous media with/without integrated latent-heat thermal energy storage (LHTES). To achieve this goal, the Lewis-Nielsen model has been calibrated to predict the insulation properties of mineralized foamed concretes. Two pore-related microstructural fitting parameters, A and Φ_m, are presented according to the available data in the literature. In this regard, new findings are implemented for the classification of pore structure and prediction of the homogenized thermal conductivity of two-phase cementitious foams with or without phase change materials. The calibration and predictive analyses have been extended to a wide range of experimental data, including variation of binder types, porosities, and latent components. The presented analytical approach appears to agree well with experimental results and can be employed in the design of two-phase mineral foam materials. Then, to assess the thermal behavior of the predicted insulating envelopes, a one-dimensional (1D) enthalpy-based model is used which combines Fourier's law of heat conduction, the first law of thermodynamics, Lewis-Nielsen conductivities, and the mixture theory for LHTES additions. The results demonstrated the importance of volumetric heat capacity for the thermal inertia of building envelopes.

Keywords: latent heat thermal energy storage (LHTES); thermal energy storage (TES); porous media; cementitious foam; conductivity prediction; Lewis-Nielsen model; Fourier's law; dynamic envelopes; energy efficiency

1. Introduction

The built environment is responsible for over 30% of global energy consumption and contributes nearly 40% of total direct and indirect greenhouse gas emissions [1–3]. In this context, the use of thermal energy storage (TES) technologies and latent-heat thermal energy storage (LHTES) systems has an enormous potential for novel applications in buildings [4–6]. Utilizing phase-change materials (PCMs), one of the most effective methods for increasing the latent heat storage in buildings, offers viable solutions for improving energy efficiency, enhancing the performance of HVAC and on-site renewable energy systems, i.e., reducing peak loads, maintaining indoor comfort levels, regulating temperature excursions and attenuating extreme thermal values [7–11].

Over the past few years, numerous articles have been dedicated to exploring the use of PCMs in construction and building materials [12–14]. More specifically, the ability to store and release large amounts of heat into PCMs mixed within insulating envelopes can effectively reduce the thermal conditioning load of the building [15–24]. In this context, a significant proportion of the research on PCMs focuses on their incorporation into porous cementitious or other binder-based composites (i.e., having a density <800 kg/m^3) to be used for thermal insulation in nearly-zero-energy buildings (nZEB) [25–28]. These composites can provide excellent heat blocks and are non-flammable [29,30]. However, the thermophysical properties of cementitious foams largely depend on the microstructural parameters, including air content as well as inner pore morphology and the amount/type of binders, aggregates, and, eventually, additional components like fillers, fibers, or rubber particles [31–38]. The correlation between microstructural parameters and related thermophysical properties of porous cementitious composites has been the object of extensive study for many decades due to its wide applicability in engineering problems and applications.

From the modelling point of view, on one hand, several theoretical and empirical models have been proposed in the literature to deal with describing the conductivity (either thermal or electrical) of composites. In this context, the most popular expressions are those due to the proposals by Maxwell [39–42] and Bruggeman [43–46]. They are often based on expressions providing the effective conductivity of a homogenized composite, and built up by combining the conductivity of the multi-phase components, such as the matrix, fillers, air-voids, fibers, etc. The weighting parameters in these formulas are commonly determined by either the volumetric or the weight fraction of each individual phase. However, the influence of the pores (i.e., amounts, sizes, and distributions) on the heat-transfer properties of porous materials represents the key factor in this field [47]. Specifically, the thermal conductivity of two-phase porous systems has been reported to be strongly influenced by their pore nature and structure [48,49]. In this field, it is worth mentioning the proposal by Lewis and Nielsen [50,51], which most closely matches the transport parameters in two-phase porous systems in different cases [52–56].

On the other hand, modelling the heat storage capacity of porous cementitious systems enhanced with LHTES components has been largely explored in the literature. Most of the models arise from solving the so-called Stefan problem for phase transformation problems [57–59]. However, in construction and building materials applications, the problem can be highly simplified, and the so-called enthalpy-based method, employable for conduction-only mechanisms, represents the most used and simplified approach followed by many authors [60–65].

This study aims to propose a simplified analytical approach for predicting the homogenized thermal conductivity of porous cementitious media. A wide range of experimental data from the literature regarding the microstructure parameters of cement-based foams influencing thermal conductivity has been compiled. The links between air inclusions and fractions, geometrical classification of the pores, and the overall homogenized thermal conductivity k_{eff} of porous cementitious foams are established. The proposed analytical model concerns the stochastic characteristic of the pore structure and allows the prediction of the homogenized thermal conductivity of cementitious foams considering the shape and packing of the pores. The methodology can be further applied to solve the energy transport equations through the complex structure of mineral foams, with/without LHTES components, without the need for costly experimental designs or expensive computer analysis for ascertaining k_{eff} [66].

The present paper is structured as follows. Section 2 provides a summary of the available experimental data in the recent literature related to low-, medium-, and high-porosity cementitious media. In Section 3, the theoretical background of the Lewis-Nielsen model is discussed, including its assumptions, applicability, and further extension of the law. Section 4 presents the analytical predictions and comparisons relative to the experimental data to assess the soundness and capabilities of the methodology. Section 5 analyzes the TES (latent and sensible) performance of the studied composites, e.g., in 1D wall systems, while also examining the direct impacts of dynamic insulation parameters and the variable effusivity and diffusivity of the media. Finally, Section 6 concludes with the main remarks and provides future research directions.

2. Experimental Database

Experimental tests on variable porous cementitious systems have been employed as a reference database to show the potential and capability of the proposed modelling procedure described within Sections 3 and 4. This section reviews and analyzes a series of data available in the literature; the data cover the entire density (porosity) range, from low densities (porosities) up to high densities (porosities). The data are those tested by Batool (2015) [67], Awang et al. (2012) [68], Wei et al. (2013) [69], Mydin (2011) [70], Davraz et al. (2016) [71], Oren et al. (2020) [72], and Jiang et al. (2017) [73] & (2016) [74]; see Supplementary Material S1 for more details.

All data collected in Table 1 are highlighting the types of binders used, possible additives/fillers, target densities ρ, water-to-binder ratios, dry densities ρ, measured conductivity values of cementitious matrices/skeleton (either pastes or mortars), dispersed air content (porosity), type of foaming agent employed, methods of foaming, and measurement methods for the foam conductivity, as well as how all of these factors are affecting the overall conductivity of the foam system (last column of the table). The review of these data, as listed in Table 1, reveals that the use of different pozzolanic additives/fillers into the matrix can affect the porosity and the related properties of the foam, and generally leads to a reduction of the thermal conductivity of the composite.

Table 1. Experimental database of foam concrete by [67–74].

Authors	Binder Type	Additive/Filler	W/b Ratio	Target Density ρ (kg/m³)	Dry Density ρ (kg/m³)	Matrix Conductivity (k/W/m)	Dispersed Air Content c	Foaming Agent	Foaming Method	Conductivity Measurnment Method	Foam Conductivity (k/W/m)
Batool, F. (2015) [1]	HE portland cement	Ref	0.69	800	724	0.458	0.49	Synthetic	Physical air-entraining	Transient plane source (TPS)	0.180
				600	492	0.458	0.64				0.125
				400	303	0.458	0.78				0.076
		FA10		800	751	0.458	0.50				0.183
				600	549	0.458	0.62				0.136
				400	360	0.458	0.76				0.081
		FA20		800	710	0.407	0.50				0.182
				600	499	0.407	0.64				0.122
				400	273	0.407	0.81				0.071
		SF10		800	755	0.415	0.50				0.173
				600	499	0.415	0.65				0.111
				400	320	0.415	0.78				0.079
		SF20		800	750	0.416	0.53				0.166
				600	474	0.416	0.70				0.100
				400	270	0.416	0.83				0.071
		MK10		800	759	0.456	0.50				0.176
				600	529	0.456	0.65				0.119
				400	295	0.456	0.81				0.074
		MK20		800	756	0.461	0.51				0.179
				600	533	0.461	0.65				0.121
				400	276	0.461	0.82				0.074
Awang, H. et al. (2012) [2]	SEM1 Portland cement/sand (1:1.5)	RefSa	0.45	600	n.a.	1.400	0.69	Protein-based	Physical air-entraining	Transient plane source (TPS)	0.190
				1000	n.a.	1.400	0.49				0.430
				1400	n.a.	1.400	0.36				0.590
		FA15		600	n.a.	n.a.	0.69				0.170
		FA30		600	n.a.	n.a.	0.70				0.160
		RL15		600	n.a.	n.a.	0.72				0.160
		RL30		600	n.a.	n.a.	0.70				0.200
		PF20		600	n.a.	n.a.	0.70				0.180
		PF40		600	n.a.	n.a.	0.71				0.180
		FA15		1000	n.a.	n.a.	0.50				0.380
		FA30		1000	n.a.	n.a.	0.51				0.360
		RL15		1000	n.a.	n.a.	0.56				0.310
		RL30		1000	n.a.	n.a.	0.55				0.310
		PF20		1000	n.a.	n.a.	0.54				0.310
		PF40		1000	n.a.	n.a.	0.55				0.320
		FA15		1400	n.a.	n.a.	0.35				0.580
		FA30		1400	n.a.	n.a.	0.35				0.610
		RL15		1400	n.a.	n.a.	0.36				0.590
		RL30		1400	n.a.	n.a.	0.37				0.530
		PF20		1400	n.a.	n.a.	0.36				0.600
		PF40		1400	n.a.	n.a.	0.35				0.560

Table 1. *Cont.*

Authors	Binder Type	Additive/Filler	W/b Ratio	Target Density ρ (kg/m³)	Dry Density ρ (kg/m³)	Matrix Conductivity (k/W/m)	Dispersed Air Content ε	Foaming Agent	Foaming Method	Conductivity Measurnment Method	Foam Conductivity (k/W/m)
Wei, S. et al. (2013) [3]	CEM I	FA20	0.40	1900	1870	0.500	0.00	Protein-based	Physical air-entraining	Rapid-K Thermal Conductivity test	0.500
				1700	1636	0.500	0.13				0.423
				1500	1461	0.500	0.22				0.363
				1300	1201	0.500	0.36				0.282
				1000	948	0.500	0.49				0.217
				800	757	0.500	0.60				0.165
				600	570	0.500	0.70				0.124
				500	453	0.500	0.76				0.091
				400	374	0.500	0.80				0.080
				300	252	0.500	0.87				0.065
Mydin, M. (2011) [4]	OPC cement/sand (2:1)	Ref	0.50	650	n.a.	n.a.	0.74	Protein-based	Physical air-entraining	Hot guarded plate test (HGP)	0.230
				700	n.a.	n.a.	0.71				0.240
				800	n.a.	n.a.	0.64				0.260
				900	n.a.	n.a.	0.57				0.280
				1000	n.a.	n.a.	0.51				0.310
				1100	n.a.	n.a.	0.47				0.340
				1200	n.a.	n.a.	0.44				0.390
Davraz, M. et al. (2016) [5]	CEM I 42.5 R Portland cement	PF/L	0.30	300	364	n.a.	0.77	Protein-based	Physical air-entraining	GHP/HFM-TS EN 12667	0.090
				400	510	n.a.	0.70				0.140
				500	563	n.a.	0.64				0.140
				600	715	n.a.	0.58				0.160
				700	837	n.a.	0.51				0.220
				800	851	n.a.	0.45				0.240
				900	965	n.a.	0.38				0.270
				1000	1100	n.a.	0.32				0.300
				1100	1272	n.a.	0.26				0.320
				1200	1321	n.a.	0.19				0.360
				1300	1429	n.a.	0.13				0.430
				1400	1531	n.a.	0.07				0.500
Oren, O. H. et al. (2020) [6]	CEM II 42.5R/sand	FA40/Sa	0.91	1000	975	n.a.	0.53	Protein-based	Physical air-entraining	Transient plane source (TPS)	0.225
			0.68	1000	1029	n.a.	0.51				0.235
			0.55	1000	1033	n.a.	0.50				0.256
		FA40/Sa/GBS	0.91	1000	949	n.a.	0.53				0.220
			0.68	1000	1004	n.a.	0.52				0.226
			0.55	1000	1030	n.a.	0.51				0.230
		FA40/GBS	0.91	1000	926	n.a.	0.53				0.208
			0.68	1000	996	n.a.	0.50				0.226
			0.55	1000	1132	n.a.	0.49				0.264
Jiang, J. et al. (2017) [7] (2016) [8]	CEM I 42.5 R Portland cement	ER0%	0.60	n.a.	n.a.	n.a.	n.a.	Protein-based	Physical air-entraining	Heat Flow Meter (HFM)	0.059
		ER3%	0.60	n.a.	n.a.	n.a.	n.a.				0.064
		ER6%	0.60	n.a.	n.a.	n.a.	n.a.				0.063
		ER0%	0.80	n.a.	n.a.	n.a.	n.a.				0.060
		ER3%	0.80	n.a.	n.a.	n.a.	n.a.				0.063
		ER6%	0.80	n.a.	n.a.	n.a.	n.a.				0.064
		ER9%	0.80	n.a.	n.a.	n.a.	n.a.				0.065
		ER9%	0.60	n.a.	n.a.	n.a.	0.89				0.069
		Ref	0.80	n.a.	n.a.	n.a.	0.92				0.052
		C/SF6	0.80	n.a.	n.a.	n.a.	0.91				0.050
		C/MK20	0.80	n.a.	n.a.	n.a.	0.92				0.054
		C/MK20/SF6	0.80	n.a.	n.a.	n.a.	0.92				0.054
		Ref	1.00	n.a.	n.a.	n.a.	0.91				0.053
		C/SF6	1.00	n.a.	n.a.	n.a.	0.91				0.051
		C/MK20	1.00	n.a.	n.a.	n.a.	0.91				0.053
		C/MK20/SF6	1.00	n.a.	n.a.	n.a.	0.91				0.055

Figure 1 provides a plot of the effective conductivity, k_{eff} (W/m/K), as a function of target density (kg/m^3), thus summarizing the foams' results with different concentrations of additives and fillers, as collected from experimental datasets. Besides the different types of components and materials employed, Figure 1 depicts a clear increase in the overall thermal conductivity, with an increase in density in all the experiments. The results have been presented in four different foam classes, as described in [75,76]: insulating ultra lightweight (UL) foams having 0–300 kg/m^3 density; insulating low-density foams, with 300–800 kg/m^3 density; moderate strength foams with 800–1350 kg/m^3 density; and structural foams with densities higher than or equal to 1350 kg/m^3 density.

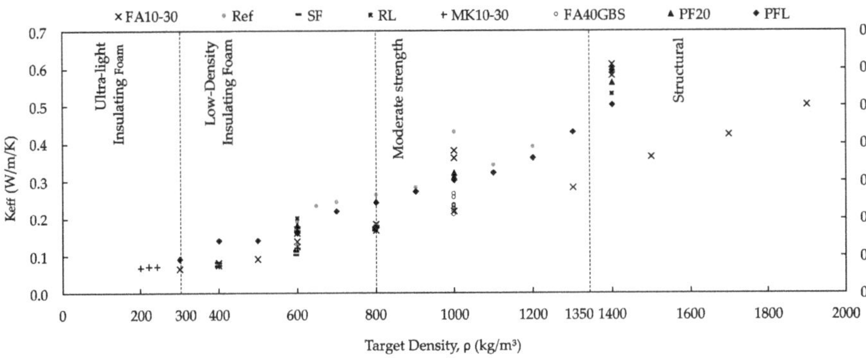

Figure 1. K_{eff} (W/m/K) vs. target density, ρ (kg/m^3) of concrete foams by [67–74]. Ref: Reference foam, FA: Fly ash, SF: Silica fume, RL/L: Limestone, MK: Metakaolin, GBS: Granulated blast furnace slag, PF: Polypropylene fiber.

3. Heat Conduction in Two-Phase Porous System

This section presents a simplified approach for the determination of the homogenized thermal conductivity, k_{eff} (W/m/K) of cementitious foam composites. The model is based on stacked porous amounts (porosity), continuous- and dispersed-phase conductivities, and geometrical classification of pores according to the comprehensive study of the aqueous foam structures presented in [77–79]. The main objective is to propose pore geometrical-fitting parameters that can be employed in designing LHTES/TES-enhanced (two phases) cementitious foams with general low-to-high porosity.

3.1. Modified Lewis-Nielsen Model for Conductivity of Porous Cementitious Composites

Cementitious foam's microstructure, considered as a two-phase system, can be modelled as a continuous phase made of cement paste (skeleton) and a dispersed phase made of air-voids (Figure 2). The main thermophysical parameters of the two-phase system which govern the homogenized heat transport phenomena include: (i) the specific heat, (ii) the density, (iii) the thermal conductivity (being i., ii., and iii. of the single phases), and (iv) possible thermal lag, i.e., interfaces between these phases. Knowing these parameters, one can describe the thermal behavior of the two-phase porous media.

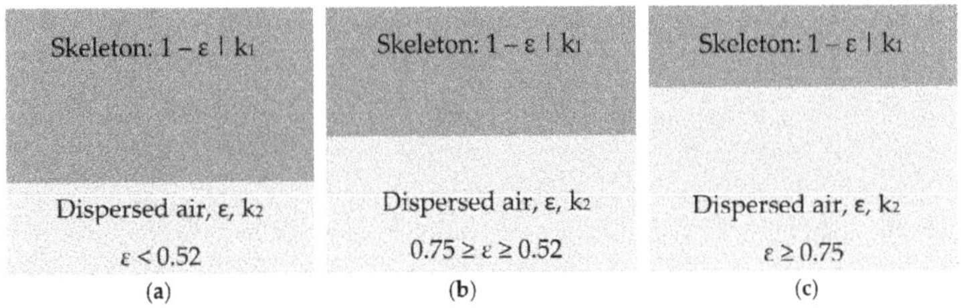

Figure 2. Schematic illustration of a two-phase cementitious foam without considering the meso-scale air-void distribution: (**a**) low- ($\varepsilon < 0.52$), (**b**) medium- ($0.75 \geq \varepsilon \geq 0.52$), and (**c**) high-porosity ($\varepsilon \geq 0.75$).

The thermal conductivity (k) in a two-phase porous system is significantly influenced by the conductivities of the continuous phase (k_1) and the dispersed phase (k_2), and by their concentrations (i.e., $1 - \varepsilon$ for the skeleton and ε for the dispersed air, as in Figure 2). In this regard, numerous analytical equations have been proposed to solve the effective (i.e., homogenized) thermal conductivity of these two-phase systems. These models are capable of accurately predicting the effective heat transfer by considering and weighting the porosity and conductivity of each single component. However, these approaches become limited when the pore volume fraction increases and, at the same time, the isolated pores become connected [69]. Nevertheless, the study conducted by Lewis and Nielsen [50,51], while originally based on the theory of the elastic moduli of composite materials, has been found to be a very good approximation for thermal conduction, if enough information about the air content and the pore structure is available.

3.2. Fitting Parameters A and ψ for Porous Cementitious Composites

In this work, the following expression by Lewis and Nielsen [50] has been used as homogenization rule for predicting the conductivity of cementitious foams:

$$\frac{K}{K_1} = \frac{1 + A\frac{K_2/K_1 - 1}{K_2/K_1 + A}\varepsilon}{1 - \frac{K_2/K_1 - 1}{K_2/K_1 + A}\psi\varepsilon} \tag{1}$$

where ε is the fraction of air inclusion, A the shape factor for the dispersed air-voids, and ψ considers the type of packing of the pores, as well as the presence of either predominantly closed or interconnected pores. In Equation (1), the accurate definition (and calibration) of A and ψ represents the key task, one which requires a fundamental knowledge and study of the pore structure of the cementitious foams. For the evolution of A and ψ, it is also important to consider the hydration process and aging time of the foam. Foam pore structures can change during the hardening process due to rheological changes of the cement matrix, with consequent volumetric reduction (shrinkage) of the paste due to drying processes. Also, local failure of individual lamellae at advanced stiffening/solidification time steps can lead to pore systems consisting of intersecting (coalescing) spherical pores. In addition to volume changes, additives with different surface characteristics (polarity) can also influence pore structure developments, leading to uncertainties in pore structure and deviation between the real (measured) and the theoretical (calculated) porosity of the composite.

In the fresh state of the foam, bubble coalescences and/or bubble coarsening are directly related to an increasing of the air inclusion amount, which leads to an increase in the mean pore diameter and a shift of the pore-size distribution towards larger pores. In this regard, pore systems with a predominance of connected pores are formed from polyhedral-shaped foams which have very high air contents ($\varepsilon \geq 0.75$) and exhibit extremely thin lamellae (ca. 20–30 µm). Due to this, highly irregular structures are associated with high-porosity foams with a dwindled cement matrix. In foams with a predominance of spherical bubbles, the minimum and maximum pore sizes, together with their distribution, can influence the maximum achievable porosity. Thus, assuming an ideal distribution of the pore sizes, a geometric maximum-possible porosity of 0.75 can be achieved for foams with spherical bubbles. A classification into polyhedral foams, spherical foams, and bubble systems can be adopted for the characterization of foamed cement pastes. This classification represents a correlation between geometry, fraction of air inclusion (porosity) and bulk density of a fictitious aqueous phase with a bulk density of 2.0 kg/dm^3 [78].

It follows that, by considering the correlation between fraction of air inclusion and void geometry, as sketched in Figure 3, the calibration factors A and ψ can be defined according to the literature. Particularly, the constant A, related to the generalized Einstein coefficient (Einstein, 1905, 1906), which represents primarily the shape of the air inclusions, for any direction of heat flow, can be calculated for aggregates of spheres as [51]:

$$A = \frac{2.5}{\Phi_m} - 1 \tag{2}$$

where Φ_m represents the maximum packing fraction of the randomly dispersed air voids, defined as the true volume of the air voids in cementitious foam divided by the volume they appear to occupy when packed to their maximum extent. In this sense, the three types of foam classes can be taken from Table 2.

(**a**) Random closed-cell bubbles $\varepsilon < 0.52$ (**b**) Spherical bubbles $0.75 \geq \varepsilon \geq 0.52$ (**c**) Polyhedral bubbles $\varepsilon \geq 0.75$

Figure 3. Correlation between the fraction of air inclusion (ε) and air void geometry in cementitious foams of (**a**) low porosity, (**b**) medium porosity, and (**c**) high porosity, based on the findings of [77,79].

Table 2. Air-void shape/packing classifications of low-, medium-, and high-porosity cementitious foam; see Supplementary Material S2 for more details.

Air Fraction	Foam Class	Shape of Air Voids	Type of Packing	Maximum Packing Fraction (Φ_m)
$\varepsilon \geq 0.75$	I	Polyhedral	Hexagonal close	0.74
$0.75 \geq \varepsilon \geq 0.52$	II	Spherical	Random close	0.64
$\varepsilon < 0.52$	III	Bubbles	Random loose	0.60

The structural parameter ψ, needed for evaluating the conductivity of cementitious foam, can be estimated as [51]:

$$\psi = 1 + \left(\frac{1 - \Phi_m}{\Phi_m{}^2}\right)\varepsilon \tag{3}$$

In this study, Equations (1)–(3) are employed for predicting the homogenized (effective) conductivity of the analyzed low-to-high porous cementitious foams. The proposed methodological approach enables precise determination of Φ_m (refer to Table 2) for solving Equations (2) and (3) within each foam class, simplifying Equation (1) for enhanced practicality in engineering applications. Consequently, the only remaining unknown parameters are porosity and matrix conductivity. This represents a significant advancement in thermal conductivity assessment, particularly addressing the impact of pore geometry within the mid-to-high-porosity range of mineral foams. This advancement greatly contributes to standardizing the thermal conductivity of mineral foams, enhancing their utility and efficiency in engineering design and applications.

It is important to emphasize that the proposed fitting parameters for mineral foams in this study are based on well-established mathematical principles and exclusively tied to the system's geometry; see Supplementary Material S2 for more details. Determining the correct parameters would be arbitrary and impractical without a comprehensive understanding of the factors influencing thermal conductivity and knowledge of mineral foam pore structures.

4. Analytical Predictions vs. Experimental Data

4.1. Classification of Test Data

In this section, the experimental data presented in Section 2 are analyzed to explore potential correlations between the fraction of air inclusion (ε), target density, and effective foam conductivity. The primary aim of this analysis is to classify the data into specific categories, such as low-, medium-, and high-porosity cementitious foams. Additionally, a categorization of the data based on air-void structure/geometry, such as random closed-cell bubbles, spherical bubbles, or polyhedral ones, is considered, as presented in Section 3.2.

This information is crucial for the accurate determination of calibration parameters "A" and "*f*" in Equation (1).

Figure 4 illustrates the relationship between the fraction of air inclusion (ε) and the target density for all datasets [67–74] considered in this study. It demonstrates a consistent linear decrease of porosity with increased density across all the data points. Furthermore, by employing the classification outlined in Section 3.2, the data can be categorized into three distinct foam classes: "Class I" for foams with polyhedral-shaped air voids (exhibiting hexagonal close packing) and a fraction of air inclusion (ε) greater than or equal to 0.75; "Class II", consisting of foams with spherical-shaped bubbles, with random close packing and a fraction of air inclusion ranging from 0.75 to 0.52; lastly, "Class III" are foams with low porosity, with random loose packing and a fraction of air inclusion (ε) below 0.52.

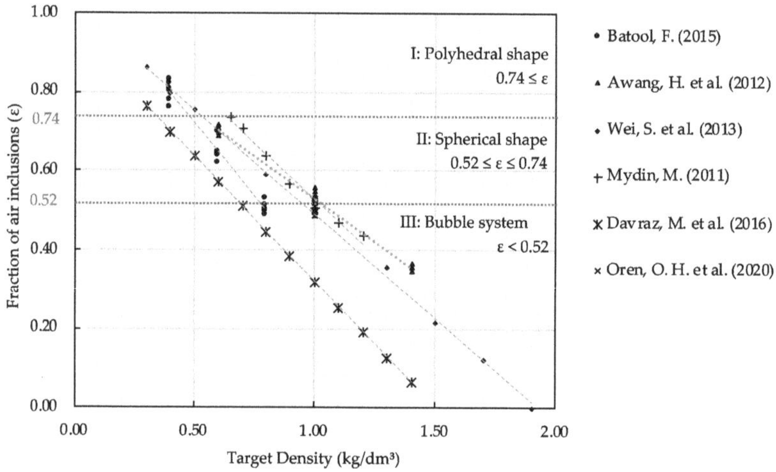

Figure 4. Fraction of air inclusion (ε) vs. target density (kg/dm^3); data by [67–74].

Figure 5 displays the correlation between the foams' homogenized conductivity (k_{eff}) and target density for collected datasets [67–74]. It exhibits a clear linear increase of conductivity against increased density across all the data points. In this case, regarding the presented classification described in Section 2, the data can be categorized into four distinct subtypes: "ultra-light insulating foams", which possess a target density that is equal to or lower than 0.30 kg/dm^3; "low-density insulating foams", which includes foams with target densities ranging from 0.30 kg/dm^3 to 0.80 kg/dm^3; "moderate-strength foams", classifying foams having a target density ranging from 0.80 kg/dm^3 to 1.35 kg/dm^3; and finally, "structural foams", comprising foams with a target density higher than 1.35 kg/dm^3. It is worth noting that as the density increases, the insulation performance of the material decreases, as indicated by a higher level of foam conductivity.

Figure 6 combines the information from Figures 4 and 5, establishing a correlation between foam effective/homogenized conductivity (k_{eff}) and the fraction of air inclusion (ε) for the presented datasets. This graph clarifies the classification of the considered foams, based on air-void structure/geometry and packing characteristics, presenting the related conductivity range (upper and lower values) expected for each foam system class. The classification in Figure 6 identifies three foam classes: the first one consists of foams with bubble types characterized by low porosity and random loose packing (ε < 0.52), with expected k_{eff} higher than 0.2 W/m/K; the second one comprises foams with predominantly non-connected spherical-shape voids, exhibiting random close packing, and having a fraction of air inclusion ranging from 0.52 to 0.75, with expected k_{eff} between 0.2 W/m/K and 0.1 W/m/K; lastly, the third class includes polyhedral-shaped foams, with hexagonal close packing and ε greater than or equal to 0.75 with expected k_{eff} less than 0.1 W/m/K.

An interesting observation arising from Figure 6 is that the type of internal structure in the foam is directly linked to the total air fraction and the target density of the system.

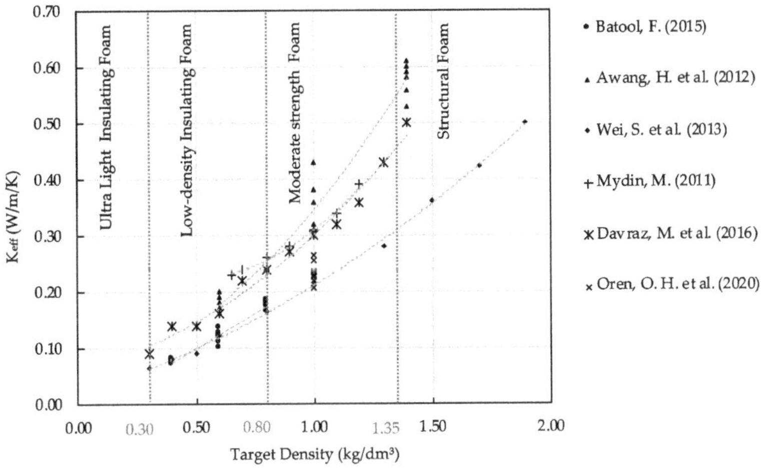

Figure 5. K_{eff} (W/m/K) vs. target density (kg/dm^3); data from [67–74].

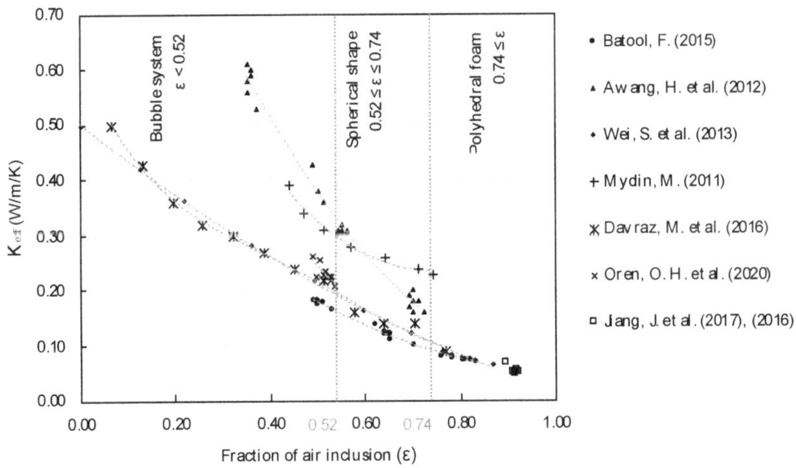

Figure 6. K_{eff} (W/m/K) vs. fraction of air inclusion (ε): data by [67–74].

The classifications, as shown between Figures 4–6, are crucial for guiding the subsequent section, which presents a unified approach for modeling heat transfer conductivity in various types of foams. The approach considers a wide range of porosity levels, extending from low to high porosity cementitious foams, and points out the effects of the different types of air-void geometry and meso-structural types. Thus, by understanding the influence of these factors, this section provides valuable insights into the heat-transfer characteristics of mineral foams, enabling a comprehensive understanding of their homogenized thermal conductivity.

4.2. Analytical Description vs. Experimental Measurements of Thermal Conductivity K_{eff}

In this section, the homogenized thermal conductivity of each foam is predicted and compared to the corresponding experimental data from Section 2, as also categorized in

Section 4.1. A total of one hundred foam cases were analyzed, and all the data were unified and calibrated by fitting the parameters "A", "Φ_m", and "ψ", as listed in Table 3.

Table 3. Calibrated "A", "Φ_m", and "ψ" parameters for the foam concretes determined by [67–74].

Air Fraction	Foam Class	A	Φ_m	ψ
$\varepsilon \geq 0.75$	I	2.37	0.74	1.370
$0.75 \geq \varepsilon \geq 0.52$	II	2.90	0.64	1.563
$\varepsilon < 0.52$	III	3.10	0.60	1.546

The equation (Equation (1)) used for the analysis is modified, as:

$$\frac{K}{K_1} = \frac{1 + A\frac{K_2/K_1 - 1}{K_2/K_1 + A}\varepsilon}{1 - \frac{K_2/K_1 - 1}{K_2/K_1 + A}\left(1 + \left(\frac{1-\Phi_m}{\Phi_m{}^2}\right)\varepsilon\right)\varepsilon} \quad (4)$$

Figure 7 illustrates all the data in terms of effective thermal conductivity (k_{eff}), based on the data from [67–74], and the predicted values obtained using the modified Lewis–Nielsen expression proposed in Equation (4).

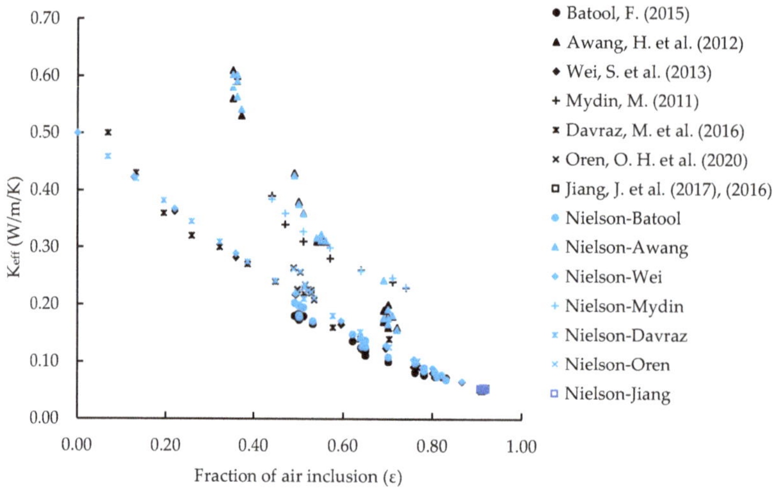

Figure 7. K_{eff} (W/m/K) [67–74] vs. fraction of air inclusion (ε); blue points indicate the predicted values and black points indicate the experimental findings.

Notably, the homogenized conductivity predictions can be achieved by solely calibrating "A" and "Φ_m" parameters, which are mainly influenced within the three distinct foam classes mentioned in Section 3.2. Specifically, the fundamental factors are the foam air-volume fraction and the dispersed air-geometry characteristics, while one can disregard other factors, such as binder variations, casting types, foaming agents, and other components/amounts. This aspect represents the significant contribution of this study, as the application of the modified Lewis-Nielsen expression to an extensive dataset of mineral foams with widely scattered formulations and thermal responses has not been attempted previously. This limitation arises from the lack of sufficient experimental data regarding the classification of pore geometry in mineral foams, a crucial factor for determining the appropriate values of parameters A and Φ_m in model implementation.

The presented results show that, by considering the shape of the dispersed air and their packing characteristics, a consistent stable response for each foam type can be achieved, i.e., class I (polyhedral-shaped air voids), class II (spherical-shaped), and class III foams (loosely packed bubble inclusion). There is substantial good agreement between the theory and the experimental data.

To demonstrate the effectiveness of this approach, Figure 8 shows the homogenized predicted conductivity values plotted against the employed data. The figure displays both the linear-regression curve and the square-root error, both of which help to identify the best-fit line that presents the relationship between the measured variables and the predicted ones.

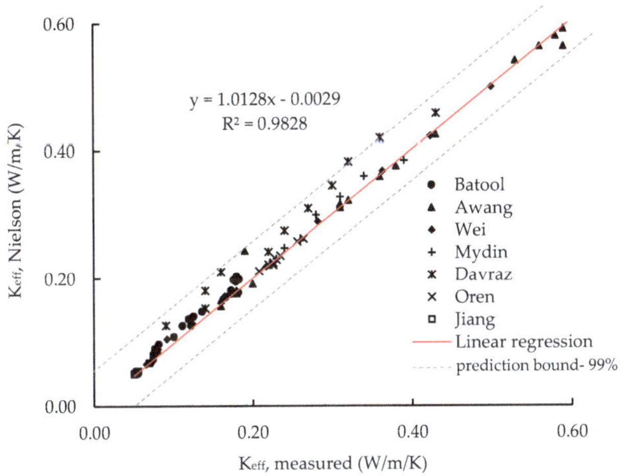

Figure 8. Lewis-Nielsen predictions: k_{eff}, Nielson (W/m/K) vs. measured conductivity k_{eff}, measured (W/m/K), for all collected data; comparison and linear regression + R2, slope >1 model overestimates the conductivity data.

The prediction results reveal promising outcomes, with a high R^2 value of 0.9828 and a regression line slope of 1.0128. In Supplementary Material S3, Figures S2–S8 show the homogenized predicted conductivity values plotted against the employed data, analyzed author by author. The agreement between experimental values and predicted ones is remarkable; see Figure 8. A prediction interval, within which, with a probability of 99%, future observations are expected to fall, is presented in Figure 8. It provides an uncertainty of 0.02 W/m/K surrounding the predicted foam homogenized conductivity value by applying the proposed approach. The scatter in the data can potentially be attributed to the experimental errors due to sampling, measurement techniques, and measurement accuracy.

It is possible to further simplify and extend Equation (4) by incorporating the parameters from Table 3. However, the predictive results obtained demonstrated remarkable accuracy, closely matching the experimental data. At this point, there is no compelling reason to extend the equation further.

4.3. Extension of the Analytical Model for Porous LHTES-Systems (Mineral Foams)

The main objective of the analytical model proposed in Sections 4.1 and 4.2 was to predict the conductivity, or the homogenized one, of cementitious foams made of various types and amounts of components and having variable properties. By knowing the conductivity of the matrix (or skeleton) and considering the amount of air inclusion (porosity), and by using the analytical formula based on air dispersion shape and packing, the conductivity of the homogenized porous system was well predicted.

Based on this, now the model can be extended to predict the conductivity of latent-heat-storage-enhanced porous systems. The goal is to facilitate simulations through simple engineering calculations, without the need to perform expensive experimental tests. The estimation of the homogenized conductivity of LHTES-enhanced foams is thus considered by employing different optimization methods, such as embedding microencapsulated phase change material (MPCM) within the skeleton at various percentages (i.e., 10%, 20%, 30%, and 40%). The investigation is limited to 40%, since higher percentages have been found to compromise foam stability. The proposed analytical formula (Equation (4)) has been employed by adopting the parameters of the LHTES-enhanced skeleton (MPCM pastes) by Sam et al. [80], including the thermal conductivity data for the MPCM paste, in both liquid and solid phases. The conductivity of the air is sourced from the literature. All these data are summarized in Table 4.

Table 4. Data used in the proposed model to predict homogenized conductivity of thermal-energy-storage mineral foam composites; see Sam et al. [80].

	Vol.- %MPCM in Paste	K_{eff} Paste (w/b = 0.4) Solid Phase MPCM (W/m/K)	K_{eff} Paste (w/b = 0.4) Liquid Phase MPCM (W/m/K)
LHTES Paste	40 20	0.600 0.754	0.464 0.759
K_{eff} Ref Paste (W/m/K)		0.89–0.93	
K Air(W/m/K)		0.025	

The graphs of Figures 9 and 10 illustrate the predicted homogenized conductivities of the LHTES/TES-enhanced foams, with and without considering 20% and 40% of MPCM pastes in comparison.

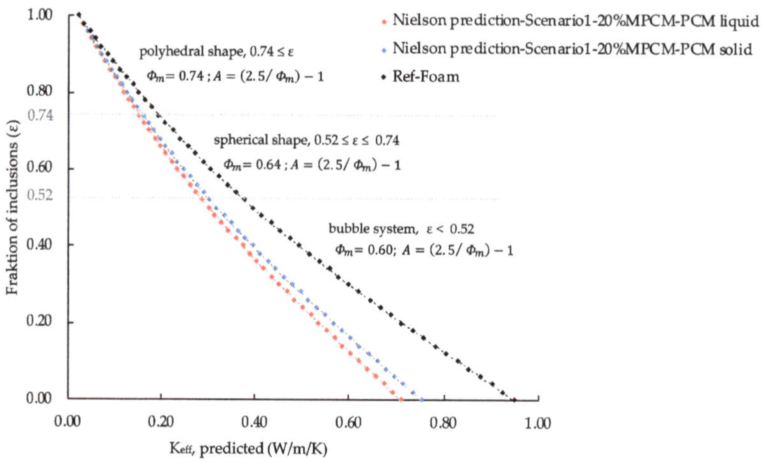

Figure 9. Foam with skeleton having 0% (Ref) and 20% MPCM: fraction of air inclusion (ε) vs. $k_{eff,}$ predicted (W/m/K).

More specifically, the skeleton of the foams consists of cement paste with a water-to-binder ratio of 0.40, which was mixed with a commercial MPCM having a melting/solidification temperature of 24 °C, with volume percentage substitutions of 0%, 20% and 40%, respectively. The experimental data were tested by the author in the previous work; see Sam et al. [80]. By analyzing the graphs, the conductivity range within each foam class and porosity level can be determined at any desired porosity. The effective homogenized values for 10% and 30% of MPCM substitutions in the paste can be smoothly interpolated from the data available in this section.

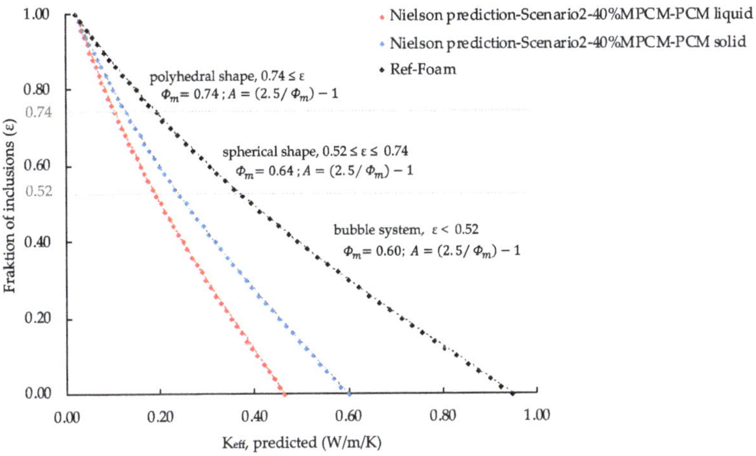

Figure 10. Foam with skeleton having 0% (Ref) and 40% MPCM: fraction of air inclusion (ε) vs. $k_{eff,}$ predicted (W/m/K).

In conclusion, Table 5 presents the predicted homogenized conductivity ranges for different foam classes and volumetric amounts of MPCM, offering insights into the conductivity of polyhedral and spherical foam systems (of both Class I and II) with porosities higher than 52%, which can be used for energy storage purposes.

Table 5. Proposed predicted conductivity range for further thermal analysis and numerical simulations. Values for specific porosity can be taken from Figures 9 and 10.

	Foam Class	Vol-. %MPCM in Paste	k_{eff} Foam Solid Phase PCM (W/m/K)	k_{eff} Foam Liquid Phase PCM (W/m/K)
TES Polyhedral Foam	I	40	≤0.14	≤0.10
		20	≤0.17	≤0.15
TES Spherical Foam	II	40	≤0.24	≤0.20
		20	≤0.30	≤0.28
TES Polyhedral Foam	I	0	≤0.2	
TES Spherical Foam	II	0	≤0.39	

This information is utilized for a 1D-wall simulation presented next, in Section 5, by incorporating porosity and effective thermal conductivity of TES-Foam with and without MPCM.

5. Integration of LHTES and Dynamic Envelopes: Application Scenario-1D Model Based on the First Law of Thermodynamics

In this section, the thermal performance of multi-layer enclosing structures consisting of expanded polystyrene (EPS) and TES-foam (foam having porosity higher than 52%) as external layers and reinforced concrete as main structural layer is assessed by using standard one-dimensional (1D) FEM simulation. More details about the applied theoretical background are presented in Supplementary Materials S4.

This chapter underscores the model's physical significance for a building envelope and illustrates how the model simplifies the intricate numerical calculations that were previously required to determine the thermal conductivity of mineral foams. These calculations include scenarios involving 20% MPCM in the skeleton and 90% porosity.

The analytical model implemented enables the straightforward representation of thermal behavior within building envelopes by the modification of input parameters such as matrix conductivity and/or porosity. This analytical capability extends to the calculation and implementation of conductivity while addressing concerns related to volumetric heat capacity through a simple one-dimensional FEM simulation.

Parametric Studies for Thermal Performance in Multi-Layer Structures (EPS vs. TES Foam Composites)

The heat flow via conduction through a two-layer enclosing structure is schematized in Figure 11: the x-axis denotes the direction of the one-dimensional heat flux, and considers a heat flow between the exterior layer, L1 (0.1 m thickness), and the interior one, L2 (0.2 m thickness), with total thickness of Δx = 0.3 m.

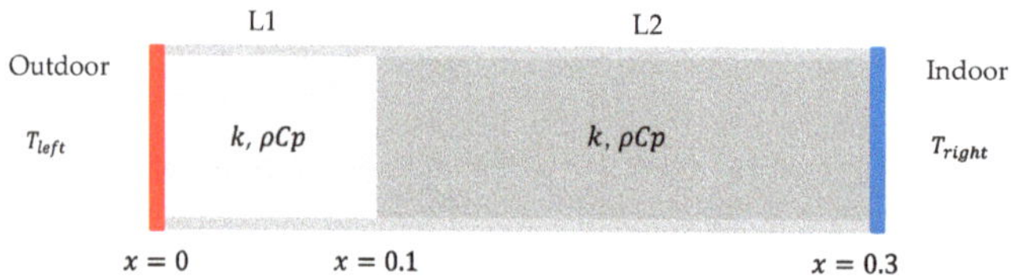

Figure 11. Schematized one-dimensional (1D) conduction model for a two-layer wall.

For one-dimensional heat flow, the temperature gradient through the structure is simulated. To solve the heat transfer problem (as shown in the Supplementary Materials), two nodes' truss finite elements (FEs) were adopted and the non-linear FEM problem was solved in MATLAB; the space discretization of the wallboard uses 100 FEs, while for the temporal one, an implicit backward Euler method with a time stepping of 5 min within 24 h was chosen. Initial boundary conditions considered an initial temperature $T_0 = 20$ °C (i.e., at t = 0), while essential (Dirichlet types) boundary conditions were fixed according to two case studies:

- "Summer" $T(x = 0, t) = T_{left} = 40$ °C $T(x = 0.3, t) = T_{right} = 20$ °C;
- "Winter" $T(x = 0, t) = T_{left} = -5$ °C $T(x = 0.3, t) = T_{right} = 20$ °C;

Two scenarios are simulated: wall section with TES-foam as insulating layer L1 containing 2% MPCM (T_m: 24); and wall section with EPS as insulating layer L1. Tables 6 and 7 provide the key thermophysical parameters of the considered EPS and TES foam with and without MPCM for the analysis of the thermal behavior of the two-layer wall sections.

Table 6. Thermophysical properties of single components relevant for calculating the volumetric heat capacity of TES foam, both with and without MPCM, using mixture theory.

	Paste [80]	Air	MPCM [80]
Density, ρ (kg/m^3)	1708	1.2	760
Sensible heat capacity, Cp (J/kg/K)	1300	1000	2100
Latent heat storage, h (kJ/kg)	0	0	195

Table 7. Thermophysical properties of composites: model input for simulating one-dimensional (1D) temperature profiles. 'TES-Foam' denotes foam with porosity higher than 52% and without integrated MPCM. * Foam with ε = 0.90, MPCM in solid phase.

	TES-Foam 0% MPCM	TES-Foam 2% MPCM	RC-Wall	EPS [81,82]
Density, ρ (kg/m^3)	200	90	2000	30
Predicted effective conductivity, k_{eff} (W/m/K)	0.07 *	0.05 *	2.1	0.04
Volumetric heat capacity, Cp, sensible (kJ/m^3/K)	223.120	210.632	1760	45
Latent heat storage, h (kJ/m^3)	0	2964	0	0

Using the measured thermophysical properties of single components shown in Table 6, the volumetric heat capacity of TES-foam with and without MPCM can be simply ascertained by applying the mixture law [61]; see Table 7. Density and effective conductivity of TES-foams with 90% porosity are two parameters derived from the presented model in Sections 3 and 4.

Figures 12 and 13 present the transient temperature distribution along the x direction across the two-layer structure, showing the differing responses of EPS and TES-Foam with 2% volumetric amount of MPCM to the heat and cold wave. The initial scenario involves both outside and inside temperatures equal to 20 °C at t_0. Subsequently, a sudden increase/decrease in the outside temperature to 40 °C/−5 °C occurs. The graphs depict the temperature distribution inside the wall at 5 min intervals. In both Figures 12 and 13, T_1 represents the temperature at location ×1, recorded 5 min after the initiation of a temperature increase or decrease, demonstrating a noticeably slower rate of temperature change in response to both heat (Figure 12) and cold (Figure 13) wave stimuli for TES-Foam.

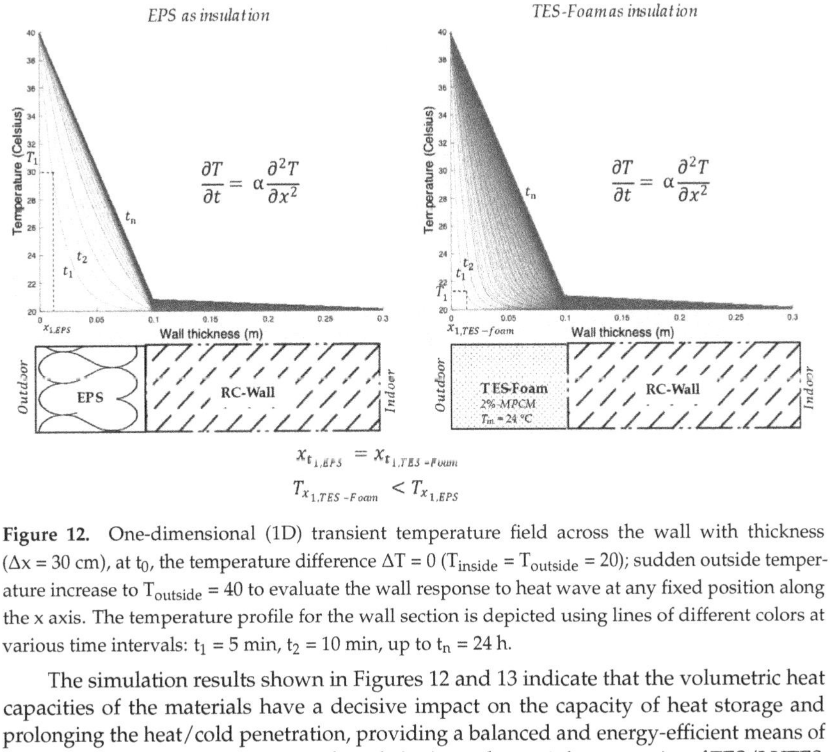

Figure 12. One-dimensional (1D) transient temperature field across the wall with thickness (Δx = 30 cm), at t_0, the temperature difference ΔT = 0 (T_{inside} = $T_{outside}$ = 20); sudden outside temperature increase to $T_{outside}$ = 40 to evaluate the wall response to heat wave at any fixed position along the x axis. The temperature profile for the wall section is depicted using lines of different colors at various time intervals: t_1 = 5 min, t_2 = 10 min, up to t_n = 24 h.

The simulation results shown in Figures 12 and 13 indicate that the volumetric heat capacities of the materials have a decisive impact on the capacity of heat storage and prolonging the heat/cold penetration, providing a balanced and energy-efficient means of maintaining indoor comfort. A significantly higher volumetric heat capacity of TES/LHTES-Foam (223.120 kJ/m^3/K/210.632 kJ/m^3/K) in comparison to EPS (45 kJ/m^3/K) reduces the required storage volume and, at the same time, the structure thickness.

TES-Foam functions as thermal mass or thermal battery, effectively moderating internal temperatures by mitigating day–night temperature fluctuations. When integrated as LHTES-Foam, TES-Foam significantly enhances its thermal capacity through the efficient latent heat storage and controlled release of heat energy. During warmer periods, it adeptly absorbs and stores excess heat, while during cooler periods, it steadily releases the stored heat, contributing to the stabilization and regulation of indoor temperatures. This dual functionality of TES Foam as both a thermal mass and an advanced LHTES material provides a robust and energy-efficient solution for maintaining a consistently comfortable indoor environment, regardless of external weather variations.

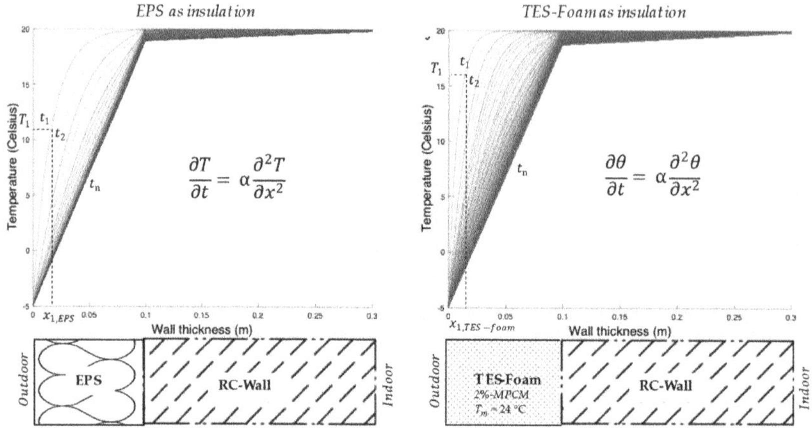

$x_{t1,EPS} = x_{t1,TES\text{-}Foam}$

$T_{x1,TES\text{-}Foam} > T_{x1,EPS}$

Figure 13. One-dimensional (1D) transient temperature field across the wall with thickness (Δx = 30 cm), at t_0, the temperature difference $\Delta T = 0$ ($T_{inside} = T_{outside}$ = 20); sudden outside temperature decrease to $T_{outside} = -5$ to evaluate the wall's response to a cold wave at any fixed position along the x axis. The temperature profile in the wall section is depicted using lines of different colors at various time intervals: t_1 = 5 min, t_2 = 10 min, up to t_n = 24 h.

Table 8 summarizes the calculated thermal diffusivity (α) and thermal effusivity (b) variables of proposed structures to assess the thermal performance under transient boundary conditions and conducts an evaluation of the energy-saving capacities of both structures. Lower thermal diffusivity of LHTES/TES-Foam (0.0237 10^{-5} m^2/s/0.0314 m^2/s), compared to EPS (0.08 10^{-5} m^2/s), provides a much slower response to temperature differences, and decreases the transient heat transfer, or the progression of a temperature wave in the system, extending the time for charging and discharging of heat/cold load and turns, all in a more thermally inert system; also see Figures 12 and 13. This is especially beneficial for reducing over-temperature degree hours in indoor spaces. A much higher thermal effusivity value for LHTES/TES-Foam (102.6 m^2/k/S$^{0.5}$/124.9 m^2/k/S$^{0.5}$ vs. 42.5 m^2/k/S$^{0.5}$ for EPS) leads to a larger amount of heat being stored, which is significant for the effectiveness of internal storage masses.

Table 8. Calculated thermophysical properties relevant for dynamic envelope evaluation; thermal diffusivity $\alpha = \lambda/\rho c_p$ and heat penetration coefficient $b = \sqrt{\lambda \rho c}_p$.

	TES-Foam 0% MPCM	TES-Foam 2% MPCM	RC-Wall	EPS
Thermal diffusivity, α 10^{-5} (m^2/S)	0.0314	0.0237	0.15	0.08
Heat penetration coefficient, b (J/(m^2/k/S$^{0.5}$)	124.9	102.6	1922.5	42.5

6. Conclusions

This paper presents an analytically based approach which combines the effective thermal conductivity of the foam skeleton with dispersed air bubbles, creating a porous media system. The objective is to provide a comprehensive description and precise prediction of the thermal behavior of TES-mineral foams with/without LHTES (incorporation of MPCM), with varying porosity levels, ranging from low to high, with the final goal of facilitating simulations through simple engineering calculations, and without the need to perform expensive experimental tests.

Considering the correlation between porosity and geometry of the dispersed air, through the porous mineral systems, six fitting values of parameters "A" and "Φ_m" were proposed for use in the modified analytical expression proposed by Lewis and Nielsen (1940), based on three distinct foam classes: Class I foams, having polyhedral-shaped air voids and a fraction of air inclusion (ε) greater than or equal to 0.75; Class II foams, having spherical-shaped bubbles and a fraction of air inclusion ranging from 0.75 to 0.52; and lastly, Class III foams, with low porosity and a fraction of air inclusion (ε) below 0.52. The calculated homogenized conductivities were validated by using the proposed fitting parameters and upon the collected experimental data, resulting in a consistently stable response for all foam classes and datasets.

By determining the effective conductivity of TES-mineral foams, with and without latent heat integration, at high porosity of 90%, and by calculating the volumetric heat capacity of the composite using the mixture theory, the potential of LHTES/TES-enhanced mineral foams to be used for thermal energy storage purposes was discussed. In this regard, thermal diffusivity and the heat penetration coefficient of the LHTES/TES-enhanced mineral foams was calculated and compared with conventional EPS. The results showed a 3.5 to 4 times higher value for thermal storage capacity and an almost 70%/60% lower thermal diffusivity for LHTES/TES-enhanced mineral foams, respectively, compared to EPS layers, without considering the latent heat storage amount (h).

Future research directions might aim at exploring new building envelopes that incorporate dynamic thermal performance. This approach differs from conventional solutions, which are based on obsolete concepts of insulation and thermal resistance (R-value) parameters. The main goal will be to synergistically integrate both insulation materials and latent heat thermal storage (LHTS), thus opening new perspectives for prospective novel envelope designs. Research studies regarding their long-term stability, durability, and performance under different environmental conditions would contribute to a comprehensive understanding of their applicability in real-world scenarios.

Supplementary Materials: The following supporting information can be downloaded at: https://www.mdpi.com/article/10.3390/en16186687/s1.

Author Contributions: Conceptualization, M.N.S.; Methodology, M.N.S.; Validation, M.N.S.; Formal analysis, M.N.S.; Investigation, M.N.S.; Data curation, M.N.S.; Writing—original draft, M.N.S.; Writing—review & editing, M.N.S., J.S. and H.V.L.; Visualization, M.N.S.; Supervision, J.S. and H.V.L.; Funding acquisition, J.S. and H.V.L. All authors have read and agreed to the published version of the manuscript.

Funding: We acknowledge support by the Deutsche Forschungsgemeinschaft (DFG, German Research Foundation) and the Open Access Publishing Fund of Technical University of Darmstadt.

Data Availability Statement: The data that support the findings of this study are available in Supplementary Materials. Further data are available from the authors upon reasonable request.

Conflicts of Interest: The authors declare that they have no known competing financial interest or personal relationships that could have appeared to influence the work reported in this paper.

References

1. Miller, S.A.; Habert, G.; Myers, R.J.; Harvey, J.T. Achieving net zero greenhouse gas emissions in the cement industry via value chain mitigation strategies. *One Earth* **2021**, *4*, 1398–1411. [CrossRef]
2. Chen, X.-S.; Li, Y.-P.; Jiang, Y.-L.; Liu, Y.-X.; Zhang, T. Theoretical research on gas seepage in the formations surrounding bedded gas storage salt cavern. *Pet. Sci.* **2022**, *19*, 1766–1778. [CrossRef]
3. Xue, W.; Wang, Y.; Chen, Z.; Liu, H. An integrated model with stable numerical methods for fractured underground gas storage. *J. Clean. Prod.* **2023**, *393*, 136268. [CrossRef]
4. Yang, T.; King, W.P.; Miljkovic, N. Phase change material-based thermal energy storage. *Cell Rep. Phys. Sci.* **2021**, *2*. [CrossRef]
5. Lu, S.; Lin, Q.; Liu, Y.; Yue, L.; Wang, R. Study on thermal performance improvement technology of latent heat thermal energy storage for building heating. *Appl. Energy* **2022**, *323*, 119594. [CrossRef]

6. Hassan, F.; Jamil, F.; Hussain, A.; Ali, H.M.; Janjua, M.M.; Khushnood, S.; Farhan, M.; Altaf, K.; Said, Z.; Li, C. Recent advancements in latent heat phase change materials and their applications for thermal energy storage and buildings: A state of the art review. *Sustain. Energy Technol. Assess.* **2021**, *49*, 101646. [CrossRef]
7. Heim, D.; Pawłowski, M. The methodology of thermal energy management for nearly zero energy buildings. *Period. Polytech. Civ. Eng.* **2019**, *63*, 499–517. [CrossRef]
8. Al-Yasiri, Q.; Szabó, M. Experimental study of PCM-enhanced building envelope towards energy-saving and decarbonisation in a severe hot climate. *Energy Build.* **2023**, *279*, 112680. [CrossRef]
9. Piselli, C.; Prabhakar, M.; de Gracia, A.; Saffari, M.; Pisello, A.L.; Cabeza, L.F. Optimal control of natural ventilation as passive cooling strategy for improving the energy performance of building envelope with PCM integration. *Renew. Energy* **2020**, *162*, 171–181. [CrossRef]
10. Saffari, M.; de Gracia, A.; Fernández, C.; Cabeza, L.F. Simulation-based optimization of PCM melting temperature to improve the energy performance in buildings. *Appl. Energy* **2017**, *202*, 420–434. [CrossRef]
11. Park, J.H.; Berardi, U.; Chang, S.J.; Wi, S.; Kang, Y.; Kim, S. Energy retrofit of PCM-applied apartment buildings considering building orientation and height. *Energy* **2021**, *222*, 119877. [CrossRef]
12. Ryms, M.; Klugmann-Radziemska, E. Possibilities and benefits of a new method of modifying conventional building materials with phase-change materials (PCMs). *Constr. Build. Mater.* **2019**, *211*, 1013–1024. [CrossRef]
13. Marani, A.; Nehdi, M.L. Integrating phase change materials in construction materials: Critical review. *Constr. Build. Mater.* **2019**, *217*, 36–49. [CrossRef]
14. Chihab, Y.; Bouferra, R.; Garoum, M.; Essaleh, M.; Laaroussi, N. Thermal inertia and energy efficiency enhancements of hollow clay bricks integrated with phase change materials. *J. Build. Eng.* **2022**, *53*, 104569. [CrossRef]
15. Sawadogo, M.; Duquesne, M.; Belarbi, R.; Hamami, A.E.A.; Godin, A. Review on the integration of phase change materials in building envelopes for passive latent heat storage. *Appl. Sci.* **2021**, *11*, 9305. [CrossRef]
16. Ben Romdhane, S.; Amamou, A.; Ben Khalifa, R.; Saïd, N.M.; Younsi, Z.; Jemni, A. A review on thermal energy storage using phase change materials in passive building applications. *J. Build. Eng.* **2020**, *32*, 101563. [CrossRef]
17. Ascione, F.; Bianco, N.; De Masi, R.F.; de' Rossi, F.; Vanoli, G.P. Energy refurbishment of existing buildings through the use of phase change materials: Energy savings and indoor comfort in the cooling season. *Appl. Energy* **2014**, *113*, 990–1007. [CrossRef]
18. Du, K.; Calautit, J.; Wang, Z.; Wu, Y.; Liu, H. A review of the applications of phase change materials in cooling, heating and power generation in different temperature ranges. *Appl. Energy* **2018**, *220*, 242–273. [CrossRef]
19. Drissi, S.; Mo, K.H.; Falchetto, A.C.; Ling, T.-C. Understanding the compressive strength degradation mechanism of cement-paste incorporating phase change material. *Cem. Concr. Compos.* **2021**, *124*, 104249. [CrossRef]
20. Bao, X.; Tian, Y.; Yuan, L.; Cui, H.; Tang, W.; Fung, W.; Qi, H. Development of high performance PCM cement composites for passive solar buildings. *Energy Build.* **2019**, *194*, 33–45. [CrossRef]
21. Kishore, R.A.; Bianchi, M.V.; Booten, C.; Vidal, J.; Jackson, R. Optimizing PCM-integrated walls for potential energy savings in U.S. Buildings. *Energy Build.* **2020**, *226*, 110355. [CrossRef]
22. Yang, Y.K.; Kim, M.Y.; Chung, M.H.; Park, J.C. PCM cool roof systems for mitigating urban heat island—An experimental and numerical analysis. *Energy Build.* **2019**, *205*, 109537. [CrossRef]
23. Abouebeid, S.; Hany, N.; Mosaad, G. Towards enhancing indoor environmental quality: A case of smart adaptive building envelopes in a coastal city. *F1000Research* **2022**, *11*, 731. [CrossRef]
24. Marszal-Pomianowska, A.J.; Johra, H.; Madsen, H.; Kazmi, H.S.; Knotzer, A.; Salom, J.; Lopes, R. Principles of Energy Flexible Buildings: Energy in Buildings and Communities Programme. 2020. Available online: https://www.annex67.org/media/1918/principles-of-energy-flexible-buildings.pdf (accessed on 15 August 2023).
25. Hekimoğlu, G.; Nas, M.; Ouikhalfan, M.; Sarı, A.; Tyagi, V.; Sharma, R.; Kurbetci, S.; Saleh, T.A. Silica fume/capric acid-stearic acid PCM included-cementitious composite for thermal controlling of buildings: Thermal energy storage and mechanical properties. *Energy* **2020**, *219*, 119588. [CrossRef]
26. Cabeza, L.F.; Navarro, L.; Pisello, A.L.; Olivieri, L.; Bartolomé, C.; Sánchez, J.; Álvarez, S.; Tenorio, J.A. Behaviour of a concrete wall containing micro-encapsulated PCM after a decade of its construction. *Sol. Energy* **2020**, *200*, 108–113. [CrossRef]
27. Omrany, H.; Chang, R.; Soebarto, V.; Zhang, Y.; Ghaffarianhoseini, A.; Zuo, J. A bibliometric review of net zero energy building research 1995–2022. *Energy Build.* **2022**, *262*, 111996. [CrossRef]
28. Posani, M.; Veiga, M.D.R.; de Freitas, V.P. Towards resilience and sustainability for historic buildings: A review of envelope retrofit possibilities and a discussion on hygric compatibility of thermal insulations. Thermal Insulations. *Int. J. Arch. Heritage* **2019**, *15*, 807–823. [CrossRef]
29. Gilka-Bötzow, A.; Folino, P.; Maier, A.; Koenders, E.A.B.; Caggiano, A. Triaxial failure behavior of highly porous cementitious foams used as heat insulation. *Processes* **2021**, *9*, 1373. [CrossRef]
30. Vaou, V.; Panias, D. Thermal insulating foamy geopolymers from perlite. *Miner. Eng.* **2010**, *23*, 1146–1151. [CrossRef]
31. Modesti, M.; Lorenzetti, A.; Besco, S. Influence of nanofillers on thermal insulating properties of polyurethane nanocomposites foams. *Polym. Eng. Sci.* **2007**, *47*, 1351–1358. [CrossRef]
32. Gonçalves, M.C.; Margarido, F. *Materials for Construction and Civil Engineering*; Springer: Cham, Switzerland, 2015.
33. Arzumanyan, A. Technological peculiarities of non autoclaved foam concrete production on the base of volcanic pumice aggregates. In Materials Science. *Mater. Sci. Forum* **2019**, *974*, 206–210. [CrossRef]

34. Puga, H.; Carneiro, V.; Jesus, C.; Pereira, J.; Lopes, V. Influence of particle diameter in mechanical performance of Al expanded clay syntactic foams. *Compos. Struct.* **2018**, *184*, 698–703. [CrossRef]
35. Gao, H.; Liu, H.; Liao, L.; Mei, L.; Lv, G.; Liang, L.; Zhu, G.; Wang, Z.; Huang, D. Improvement of performance of foam perlite thermal insulation material by the design of a triple-hierarchical porous structure. *Energy Build.* **2019**, *200*, 21–30. [CrossRef]
36. Rugele, K.; Lehmhus, D.; Hussainova, I.; Peculevica, J.; Lisnanskis, M.; Shishkin, A. Effect of fly-ash cenospheres on properties of clay-ceramic syntactic foams. *Materials* **2017**, *10*, 828. [CrossRef] [PubMed]
37. Shishkin, A.; Bumanis, G.; Irtiseva, K.; Ozolins, J.; Korjakins, A. Clay Ceramic Hollow Sphere—Cement Syntactic Foam Composite for Building Applications. *Key Eng. Mater.* **2019**, *800*, 228–234. [CrossRef]
38. Fernando, P.; Jayasinghe, M.; Jayasinghe, C. Structural feasibility of Expanded Polystyrene (EPS) based lightweight concrete sandwich wall panels. *Constr. Build. Mater.* **2017**, *139*, 45–51. [CrossRef]
39. Maxwell, J.C. *A Treatise on Electricity and Magnetism Dover Publications*, Unabriged 3rd ed.; Dover Publications, Inc.: New York, NY, USA, 1954; Volume 1.
40. Wellander, N. Homogenization of the Maxwell Equations: Case II. Nonlinear Conductivity. *Appl. Math.* **2002**, *47*, 255–283. [CrossRef]
41. Angell, T.S.; Kirsch, A. The conductive boundary condition for Maxwell's equations. *SIAM J. Appl. Math.* **1992**, *52*, 1597–1610. [CrossRef]
42. Nemilov, S.V. The review of possible interrelations between ionic conductivity, internal friction and the viscosity of glasses and glass forming melts within the framework of Maxwell equations. *J. Non-Crystalline Solids* **2011**, *357*, 1243–1263. [CrossRef]
43. Goncharenko, A.V. Generalizations of the Bruggeman equation and a concept of shape-distributed particle composites. *Phys. Rev. E* **2003**, *68*, 041108. [CrossRef]
44. Tjaden, B.; Cooper, S.J.; Brett, D.J.; Kramer, D.; Shearing, P.R. On the origin and application of the Bruggeman correlation for analysing transport phenomena in electrochemical systems. *Curr. Opin. Chem. Eng.* **2016**, *12*, 44–51. [CrossRef]
45. Liu, H.; Zhang, G.; Li, D.; Wang, C.; Bai, S.; Li, G.; Wang, G. Three-dimensional multi-phase simulation of cooling patterns for proton exchange membrane fuel cell based on a modified Bruggeman equation. *Appl. Therm. Eng.* **2020**, *174*, 115313. [CrossRef]
46. Myers, M.T. Pore Modeling: Extending the Hanai-Bruggeman Equation. In Proceedings of the SPWLA 30th Annual Logging Symposium, Denver, CO, USA, 11–14 June 1989.
47. Sanjaya, C.S. Pore Size Effect on Heat Transfer through Porous Medium. Ph.D. Thesis, National University of Singapore, Singapore, 2011.
48. Skibinski, J.; Cwieka, K.; Ibrahim, S.H.; Wejrzanowski, T. Influence of pore size variation on thermal conductivity of open-porous foams. *Materials* **2019**, *12*, 2017. [CrossRef] [PubMed]
49. Chen, J.; Wang, H.; Xie, P.; Najm, H. Analysis of thermal conductivity of porous concrete using laboratory measurements and microstructure models. *Constr. Build. Mater.* **2019**, *218*, 90–98. [CrossRef]
50. Lewis, T.B.; Nielsen, L.E. Dynamic mechanical properties of particulate-filled composites. *J. Appl. Polym. Sci.* **1970**, *14*, 1449–1471. [CrossRef]
51. Nielsen, L.E. The thermal and electrical conductivity of two-phase systems. *Ind. Eng. Chem. Fundamentals* **1974**, *13*, 17–20. [CrossRef]
52. Pal, R. On the Lewis-Nielsen model for thermal/electrical conductivity of composites. *Compos. Part A Appl. Sci. Manuf.* **2008**, *39*, 718–726. [CrossRef]
53. Kochetov, R.; Korobko, A.V.; Andritsch, T.; Morshuis, P.H.F.; Picken, S.J.; Smit, J.J. Three-phase lewis-nielsen model for the thermal conductivity of polymer nanocomposites. In Proceedings of the 2011 Annual Report Conference on Electrical Insulation and Dielectric Phenomena, Cancun, Mexico, 16–19 October 2011; pp. 338–341. [CrossRef]
54. Montes-De-Oca, L.M.; Medina-Esquivel, R.; Zambrano-Arjona, M.; Martínez-Torres, P. Thermal detection of second critical micelle concentration in SDS and CTAB aqueous solutions using a modified Lewis-Nielsen effective thermal model. *J. Mol. Liq.* **2022**, *363*, 119864. [CrossRef]
55. Pietrak, K.; Wiśniewski, T.S. A review of models for effective thermal conductivity of composite materials. *J. Power Technol.* **2015**, *95*, 14–24.
56. Wang, X.J.; Niu, X.H.; Qiu, X.W.; Istikomah, N.; Wang, L.B. Thermal conductivity of porous polymer materials considering pore special-shape and anisotropy. *Express Polym. Lett.* **2021**, *15*, 319–328. [CrossRef]
57. Javierre, E.; Vuik, C.; Vermolen, F.; van der Zwaag, S. A comparison of numerical models for one-dimensional Stefan problems. *J. Comput. Appl. Math.* **2006**, *192*, 445–459. [CrossRef]
58. Visintin, A. *Models of Phase Transitions*; Springer Science & Business Media: Berlin/Heidelberg, Germany, 1996; Volume 26.
59. Gupta, S.C. *The Classical Stefan Problem: Basic Concepts, Modelling and Analysis with Quasi-Analytical Solutions and Methods*; Elsevier: Amsterdam, The Netherlands, 2017; Volume 45.
60. Caggiano, A.; Mankel, C.; Koenders, E. Reviewing theoretical and numerical models for PCM-embedded cementitious composites. *Buildings* **2018**, *9*, 3. [CrossRef]
61. Mankel, C.; Caggiano, A.; König, A.; Schicchi, D.S.; Sam, M.N.; Koenders, E. Modelling the thermal energy storage of cementitious mortars made with PCM-recycled brick aggregates. *Materials* **2020**, *13*, 1064. [CrossRef] [PubMed]
62. Voller, V.R.; Swaminathan, C.R.; Thomas, B.G. Fixed grid techniques for phase change problems: A review. *Int. J. Numer. Methods Eng.* **1990**, *30*, 875–898. [CrossRef]

63. Šavija, B.; Zhang, H.; Schlangen, E. Influence of microencapsulated phase change material (PCM) addition on (micro) mechanical properties of cement paste. *Materials* **2017**, *10*, 863. [CrossRef]
64. Dobri, A.; Tsiantis, A.; Papathanasiou, T.; Wang, Y. Investigation of transient heat transfer in multi-scale PCM composites using a semi-analytical model. *Int. J. Heat Mass Transf.* **2021**, *175*, 121389. [CrossRef]
65. Šavija, B.; Schlangen, E. Use of phase change materials (PCMs) to mitigate early age thermal cracking in concrete: Theoretical considerations. *Constr. Build. Mater.* **2016**, *126*, 332–344. [CrossRef]
66. Fachinotti, V.; Peralta, I.; Toro, S.; Storti, B.; Caggiano, A. Automatic Generation of High-Fidelity Representative Volume Elements and Computational Homogenization for the Determination of Thermal Conductivity in Foamed Concretes. 2022. Available online: https://ssrn.com/abstract=4286262 (accessed on 15 August 2023).
67. Batool, F. Effect of Microstructure on Thermal Conductivity of Cement-Based Foam. Ph.D. Thesis, University of Alberta Libraries, Edmonton, AB, Canada, 2015.
68. Awang, H.; Mydin, M.A.O.; Roslan, A.F. Effect of additives on mechanical and thermal properties of lightweight foamed concrete. *Adv. Appl. Sci. Res.* **2012**, *3*, 3326–3338.
69. Wei, S.; Yiqiang, C.; Yunsheng, Z.; Jones, M. Characterization and simulation of microstructure and thermal properties of foamed concrete. *Constr. Build. Mater.* **2013**, *47*, 1278–1291. [CrossRef]
70. Mydin, M.A.O. Effective thermal conductivity of foamcrete of different densities. *Concr. Res. Lett.* **2011**, *2*, 181–189.
71. Davraz, M.; Kilinçarslan, Ş.; Koru, M.; Tuzlak, F. Investigation of relationships between ultrasonic pulse velocity and thermal conductivity coefficient in foam concretes. *Acta Phys. Pol. A* **2016**, *130*, 469–470. [CrossRef]
72. Oren, O.H.; Gholampour, A.; Gencel, O.; Ozbakkaloglu, T. Physical and mechanical properties of foam concretes containing granulated blast furnace slag as fine aggregate. *Constr. Build. Mater.* **2019**, *238*, 117774. [CrossRef]
73. Jiang, J.; Lu, Z.; Niu, Y.; Li, J. Investigation of the properties of high-porosity cement foams containing epoxy resin. *Constr. Build. Mater.* **2017**, *154*, 115–122. [CrossRef]
74. Jiang, J.; Lu, Z.; Niu, Y.; Li, J.; Zhang, Y. Investigation of the properties of high-porosity cement foams based on ternary Portland cement–metakaolin–silica fume blends. *Constr. Build. Mater.* **2016**, *107*, 181–190. [CrossRef]
75. Neville, A.M.; Brooks, J.J. *Concrete Technology*; Longman Scientific & Technical: New York, NY, USA, 1987; Volume 438.
76. Tran, N.P.; Nguyen, T.N.; Ngo, T.D.; Le, P.K.; Le, T.A. Strategic progress in foam Stabilization towards high-performance foam concrete for building sustainability: A state-of-the-art review. *J. Clean. Prod.* **2022**, *375*, 133939. [CrossRef]
77. Kern, T. Neuses Verfahren zur experimentellen Untersuchung wässriger Schäume. Ph.D Thesis, University of Paderborn, Paderborn, Germany, 2002.
78. Lohaus, L.; Pott, J.U. *Konstruktionsleichtbeton unter Verwendung Vorgeschäumter Luftporen*; Forschungsbericht; Institut für Baustoffe, Universität Hannover: Hanover, Germany, 2005; Available online: https://www.baustoff.uni-hannover.de/de/forschung/forschungsprojekte/forschungsprojekte-detailansicht/projects/konstruktionsleichtbeton-unter-verwendung-vorgeschaeumter-luftporen/ (accessed on 15 August 2023).
79. Pott, J.U. Entwicklungsstrategien für zementgebundene Schäume. IfB. 2006. Available online: https://www.baustoff.uni-hannover.de/fileadmin/baustoff/publications/Heft05-Pott-ZemSchaum.pdf (accessed on 15 August 2023).
80. Sam, M.; Caggiano, A.; Dubyey, L.; Dauvergne, J.-L.; Koenders, E. Thermo-physical and mechanical investigation of cementitious composites enhanced with microencapsulated phase change materials for thermal energy storage. *Constr. Build. Mater.* **2022**, *340*, 127585. [CrossRef]
81. Available online: https://issuu.com/detail-magazine/docs/978-3-920034-18-8_bk_de_daemmstoffe (accessed on 15 August 2023).
82. Schild, K. *Wärmeschutz: Grundlagen-Berechnung-Bewertung*; Springer: Berlin/Heidelberg, Germany, 2013.

energies

MDPI

Article

Examining the Impact of Multilevel Courtyards in Hot-Dry and Humid Climates

Eleazar Chidiadi and Ahmad Taki *

Leicester School of Architecture, De Montfort University, Leicester LE1 9BH, UK; p2772882@alumni365.dmu.ac.uk
* Correspondence: ahtaki@dmu.ac.uk

Abstract: Urbanisation has significantly transformed human settlements, presenting sustainability challenges, particularly in hot-dry and humid climates. The urban heat island effect and increased energy consumption exacerbate reliance on mechanical cooling and fossil fuels. As climate change escalates, developing sustainable architectural solutions that improve thermal performance and energy efficiency becomes crucial. This study examines the effects of various multilevel courtyard designs on building performance in Abuja, Nigeria, highlighting gaps in applying traditional principles to these models. A mixed-method approach, combining quantitative and qualitative techniques, assesses user perceptions, thermal performance, energy efficiency, and daylighting in multilevel courtyards. Findings indicate that optimised multilevel courtyard configurations yield a 2.15 °C reduction in temperature, enhancing indoor thermal comfort and improving natural ventilation. Users favour multilevel courtyard housing; however, challenges include inadequate daylighting on lower levels and the need for shading solutions. Compressed earth blocks exhibit better thermal performance, reducing peak temperatures by 1.19 °C compared to hollow concrete blocks. Guidelines for architects and urban planners are provided, as well as recommendations for future research on policy incentives to promote multilevel courtyard models.

Keywords: courtyards; multilevel typology; hot-dry and humid; Abuja; simulations; energy efficiency; thermal comfort; case studies; design guidelines; sustainability

1. Introduction

Courtyards are integral open spaces, either entirely or partially enclosed by surrounding structures, designed to optimise sunlight, natural ventilation, and environmental connectivity [1]. They facilitate privacy while maintaining a bond with the external environment [2] and are particularly prevalent in hot, arid climates, where they serve to mitigate harsh outdoor conditions [3]. Historical evidence indicates that courtyards have been in use for over 4000 years, with origins traced back to ancient Egyptian and Mesopotamian settlements [2]. Architectural forms, such as the light well and courtyard of Chinese homes, illustrate philosophical ideas regarding the relationship between the celestial and the terrestrial. The modern design of courtyard houses has roots in diverse cultures, including the early rectangular dwellings in Morocco and nomadic shelters in West Africa [4]. In contexts like Nigeria, courtyard typology reflects not only climatic conditions but also cultural and religious imperatives [5]. The preservation of Nigeria's architectural heritage necessitates the incorporation of climate-responsive and passive design strategies in modern courtyard developments [5]. Courtyards promote familial interactions and connections to nature [6] and serve as secure spaces for vulnerable populations, especially in conservative societies [7]. Their design often includes elements such as fountains and trees, which provide sensory relief from urban congestion and underscore the significance of courtyards in urban life. Additionally, the therapeutic aspects of courtyards contribute to health benefits [5]. Historically, they have supported various family activities, including cooking, dining, and gardening [8]. In conclusion, contemporary architectural practices should reinforce the integration of courtyards, emphasising climate-responsive designs while honouring cultural heritage, given their historical relevance and potential to enhance well-being.

In hot and humid climates, courtyards are essential for enhancing building sustainability and performance. Their flexible design allows them to harness natural elements for comfort and efficiency, enabling more effective environmental solutions in these regions.

Courtyards play a vital role in enhancing outdoor comfort and facilitating natural ventilation, which can diminish reliance on air conditioning. Research on traditional courtyard houses in the Tunisian Sahara [9] revealed that indoor temperatures remained around 27 °C despite external temperatures reaching 49 °C, owing to high thermal mass and small windows. The stack effect, as shown in Figure 1, contributes to thermal comfort by enabling cool breezes, particularly at night when courtyards retain cool air, which then circulates into surrounding rooms. The effectiveness of a courtyard is influenced by its depth, elongation, and sky exposure. Ref. [10] introduces the "malqaf" (wind catcher) as a passive ventilation method that captures cooler winds from upper layers and directs them into buildings, also mitigating dust intrusion and serving as an emergency exit during high winds. While night-time cooling strategies, as shown in Figure 1, particularly for buildings with high thermal mass, are well supported [11], the ideal ventilation methods for hot and humid regions remain contentious.

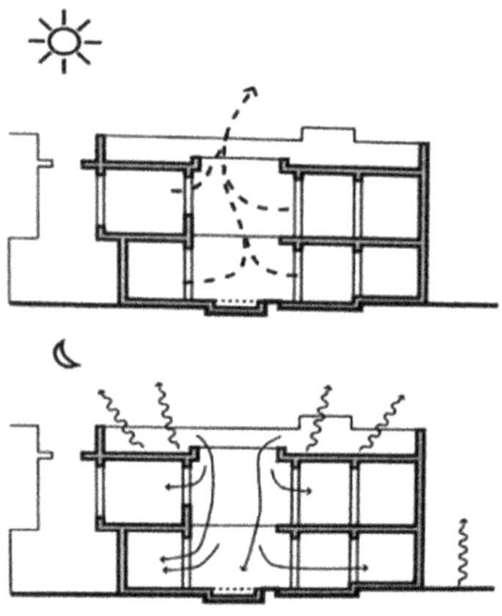

Figure 1. Stack effects during daytime and night-time ventilation—dashed and curved arrows indicate air flow, wavy or zig-zag arrows indicate radiative cooling. (source: [11]).

While some studies question the necessity of daytime ventilation, the authors of [11] argue that integrating night ventilation and sloped roofs can effectively lower indoor temperatures. In [12], computational fluid dynamics (CFD) are employed to analyse courtyard thermal performance, introducing the night-time effectiveness ratio (NTER) to evaluate the interplay between solar radiation and night cooling. Their findings suggest that courtyard width and shading elements significantly affect night ventilation effectiveness, with homes featuring galleries experiencing 30–60% improvements in thermal performance. Nevertheless, while the research provides recommendations for optimal courtyard dimensions and component ratios, it does not account for other influential factors such as wind catchers, vegetation, and evaporative cooling.

Courtyards have endured over millennia due to their ability to create thermally comfortable living spaces by mitigating peak temperatures, channelling breezes, and adjusting humidity levels [9]. The concept of thermal comfort is intertwined with architectural design, microclimatic performance, and traditional housing, particularly in hot and dry or humid climates. Ref. [8] explores the transition from traditional to non-traditional housing in Iran, emphasising the microclimatic advantages of central courtyards featuring gardens, paved areas, and water pools that foster cooler, more humid microenvironments. The vegetation and water features contribute to thermal comfort by reducing local temperatures through evapotranspiration and shading from deciduous trees. Research indicates that shaded areas in courtyards can maintain temperatures up to 3 K cooler [8]. However, the introduction of evaporative coolers in the 1960s altered thermal comfort dynamics, providing efficient cooling at a lower energy cost, albeit with drawbacks like elevated humidity and noise. Ref. [13] elucidates the significance of courtyards in regulating microclimatic conditions, revealing that orientation and enclosure height critically influence air temperature and humidity levels. While this study demonstrates the effectiveness of simulation tools such as IES<VE> for optimising thermal performance, it inadequately addresses boundary conditions and fenestration. Supporting this perspective, Refs. [14,15] recognise that variables like courtyard proportions, wall height, and surface-to-volume ratios impact thermal performance. Effective design is paramount for enhancing thermal comfort in unshaded courtyards, particularly in hot and humid climates, with simulations utilising software like ENVI-met providing insights into physiological equivalent temperature (PET) and other thermal conditions. Emphasising the integration of vegetation, water, shading devices, and natural ventilation appears crucial for improving outdoor comfort in urban courtyards. Additionally, an inclusive review of diverse courtyard configurations and climates is necessary for advancing the understanding of thermal behaviour in these spaces, with findings indicating that unshaded courtyards typically offer comfort only during early morning and evening hours [15].

The current body of literature on multilevel courtyard typologies in hot-dry and humid climates is markedly insufficient, particularly concerning the comparative performance metrics of multilevel courtyards relative to single-level counterparts. Existing research has not adequately examined the effects of multilevel designs on essential performance indicators such as lighting, shading, and social interaction. Additionally, there is a significant lack of proposed design guidelines specifically catering to multilevel courtyards in these regions, and few empirical studies have explored user perceptions and the quantitative benefits compared to conventional configurations. Consequently, there is a pressing need for comprehensive research focused on physical prototyping, innovative construction methods, and cost analyses tailored to the unique requirements of multilevel courtyard developments. Courtyards have long been a bioclimatic design strategy in hot and humid regions, enhancing thermal performance and outdoor comfort. Key strategies include integrating vegetation, water features, shading devices, and natural ventilation. They reduce noise pollution, improve safety and security, and provide suitable lighting, all contributing to social sustainability. Furthermore, courtyards encourage natural healing and should be integrated into modern architecture to promote environmental sustainability. However, there is limited knowledge on adapting enduring passive design principles to next-generation, multilevel courtyard models that address contemporary density and climate change challenges. Future research should focus on optimising configurations that prioritise human-centred factors for sustainable cooling and ventilation in urbanised tropical regions.

Therefore, the following specific objectives can be identified:

- To assess user perception and the feasibility of multilevel courtyards as a next-generation sustainable building model for increasingly dense urban areas in hot-dry and humid contexts;
- To analyse the impact of multilevel courtyard configuration variables (e.g., shading, proportions, height, and building material) on daylighting, energy efficiency, and thermal comfort in hot-dry/humid climates compared to single-level courtyards;
- To provide suitable climate-specific multilevel courtyard design guidelines and recommendations that balance environmental performance with cultural preferences.

2. Context and Background Knowledge

The courtyard is a vital passive architectural element in traditional buildings, reflecting the socio-cultural needs of its users. Its design varies across geographical locations while sharing common characteristics. In the Yoruba community of southwestern Nigeria, courtyards serve as central spaces for social interactions, mentorship, and community activities, significantly contributing to social cohesion and crime mitigation [16]. Similarly, in Indian courtyard houses, courtyards are essential for hosting ceremonies and festivals, providing an open environment for social rituals [17]. Social hierarchies and gender roles also influence courtyard arrangements. In Chinese courtyard houses, the orientation of east and west wing halls correlates with cardinal directions and social hierarchy, with specific room rankings based on cultural implications and thermal comfort [17]. Additionally, enclosed courtyards mitigate noise pollution and enhance security by centralising building openings [9]. Research on Iranian traditional four-season courtyard houses in arid regions highlights the relationship between daylighting design strategies and social sustainability. It suggests that well-designed central spaces facilitate social interaction by providing quality environments, thereby promoting communal living [18]. Despite their success in fostering social cohesion, traditional structures may struggle to meet contemporary construction demands [19].

2.1. Daylighting and Shading

In hot climates, larger southern courtyards enhance natural lighting, though they increase solar radiation, which raises temperatures [9]. A case study evaluating various courtyard shapes revealed that a rectangular design with a 2:1 ratio optimises solar protection in summer and heat gain in winter. Ref. [20] investigated how sunlight influences residential occupancy in three regions of Algeria—M'chounech, El-Oued, and Ghardaia—considering climate, culture, and architecture. Findings indicate that natural light significantly influences domestic activities and spatial organisation, particularly in El-Oued, where traditional one-storey, rectangular homes feature sandy courtyards and rooms orientated inward, relying solely on courtyard access for natural light. The spaces in El-Oued are configured based on sunlight exposure, with multipurpose areas on the southern and northern sides, while eastern and western sides house utilities like toilets and stables. The use of shaded areas, or sabats, varies seasonally. While the authors of [20] argue that traditional homes were designed to harmonise with sunlight, contemporary buildings often overlook this relationship, focusing instead on maximising daylight [11]. In hot-dry climates, building orientation prioritises shading, whereas in hot-humid climates, controlling sky exposure and courtyard roofs is more effective.

2.2. Vegetation and Water

Courtyards can effectively moderate indoor temperatures in hot climates by incorporating water and vegetation and utilising shading and evaporative cooling [10,21]. In contrast, temperate climates rely less on solar energy for temperature regulation, leading to similar courtyard dimensions predominantly used for aesthetic purposes [9]. Ref. [22] demonstrated that greenery provides significant environmental benefits, reducing indoor temperatures by up to 3.4 °C through shade and evapotranspiration, and with combined strategies like window replacement and plant integration, the reduction can reach 4.8 °C. Ref. [23] further indicated that plant evapotranspiration mitigates temperature more effectively than suggested by simulations. Building on this, Ref. [24] analysed various vegetation scenarios in a tropical urban area, finding that interventions such as curbside trees and green roofing/walls can lead to temperature reductions of up to 1.9 °C compared to traditional urban heat island conditions. This underscores the critical role of vegetation in enhancing thermal comfort in urban environments.

2.3. Construction Materials

Vernacular architecture employs local building materials like stone, brick, wood, and palm trunks, which are environmentally sustainable due to their low embodied energy and toxicity levels [25]. These materials adapt well to local climates, providing comfortable indoor environments and supporting energy efficiency through potential recycling. Thick walls constructed from brick and stone serve as effective thermal insulators, ensuring internal privacy and enhancing family protection, particularly on ground floors. This choice of materials also embodies principles of modesty and social equality [26]. Ref. [27] notes that traditional vernacular buildings utilised thick walls and roofs, often covered in mud and mortar from local limestone, which enhanced heat resistance. The post-pandemic era has prompted an intensified focus on research pertaining to the incorporation of biobased resources within the construction sector [28]. This trend is facilitating significant advancements in the repurposing of waste materials into sustainable building materials, aligning with contemporary goals for reducing environmental impact and enhancing resource efficiency in construction practices. Ref. [29] explores the use of cardboard as a sustainable construction material by highlighting its potential for structural applications. This approach not only aids in waste reduction but also fosters the creation of innovative and eco-friendly building materials that benefit both the environment and the industry. However, issues such as cardboard deformation arose four months after construction, necessitating repairs to the support tubes. Additionally, the material's sensitivity to atmospheric moisture was noted, prompting a recommendation for the installation of a waterproof barrier. Ref. [30] investigated bio-based building materials created using manufacturing waste binder and filler materials derived from the production of wood-wool cement boards. The findings indicated that the developed bio-building materials may be suitable for non-load-bearing applications, such as interior wall partitions and ceiling tiles, making them a potential option for low-stress construction projects. Additionally, the bio-based building material can function as an intermediary layer consisting of reactivated binder and recovered wood fibres for insulation in walls, roofs, and floors. Further testing and evaluation are recommended to assess the material's suitability for specific applications. In Nigeria's hot and dry climates, Refs. [21,31] highlight the significance of mud bricks, where they provide a favourable indoor climate without mechanical cooling, thus lowering operating costs while improving air quality and thermal comfort. Nonetheless, research by [32] underscores the limited durability of mud bricks, which typically have a lifespan of only seven years due to vulnerability from nearby quarry operations, necessitating frequent replacement. Ref. [10] further supports this view, indicating that mud is a traditional building material, but it may not always meet modern durability expectations. Ref. [31] emphasises the significance of utilising earth materials to attain thermal comfort in buildings and underscores the potential of compressed stabilised earth blocks (CSEBs) in promoting sustainable construction practices from a quantitative perspective. The particle size distribution analysis indicates that the average coefficient of thermal conductivity (k avg) for the NBRRI CSEB is 0.4841 W/mK. This points to a need for more comprehensive quantitative analyses to assess CSEB materials' energy efficiency and thermal performance.

2.4. Energy Efficiency

The energy efficiency of buildings and the mitigation of climate change are critical global issues, underscoring the importance of sustainable design practices. Courtyard designs, characterised by open-air areas enclosed by walls or buildings, can significantly reduce energy consumption in residential structures, thus lowering CO_2 emissions [9]. A recent study by [33] emphasises the influence of building geometry on cooling energy demand, particularly in the context of climate change and the urban heat island effect. The study conducted simulations for three case studies in Seville, Spain, known for extreme summer temperatures. Simulation A treated the courtyard as an outdoor space, ignoring its effects. Simulation B accounted for the shading effect from the courtyard's geometry, while Simulation C included the thermal values within the courtyard. The findings indicated potential cooling demand reductions of 8–18%, depending on the courtyard design, with additional impacts from the floor level and orientation of adjacent rooms.

In the context of the Hebei region in China, Ref. [34] explored the emerging trend of courtyard roofing technologies (CBRTs). Their analysis showed that T-CBRT and NT-CBRT could decrease energy consumption in open courtyards by approximately 19.97% and 15.41%, respectively, without necessitating upgrades to the existing building envelope. CBRT enhances solar energy collection and heat storage while reducing heat loss and shortening heating duration. However, it compromises traditional passive cooling benefits, emphasising the need for a balance between reducing heating demands and mitigating cooling burdens. Courtyards and atriums, integral to architectural design, provide private, tranquil spaces for communal interaction. The debate over whether courtyards should be covered or left open is ongoing. Ref. [35] compared the energy efficiency of central atriums and courtyards of equivalent proportions through computer simulations, assessing their respective energy demands and consumption. Further research is necessary to evaluate these energy-efficient technologies' economic feasibility, maintenance needs, and payback periods.

The research presented in [36,37], as cited by [38], advocates for the potential of AI-driven generative design to swiftly assess a range of design alternatives while considering various factors. This approach aims to reduce embodied carbon through the iterative process of generating and evaluating designs. AI methodologies can identify optimal solutions that significantly mitigate environmental impacts compared to traditional methods. Additionally, AI simulation tools are vital in evaluating building performance during the initial stages of the design process. By modelling key aspects such as energy consumption and indoor environmental quality, these tools empower designers to make well-informed decisions that emphasise sustainability while maintaining functionality and comfort. Ref. [39] conducted a comparative analysis of energy efficiency in buildings with a focus on atriums and courtyards, utilising EnergyPlus and Design Builder to assess the performance of a three-story courtyard structure across three distinct Turkish climates: cold (Erzurum), mild and humid (Istanbul), and hot and dry (Diyarbakir). The study examined factors such as weather conditions, courtyard form, building height and depth, construction layers, glazing types, and window-to-wall ratios. Simulation results indicated that a courtyard building with triple glazing achieved optimal performance, with both double and triple glazing contributing to significant energy savings; however, maximum energy waste for atriums occurred in June. Although atriums demonstrated superior energy conservation in winter, courtyards outperformed them in energy efficiency during the months of May to October due to the greenhouse effect. The findings suggest that employing a hybrid approach—utilising courtyards in summer and atriums in colder months—could enhance energy savings across varying climates, highlighting the potential need for retractable building designs. Further research is advised to validate these outcomes.

2.5. Abuja's Climate

Residential buildings in urban Nigeria, particularly Abuja, face significant thermal discomfort due to high indoor temperatures [40]. The urgent demand for housing has prioritised quantity over quality, leading to substandard conditions and a reliance on mechanical cooling systems, which account for approximately 53.3% of national electricity usage [41]. This situation is exacerbated by intermittent power supply and the increasing use of generators, thus compounding sustainability challenges. The implications of climate change necessitate a reassessment of architectural practices and the exploration of sustainable solutions. Abuja, spanning approximately 8000 km^2 at latitude 9°03′ N and longitude 7°29′ E, has a Köppen climate classification, as adapted from [42] and shown in Table 1, which reveals Abuja's tropical wet and dry climate, with average temperatures around 30 °C in March and 25 °C in August, and annual rainfall of 122.12 mm (4.81 inches) distributed over 145.37 days [42]. The city's population was 4,026,000 in 2024, reflecting a 4.84% increase from the previous year, and the city is characterised by a rich multicultural environment that integrates traditional and modern influences [43].

Table 1. Climate data of Abuja, Nigeria—the Köppen–Geiger classification (adapted from: [42]).

Month	Jan	Feb	Mar	Apr	May	Jun	Jul	Aug	Sep	Oct	Nov	Dec	Year
Record high °C	42.0	45.0	45.0	45.0	42.0	39.0	36.0	34.0	35.0	37.0	40.0	41.0	45.0
Average high °C	37.77	39.47	39.84	38.57	35.43	32.84	29.82	28.49	29.68	32.53	36.24	36.58	34.77
Daily mean °C	30.67	32.39	33.52	33.22	30.96	28.81	26.56	25.53	26.33	28.31	30.63	30.28	29.77
Average low °C	20.81	22.5	24.61	25.77	24.8	23.17	22.07	21.52	21.93	22.62	22.76	21.39	22.83
Record low °C	15.0	16.0	17.0	20.0	22.0	18.0	19.0	19.0	20.0	19.0	19.0	17.0	15.0
Average precipitation mm	0.22	4.7	18.33	68.03	167.49	133.23	315.46	376.68	280.07	98.99	2.18	0.0	122.12
Average precipitation days (≥1.0 mm)	0.0	1.0	3.36	9.55	17.64	18.27	26.0	27.18	26.82	15.0	0.55	0.0	12.11
Average relative humidity (%)	20.04	27.1	36.99	50.41	64.67	73.27	82.47	86.71	84.7	72.78	41.74	22.91	55.32
Mean monthly sunshine hours	11.6	11.57	11.49	11.38	11.15	10.41	9.3	8.9	9.53	10.8	11.58	11.59	10.78

Courtyards are vital architectural components in Abuja, enhancing natural ventilation and thermal comfort. Effective building design and orientation significantly influence their ventilation efficacy. However, there has been a decline in traditional courtyard utilisation attributed to compact housing demands, lifestyle changes, and ecological pressures [44]. This trend raises concerns regarding increased energy consumption and costs related to mechanical cooling [45]. Retaining courtyards could reduce reliance on artificial cooling and promote eco-friendly strategies by creating microclimatic buffer zones. Research suggests that while traditional approaches have merits, innovative solutions are necessary to overcome their limitations [46]. Abuja's unique climate offers an important context for studying multilevel courtyards in hot-dry and humid conditions, influenced by geographical factors such as wind patterns and solar exposure, as well as demographic considerations. Aligning multilevel courtyard designs with urban planning strategies can foster sustainability in the built environment. This study aims to address the research gap regarding multilevel courtyards in similar climates, providing insights for architects, urban planners, and policymakers, with an emphasis on collecting climatic data to evaluate courtyard performance and promote sustainable urban development in Abuja.

3. Methodology

The research methods commence with an identification of the research philosophy, which forms the basis for knowledge generation and understanding the research process, in line with the foundational ideas posited by [47]. The interplay between the researcher and the study subject is highlighted as critical [48], with philosophical considerations deemed essential for effective organisational research design [49]. The research design flowchart in Figure 2 provides a structured visualisation of the research process, starting with identifying the research gap and purpose through literature and context studies.

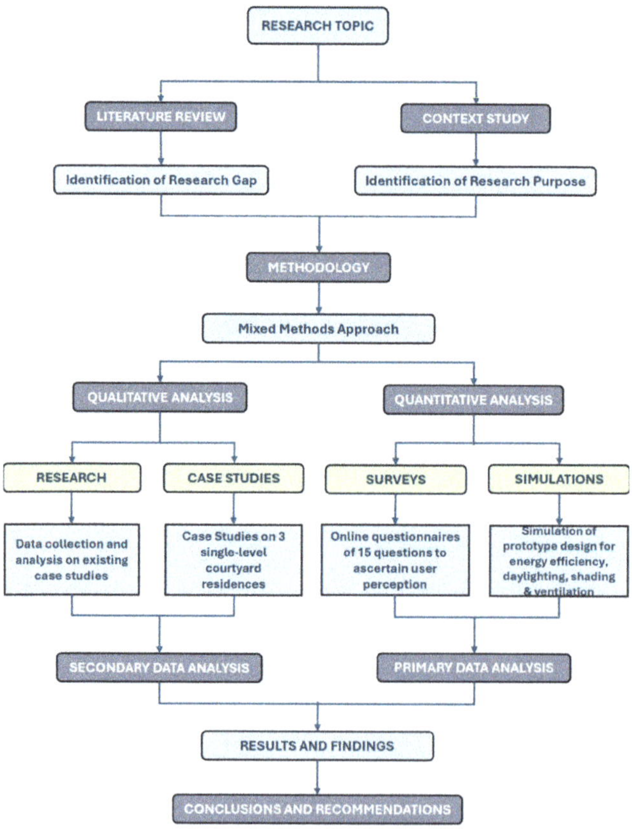

Figure 2. Research design flowchart (source: author).

It employs a mixed-methods approach, splitting into qualitative (case studies and existing studies) and quantitative (surveys and simulations) branches, followed by respective data analyses to address the identified research gap. The results are synthesised into findings, culminating in conclusions and recommendations. This organisation reflects a comprehensive study of multilevel courtyards in hot-dry/humid regions, ensuring a balanced and thorough investigation through integrated qualitative and quantitative insights. The quantitative component includes simulations conducted with DesignBuilder software version 7.0 to assess thermal performance, daylighting, ventilation, and energy efficiency. Furthermore, surveys conducted with 73 participants assess occupants' comfort perception and satisfaction, employing Google Forms for data collection. The qualitative aspect consists of three case studies of architectural projects featuring courtyards, which provide detailed insights into design characteristics and user experiences, as well as two semi-structured interviews with experts in architecture and sustainable design to gather their insights. The interviews were conducted online, and audio recordings were securely stored until and after they were transcribed. The research approach was meticulously designed to minimise any risk to the participants, and during the analysis phase, all excerpts from interviews were handled anonymously. Primary research methods entail these simulations, surveys, case studies, and interviews, collecting original data directly from participants and relevant sources. Secondary research methods involve a literature review, establishing a theoretical framework and contextual background. This multifaceted strategy ensures a thorough exploration of the research objectives, effectively blending objective performance data with subjective experiential insights. Ethics approval was secured from the De Montfort ethics committee, underscoring the commitment to ensuring ethical standards and the integrity of the research endeavour.

4. Case Studies

4.1. Case Study 1: More than a House

The "More than a House" project was designed by Lequang-Architects in Hội An, Vietnam, covering 250 m^2 and completed in 2022. The project features a spiral design that connects three households around a central courtyard, aimed at addressing microclimate challenges. This design enhances natural light and ventilation, which are limited by surrounding structures [50]. The three courtyards mitigate traffic noise and foster communal living by orienting essential spaces toward the central area, reflecting Vietnamese cultural values of social interaction. Utilising traditional ceramic bricks from Quang Nam, the design promotes thermal regulation and permeable landscaping. The project repurposes construction debris for landscaping, enhancing its eco-friendliness. The courtyard is crucial for regulating indoor thermal comfort, offering natural ventilation and daylight access in a densely built environment. By employing ceramic bricks with superior thermal properties, the project aids in passive cooling and sustainability. Overall, it exemplifies the development of energy-efficient, durable living spaces that minimise carbon footprints and conserve resources.

4.2. Case Study 2: The ReYard House

The ReYard House is a contemporary Moroccan courtyard residence that utilises waste materials in creating garden furniture, reflecting a commitment to sustainability. Key features include a microalgae-based wastewater treatment system that processes domestic wastewater within a membrane bioreactor, producing treated water for reuse and organic fertilisers. This system not only irrigates the landscape but also enhances oxygen production and sequesters CO_2. The design incorporates advanced HVAC systems to optimise energy efficiency while preserving traditional aesthetics [51]. Analytically, the separator walls serve dual functions: enabling natural ventilation and enhancing privacy by restricting external visibility, especially on the south and west facades. The wooden canopy in the yard provides shade and facilitates cooling, while the algae pond contributes to the courtyard's microclimate through evaporation. Overall, the cultivation of microalgae exemplifies a sustainable solution for carbon emissions reduction, producing oxygen at rates significantly higher than terrestrial plants and fostering an ecological balance within the courtyard.

4.3. Case Study 3: Villa LL

The residential project in northern Casablanca, Morocco, designed by Mohamed Amine Sian, spans a 1000 square meter plot and features 290 m^2 of buildable area. Drawing from traditional Moroccan architecture, the design centres on an introspective courtyard that mitigates urban noise while enhancing natural light and ventilation, effectively addressing local climatic conditions [52]. Key elements include blind façades that maintain privacy, sculpted walls for visual connectivity, and a circular swimming pool that serves as the courtyard's focal point. The incorporation of light green ceramic tiles links water and sky, harmonising with the building's palette. This architecture prioritises sustainability and energy efficiency, utilising the courtyard to improve indoor air quality through natural ventilation. Analytically, the courtyard acts as a microclimate modifier, enhancing indoor air quality and reducing HVAC energy consumption. By maximising natural light and cross-ventilation, it becomes essential to the architectural approach. Blind façades ensure privacy while maintaining a vibrant atmosphere, and structured walls provide connectivity to the exterior. Furthermore, the integration of vegetation in the courtyard supports environmental sustainability by enhancing microclimate conditions and reducing carbon dioxide levels.

4.4. Comparative Analysis of Case Studies

Table 2 below outlines a detailed comparative analysis of three case studies to evaluate the effectiveness of integrating courtyards as a passive design method. This aims to explore the strategies used in integrating courtyard designs in all cases and how it improves the overall functionality of multilevel courtyards.

The comparative analysis of three case studies—"More than a House" (250 m^2), "Re-Yard House" (140 m^2), and "Villa LL" (290 m^2)—demonstrates that courtyards play a crucial role in enhancing both environmental and social functionality within residential designs of varying scales. Environmentally, all three designs incorporate courtyards to facilitate natural ventilation, thereby minimising dependence on mechanical systems. However, their approaches to daylighting, shading, and thermal comfort differ: Case Study 1 utilises a lower southern block and a deciduous tree for shading, complemented by ceramic brick for passive cooling; Case Study 2 features a canopy and separator walls to optimise light and airflow; and Case Study 3 takes advantage of a swimming pool for evaporative cooling, with limited shading provided by eaves. Socially, the courtyards act as communal hubs, each designed to meet specific needs—supporting community activities in Case Study 1, fostering family interaction in Case Study 2, and offering private relaxation in Case Study 3. These findings underscore the adaptability of courtyards as a passive design element, seamlessly integrating diverse architectural approaches and user requirements.

Table 2. Comparative analysis of case studies (source: author).

Factors	Case Study 1: More than a House	Case Study 2: Reyard House	Case Study 3: Villa LL
Building Area	250 m^2	140 m^2	290 m^2
Material	Brick, ceramic brick, rubble	Rammed earth, wooden-framed panels, bamboo sticks	Concrete, ceramics
Daylight and Shading	The southern block of the building's design incorporates a lower height to facilitate natural light penetration into the courtyard, thereby illuminating the internal spaces. The shading mechanism provided within the courtyard is limited to a deciduous tree.	The courtyard's design was optimized to facilitate ample daylight penetration into the indoor spaces, while providing shading from the intense sun and maintaining indoor temperatures through a canopy.	The courtyard lacks shading components other than the eaves of the building. However, it provides ample illumination to the interior.
Ventilation	The courtyard functions as a primary means for facilitating natural ventilation in the building. No HVAC systems have been detected on the premises.	The canopy is engineered to efficiently shield the yard from direct sunlight while also incorporating a wind redirection system that captures and channels wind into the yard space.	The courtyard serves as the central point of the project and facilitates cross ventilation, which helps in the penetration of fresh air into the surrounding rooms. This feature enhances the ventilation system and improves the overall air quality within the building.
Thermal Comfort and Cooling	The indigenous construction material of Quang Nam, ceramic brick, when combined with the courtyard's ventilation, plays a pivotal role in regulating the indoor temperature of buildings. This, in turn, results in a cooler indoor environment and facilitates the release of excess heat.	The separator walls have been engineered to facilitate the free flow of wind through the yard, thereby promoting convective heat transfer and effectively reducing the ambient temperature.	The swimming pool plays a crucial role in the design by providing indoor thermal comfort through evaporative cooling.
Socio-Cultural	The public spaces have been designed with a central garden serving as a hub for community activities. This garden comprises a living room, two side-by-side kitchens, a laundry area, and a dedicated agricultural harvest space. The garden is also a welcoming spot for the residents to relax and socialize.	The courtyard serves as a communal space facilitating social interaction among families. It provides a platform for spending quality time with loved ones while also enhancing the overall aesthetic appeal of the property.	The courtyard is designed to offer a secluded and contemplative seating area that encourages familial interactions by providing a respite from the bustling urban environment. Its layout and construction are optimized to ensure an intimate ambiance that facilitates relaxation and introspection.

5. Survey

Descriptive statistics, as defined by [53], involves summarising dataset features, particularly frequency counts of categories. Standard formats for displaying frequency data include tables, percentiles, bar charts, and histograms, which are essential tools in various fields like social sciences and market research. This employs Google Forms for data collection and presents residents' views and experiences in contemporary single-level courtyard houses. A total of 100 questionnaires were distributed to residents of courtyard buildings in Abuja, resulting in 73 completed responses.

5.1. Gender Analysis and Age Distribution

A survey of courtyard home occupants produced 73 completed responses out of 100 distributed questionnaires, resulting in a 73% response rate. The gender analysis (Figure 3) revealed that 57.5% of respondents were male and 42.5% were female, ensuring representative household input. In terms of age demographics, 57.5% of respondents were aged 26–35 (Figure 4), highlighting the presence of young professionals and family-raisers critical to urban housing needs. Additionally, 19.2% were aged 18–25, while 16.4% were between 36–45, with fewer respondents in the older age brackets. This age distribution indicates a balanced mix of perspectives, particularly from those in key household formation stages, relevant to design and policy considerations.

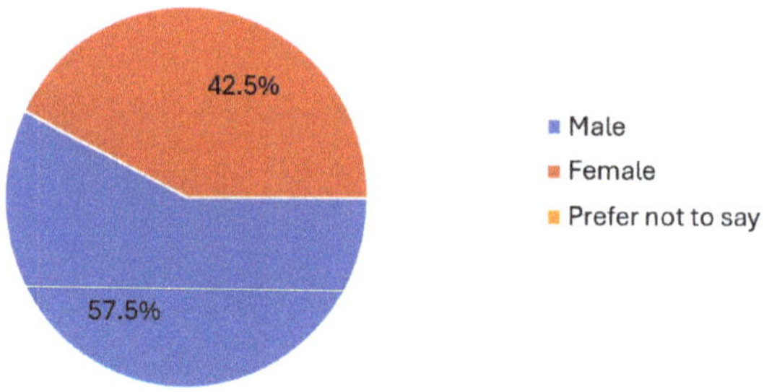

Figure 3. Gender analysis chart (source: author via Google Forms).

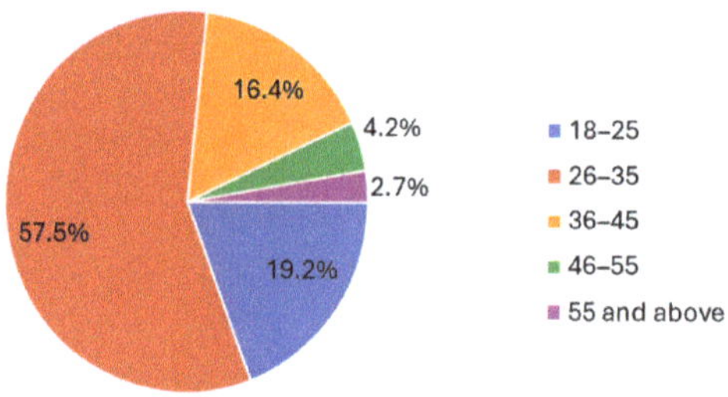

Figure 4. Age group analysis chart (source: author via Google Forms).

5.2. Courtyard Size and Type Analysis

Most participants (61.6%) have average-sized courtyards, with 23.3% having small and 15.1% having big courtyards (Figure 5). The prevalence of average-sized courtyards suggests they are common in the residential area, but the small sample size limits conclusions about big courtyards. A vast majority (97.3%), as shown in Figure 6, reported having single-level courtyards, while only 2.7% lived in multilevel courtyards. The data indicates single-level courtyards are the norm, making it challenging to draw conclusions from the small number of multilevel residents. Understanding the preferences and perceptions of both groups could provide valuable insights for consumer research.

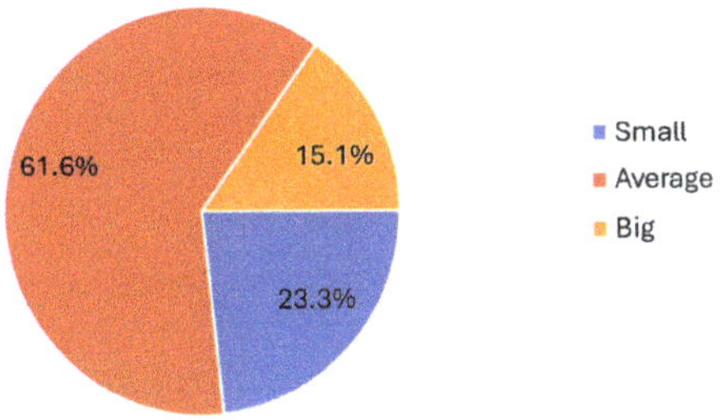

Figure 5. Courtyard size analysis chart (source: author via Google Forms).

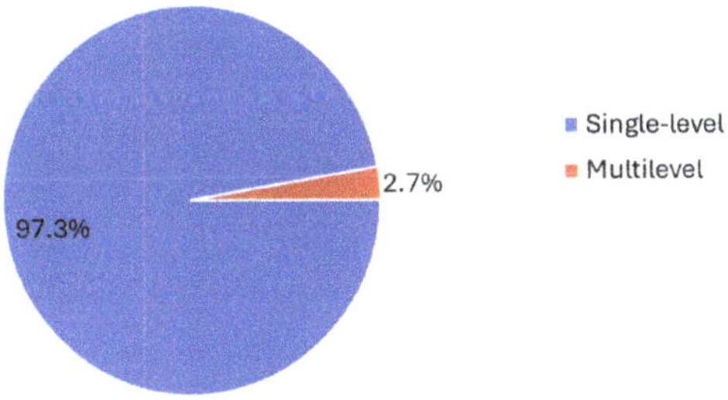

Figure 6. Courtyard type analysis chart (source: author via Google Forms).

5.3. Primary Courtyard Purpose and Comfort Analysis

The survey assessed the primary purposes of existing courtyards (Figure 7), with 28.8% of respondents identifying natural ventilation as the primary function, highlighting its role in passive cooling. Social activities were chosen by 41.1%, indicating the courtyard's significance for community interactions. Daylighting was noted by 23.3%, and privacy was identified by 19.2%. Nearly 48% of respondents, as shown in Figure 8, felt comfortable in their courtyards, with 34.2% reporting they felt very comfortable. A small percentage (4.3%) felt slightly comfortable, while 6.8% were slightly or very uncomfortable. This suggests most courtyards serve their purpose but indicates potential for improvements in design elements like temperature regulation, seating, greenery, and aesthetics to enhance comfort.

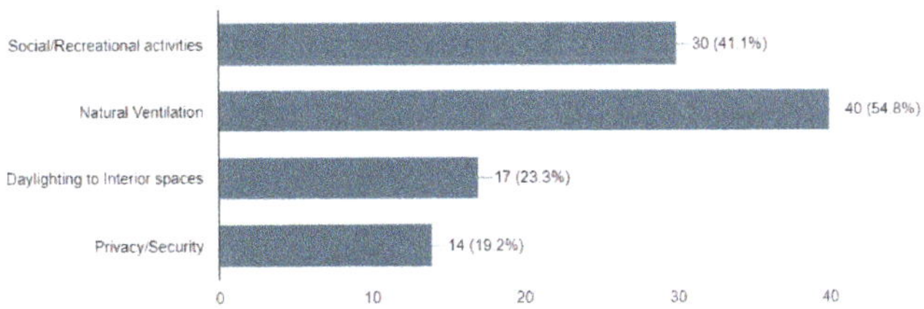

Figure 7. Courtyard purpose analysis chart (source: author via Google Forms).

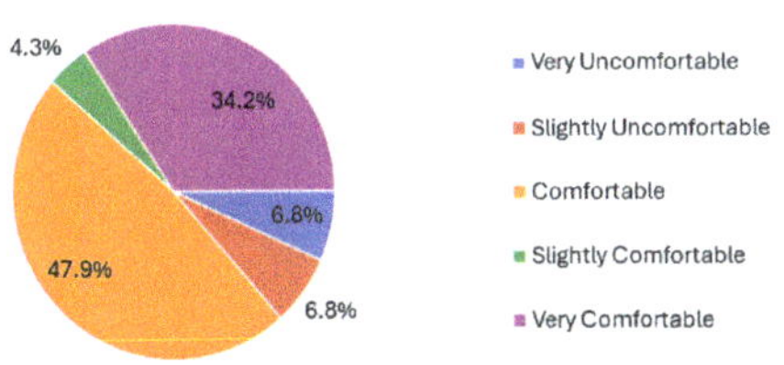

Figure 8. Comfort level analysis chart (source: author via Google Forms).

5.4. Indoor Air Quality and Social Activity Analysis

In total, 79.2% of participants believe courtyard buildings have cooler indoor air quality (Figure 9) compared to conventional buildings, supporting the benefits of natural ventilation; 8.3% perceive no difference in air quality, indicating potential neutral views or a lack of awareness; and 12.5% feel courtyard buildings are warmer, suggesting a need for design improvements and education on effective courtyard configurations. Regarding activities, 50.7% of participants use courtyards for family time and gatherings, while 52.1% indicated they used the courtyard for partying and connecting with each other (Figure 10), highlighting their role as cultural and communal spaces, respectively. In addition, 30.1% reported using courtyards for drying clothes, and only 11% for cooking, reflecting a shift in traditional uses. The main social function of courtyards is to facilitate family gatherings and community events, emphasising the need for multilevel courtyard designs to support these activities.

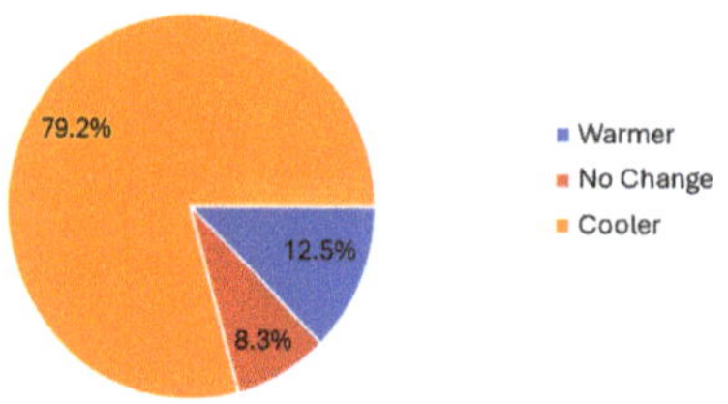

Figure 9. Indoor air analysis chart (source: author via Google Forms).

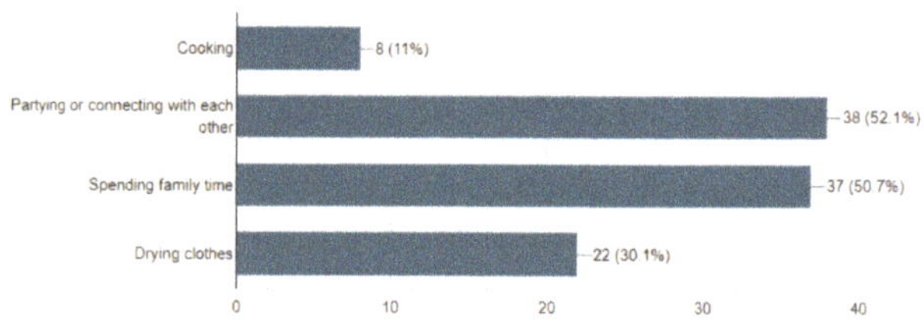

Figure 10. Activity analysis chart (source: author via Google Forms).

5.5. Multilevel Courtyard Analysis

Participants were asked about the impact of multilevel courtyards on ventilation, with 83.6% (Figure 11) believing it would enhance airflow compared to single-level courtyards, while 12.3% responded "maybe", citing a lack of local examples, and only 4.1% disagreed, expressing scepticism about the benefits. The results indicate general support for the ventilation concept but emphasise the need for simulations and prototypes to convert sceptics, with measurable improvements in airflow necessary to bolster the case for multilevel courtyards.

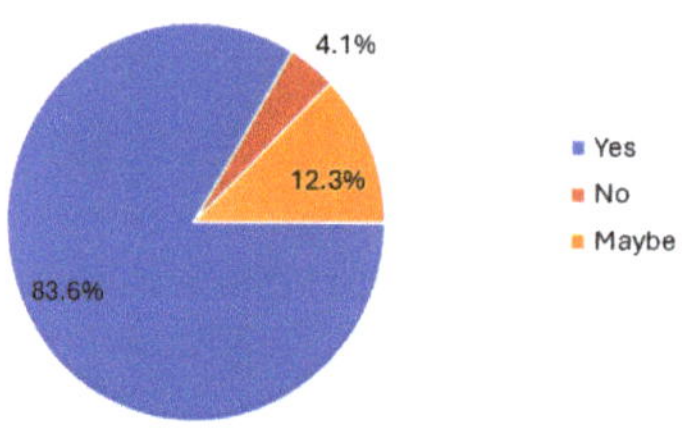

Figure 11. Ventilation analysis chart (source: author via Google Forms).

The survey on multilevel courtyard architecture assessed participants' openness to the concept. Results (Figure 12) showed that 38.4% rated their receptiveness as high (4 or 5), while 19.2% rated it moderately (3). Only a small percentage (1.4% at 2 and 2.7% at 1) showed low receptiveness, indicating overall positive reception. Regarding interest in living in multilevel courtyard buildings, as shown in Figure 13, 72.6% expressed strong interest (yes), with 16.4% responding maybe, suggesting potential demand. Only 11% answered no. Combined yes/maybe responses indicate 89% interest, highlighting the need for design refinement based on user feedback to expand market appeal.

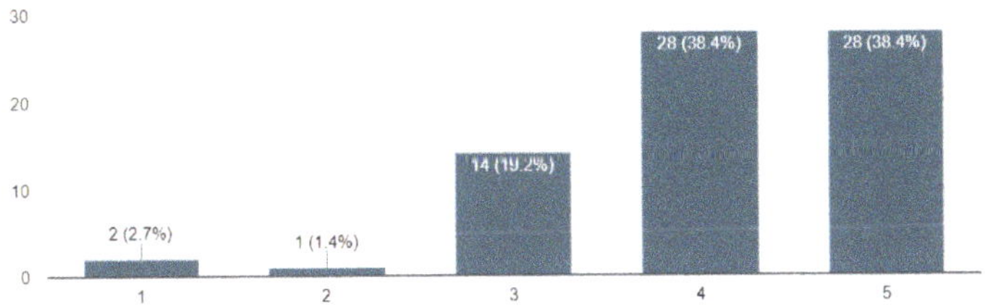

Figure 12. Receptiveness analysis chart (source: author via Google Forms).

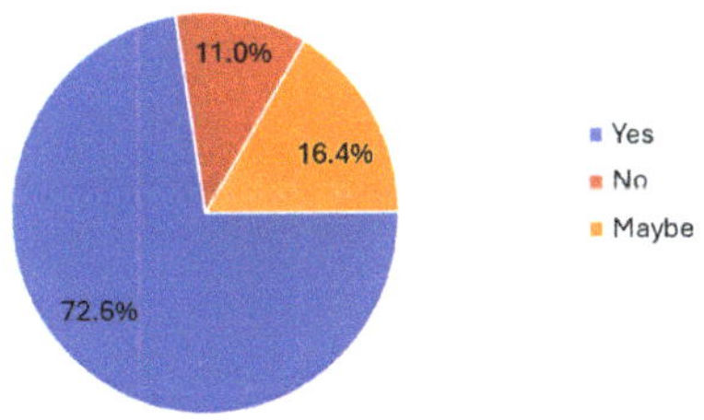

Figure 13. Multilevel interest analysis chart (source: author via Google Forms).

A survey of Abuja residents indicates that courtyards provide benefits such as ventilation, temperature regulation, and aesthetic appeal. There is notable interest in multilevel courtyard housing, although concerns about accessibility for disabled residents were highlighted. The challenge is to ensure that individuals with mobility impairments can effectively use these spaces. The findings suggest that while courtyards are culturally accepted, modernisation and inclusivity in design are crucial for the success of multilevel courtyards.

6. Simulations

The environmental performance of multilevel courtyard buildings is analysed, focusing on how design variables like shading, proportions, height, and materials impact daylighting, energy efficiency, and thermal comfort compared to single-level courtyards. The methodology began by modelling the case study using DesignBuilder software version 7 and inputting contextual data, such as location, materials, lighting, and ventilation, into the software. This process aims to determine the building's comfort and sustainability, identify areas for improvement, and contribute to a set of guidelines for future applications. The initial step in the methodology involved modelling the reference project using comprehensive data to create accurate two-dimensional models (Figures 14–19). The location and internal space characteristics are crucial for precise climatological analysis and simulations.

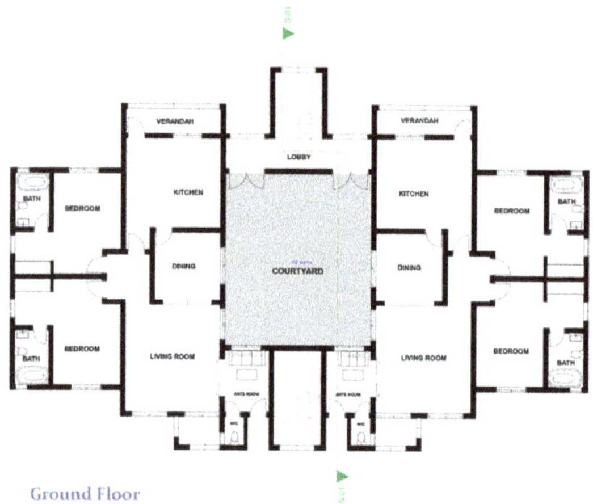

Figure 14. Test model—ground floor (source: author).

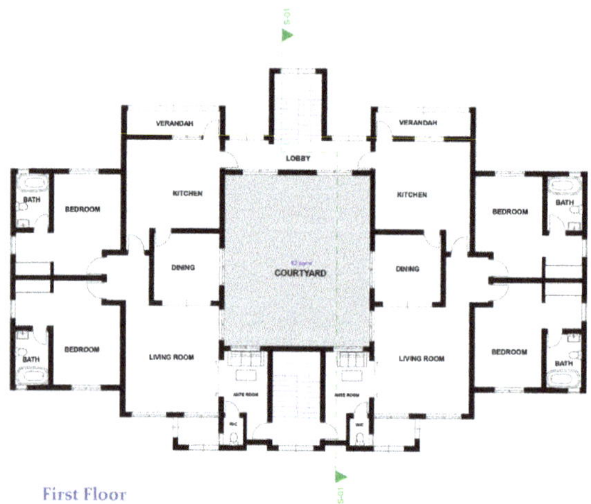

Figure 15. Test model—first floor (source: author).

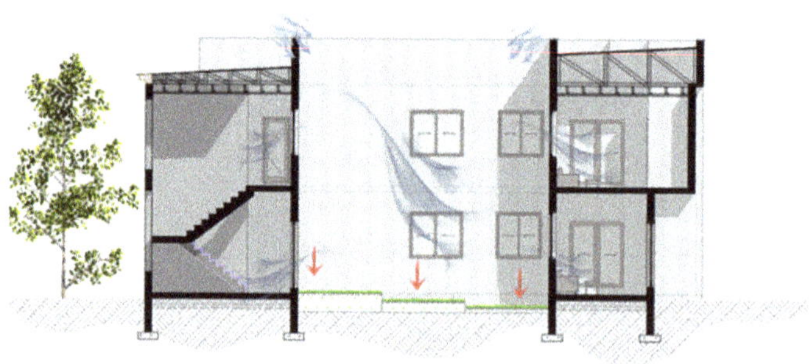

Figure 16. Multilevel test model 1 section—tiered levels—directional arrows indicate multilevel design variation (source: author).

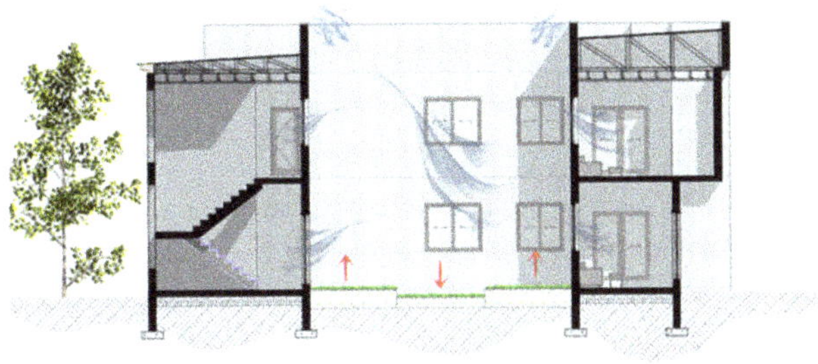

Figure 17. Multilevel test model 2 section—alternated levels—directional arrows indicate multilevel design variation (source: author).

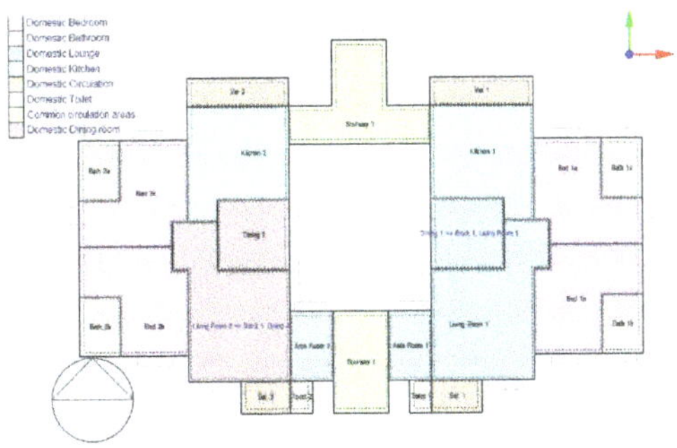

Figure 18. Activity template—ground floor (source: author via DesignBuilder).

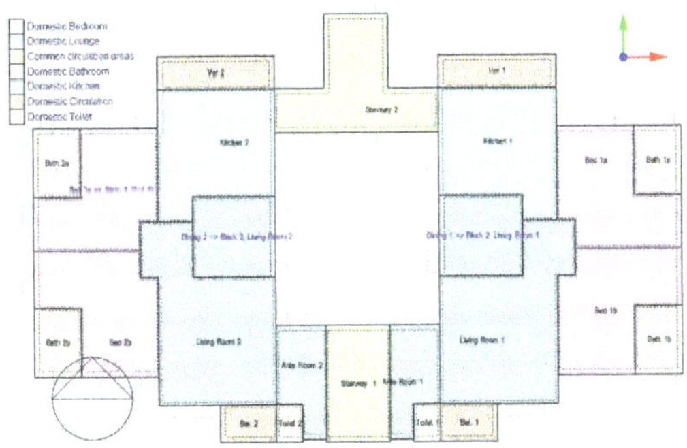

Figure 19. Activity template—first floor (source: author via DesignBuilder).

A series of simulations tested various configurations, including courtyard dimensions and construction materials. The outputs quantified lighting, cooling loads, and comfort levels, benchmarked against established standards and single-level models. The study aims to identify optimal multilevel courtyard designs that enhance passive climate control and energy efficiency in Nigeria's hot-humid context, ultimately providing guidelines for integrating these structures as a sustainable strategy in the region.

6.1. Single-Level Courtyard with Hollow Concrete Blocks and Compressed Earth Blocks

A comparative study was conducted using DesignBuilder software to evaluate the thermal performance of concrete blocks and compressed earth blocks (CEB) for a courtyard building in Abuja, Nigeria, focusing on natural ventilation without mechanical cooling. The results indicated that HCB achieved an operative temperature of 31.45 °C (Figure 20), while CEB outperformed, achieving an operative temperature of 30.26 °C (Figure 21). This 1.19 °C difference is attributed to the lower U-value of CEB (1.040 W/m^{2}K) relative to HCB (1.446 W/m^{2}K), indicating superior insulation and thermal mass properties in CEB. The findings suggest that CEB can more effectively reduce cooling loads in hot climates than concrete blocks. The findings indicate that compressed earth blocks (CEBs) exhibit a superior capacity to mitigate cooling loads in hot climates compared to conventional concrete blocks. This observation corroborates the assertions made by [31], who advocate that compressed stabilised earth blocks (CSEBs) represent a more environmentally sustainable and energy-efficient alternative to traditional concrete.

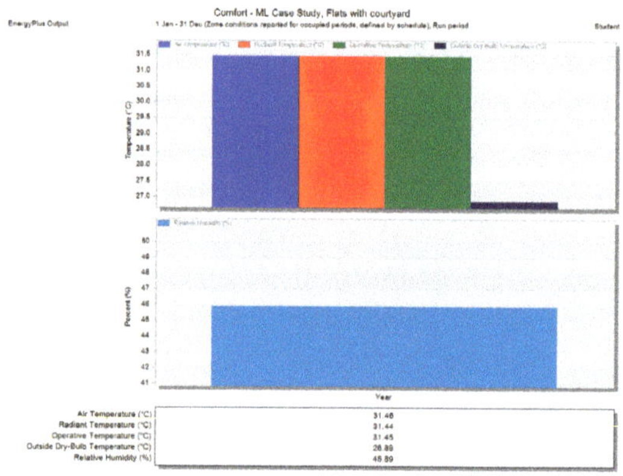

Figure 20. Operative temperature with HCB (source: author via DesignBuilder).

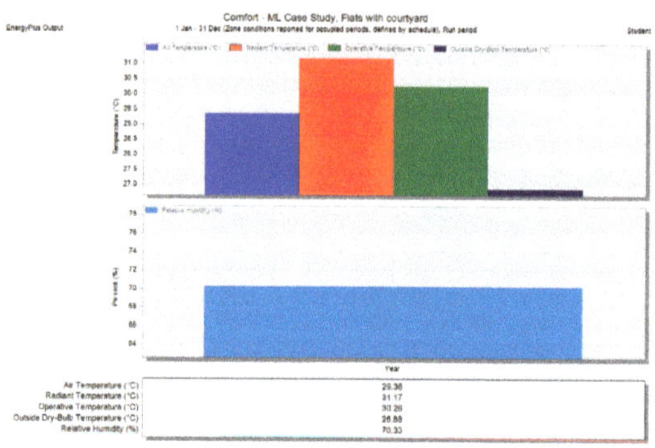

Figure 21. Operative temperature with CEB (source: author via DesignBuilder).

6.2. Comparative Thermal Analysis: Single-Level vs. Multilevel Courtyard

A thermal comfort simulation was conducted on a multilevel courtyard building model to assess the influence of varying ground levels on indoor temperatures. The analysis focused on concrete paving blocks and mud bricks. The initial simulation on a single ground level yielded an indoor operative temperature of 30.26 °C (Figure 22). Altering the ground levels reduced the operative temperature to 28.11 °C (Figure 23), a reduction of 2.15 °C. This decrease can be attributed to enhanced airflow and ventilation provided by the varied ground levels, aligning with findings by [54] that increased soil depth reduces temperature amplitude and the chimney effect facilitates improved stack ventilation. Additionally, substituting concrete with mud bricks mitigates the urban heat island effect due to the former's high thermal conductivity, while mud bricks offer lower thermal conductivity, thereby reducing heat transfer indoors. Overall, this thermal simulation suggests that replacing concrete paving with mud brick and optimising airflow through the multilevel courtyard design can passively lower operative temperatures by over 2 °C, enhancing thermal comfort through high-thermal-mass materials.

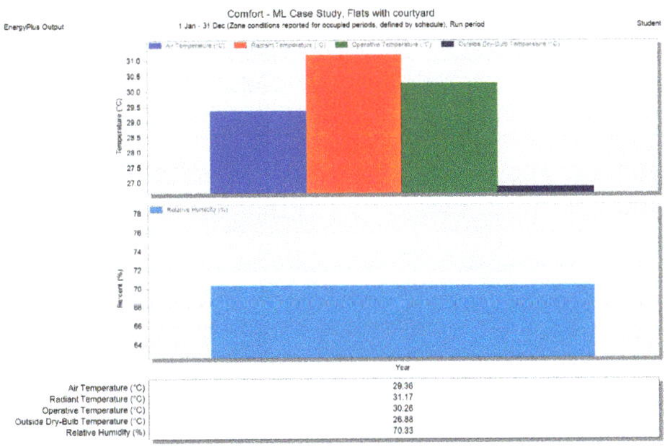

Figure 22. Operative temperature with single level (source: author via DesignBuilder).

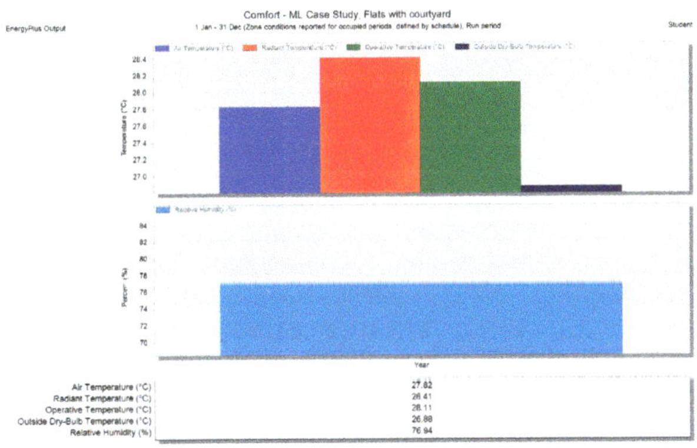

Figure 23. Operative temperature with multiple levels (source: author via DesignBuilder).

6.3. Daylighting Performance

Daylighting simulations were conducted for the courtyard building design using DesignBuilder software to evaluate daylight adequacy in accordance with BREEAM standards. Various parameters, including building geometry, room layouts, courtyard proportions, window specifications, and site location, were modelled, and multiple design iterations were tested for optimisation. BREEAM stipulates that at least 80% of net lettable floor area in occupied spaces must achieve adequate daylight at a working plane height of 0.7 m. Findings from the multilevel case study revealed that the building attained natural lighting levels of 49.1% (as shown in Table 3, row 1), with the ground floor, as illustrated in Figure 24, and the first floor, depicted in Figure 25, exhibiting deficiencies in lighting levels. Notably, spaces on the first floor, including the kitchen and living room, received more daylight than their ground-floor counterparts. These results indicate that elevation significantly influences daylighting, yet the current design fails to meet BREEAM's minimum daylight standards for certain occupied areas, particularly on lower floors.

Table 3. Annual daylighting factor report analysis (source: author).

Simulation	Percentage of Area Meeting BREEAM Requirements	Status	Analysis
Annual Daylighting Report on Multilevel Courtyard	49.1%	Fail	Despite not achieving BREEAM's minimum daylight standards, the first-floor areas demonstrated a marked increase in daylight penetration compared to those on the ground floor. These findings underscore the critical role that elevation plays in daylight distribution within built environments.
Annual Daylighting Report with CBRT	41.6%	Fail	While polycarbonate permits some diffuse light penetration, it significantly hinders direct sunlight, which is essential for fostering a well-lit and welcoming atmosphere. The courtyard's heavy reliance on artificial lighting highlights a considerable drawback regarding energy consumption and the overall ambiance of space.
Annual Daylighting Report with Courtyard Size Increment	53.5%	Fail	The expansion of the courtyard dimensions resulted in a notable enhancement of daylight penetration, increasing from 49.1% to 53.5%. This underscores the correlation between spatial volume and interior illumination levels.
Annual Daylighting Report with CBRT and Size Increment	45.7%	Fail	The interplay of light and shade is vital in outdoor spaces like courtyards. The enlarged courtyard's baseline model shows a daylight factor of 53.3%, highlighting the importance of natural light for a vibrant atmosphere. However, the CBRT shading system reduces this factor by 7.6% to 45.7%, offering a thoughtful approach that enhances usability and comfort. This reduction can soften direct sunlight, creating cooler areas that encourage relaxation and social interaction.

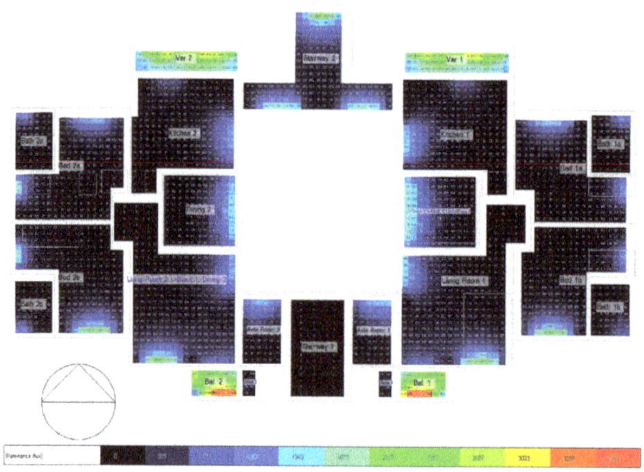

Figure 24. Ground floor—annual daylight factor (source: author via DesignBuilder).

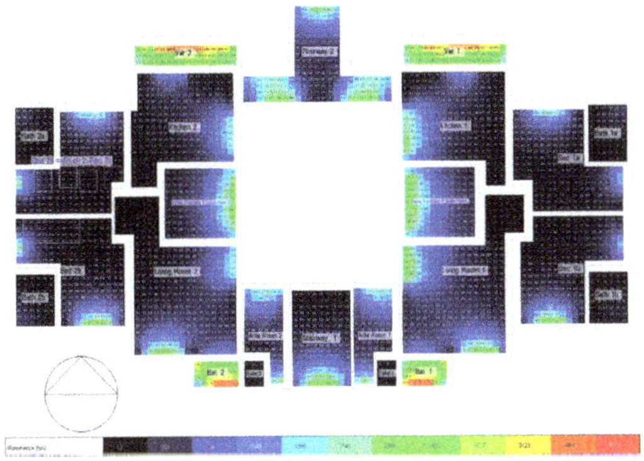

Figure 25. First floor—annual daylight factor (source: author via DesignBuilder).

The modification of the courtyard building, which involved reducing its height by one floor, prompted further daylighting simulations to evaluate its impact on interior daylight levels. This adjustment aimed to improve previously substandard daylight levels on the lower floors, as per BREEAM guidelines. The height reduction enhanced sunlight accessibility (Figure 26), particularly in winter mornings and evenings, resulting in a significant increase in average daylight factors (ADFs) for the ground floor, from 29.9% to 56.2%. However, this improvement still did not meet BREEAM's minimum residential standard of 80%. The second-floor spaces remained compliant with the standards. Although the height reduction yielded quantifiable benefits for the lower levels, it did not fully address the shortcomings, indicating that further modifications, such as enlarging the courtyard, may be needed to achieve compliance.

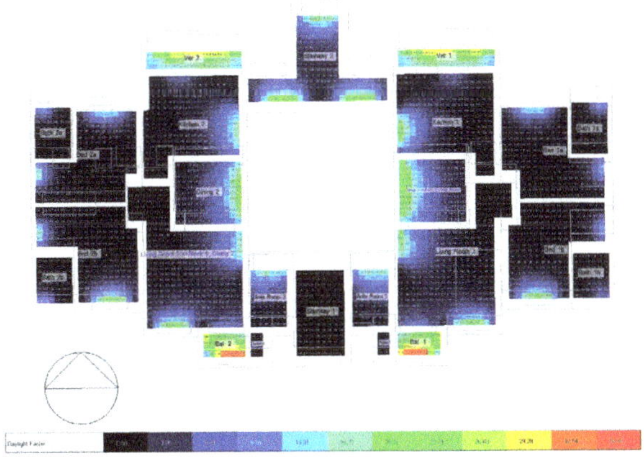

Figure 26. Annual daylight factor with height reduction (source: author via DesignBuilder).

6.4. Shading Analysis: Polycarbonate CBRT

Further daylighting simulations evaluated the impact of polycarbonate roofing on the central courtyard, informed by [34]. The incorporation of polycarbonate, as shown in Figure 27, transformed the open courtyard into an enclosed atrium-like space, yet the simulations revealed that this configuration significantly impaired daylighting in adjacent interior areas (Figure 28). The average daylight factor (ADF) within the courtyard building decreased from 49.1% to 41.6% (Table 3, row 2), a reduction of 7.5%, which falls below the BREEAM-recommended minimum ADF of 80% for residential spaces. Although polycarbonate allows some diffuse light, it obstructs direct the natural light that was previously enjoyed in the open courtyard. As a result, the enclosed condition relies solely on artificial lighting and fails to achieve sufficient daylighting. These findings suggest that utilising polycarbonate roofing in this context is inadvisable, highlighting the necessity of maintaining an open courtyard to satisfy daylighting requirements.

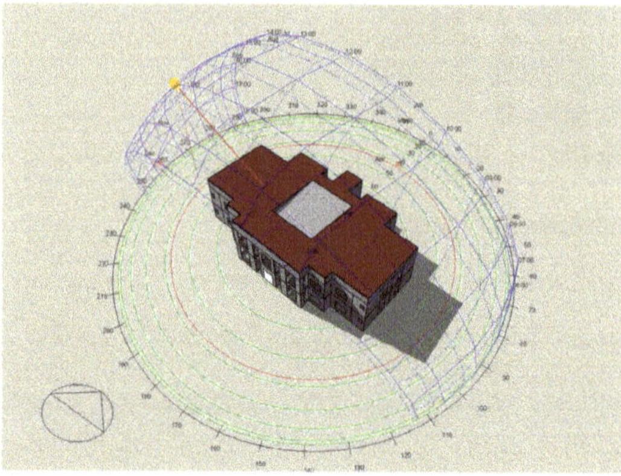

Figure 27. Multilevel with courtyard roofing (source: author via DesignBuilder).

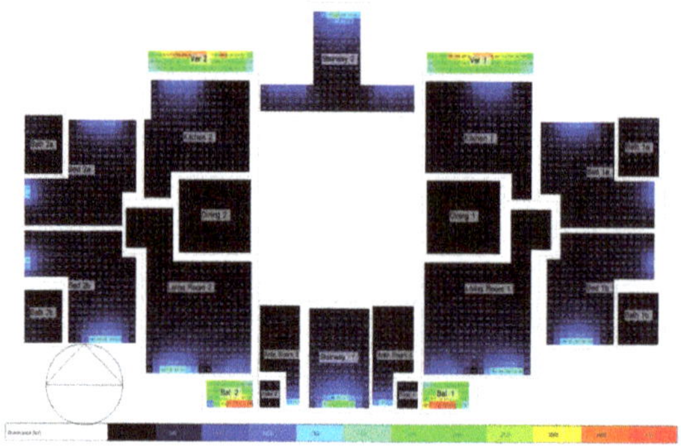

Figure 28. Annual daylight factor with CBRT (source: author via DesignBuilder).

6.5. Energy Efficiency Analysis: Multilevel Courtyard

An energy simulation was performed to assess the electricity consumption of a multilevel courtyard building model, which totalled 25,813.35 kWh annually, as shown in Figure 29. The predominant energy use was attributed to domestic hot water at 19,788.92 kWh, constituting approximately 77% of total consumption. Conversely, lighting accounted for only 5002.79 kWh, or about 19% of total usage (Figure 30). This indicates that the multilevel courtyard design provides substantial energy conservation benefits by facilitating passive daylighting, which minimises the reliance on artificial lighting during daylight hours. Consequently, the optimisation of daylight penetration in this design effectively reduces lighting energy consumption to 19% of overall electricity demand.

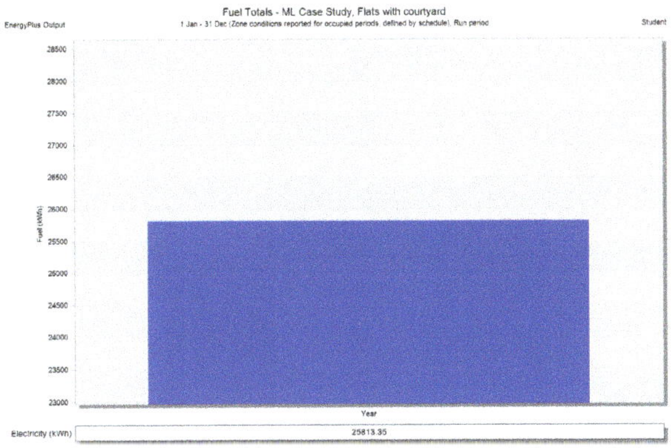

Figure 29. Fuel totals (source: author via DesignBuilder).

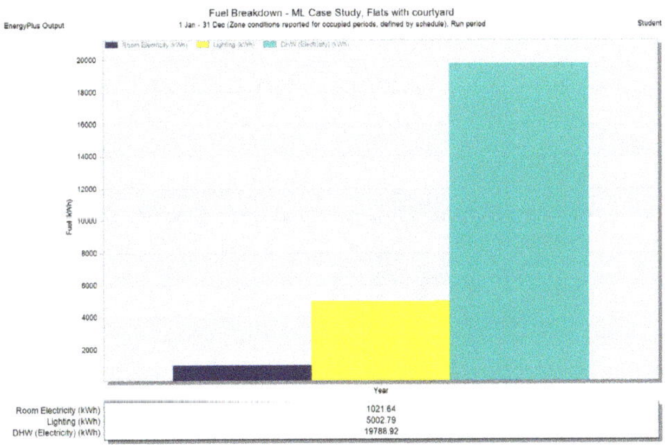

Figure 30. Fuel breakdown (source: author via DesignBuilder).

6.6. Thermal Comfort Analysis—HVAC Integration

Further thermal comfort simulations were conducted to evaluate the impact of different cooling strategies in the courtyard building. The focus was on comparing natural ventilation alone with the integration of a split-unit HVAC system. Initial simulations revealed that relying solely on passive strategies resulted in operative temperatures exceeding the comfort threshold of 28.11 °C. This suggests that passive cooling is inadequate for the hot-humid climate of Abuja. Incorporating a split-unit system with chilled water cooling alongside operable windows and courtyard ventilation reduced operative temperatures to 24.95 °C (Figure 31). This hybrid strategy capitalises on the benefits of natural ventilation while providing active cooling, when necessary, consistent with adaptive comfort standards recommending mechanical cooling for temperatures above 28 °C [55]. The observed temperature reduction of 3.16 °C emphasises the advantages of integrating natural ventilation with mechanical cooling. However, this approach resulted in a significant increase in energy demand for cooling, which reached 112.71 MWh, supplemented by 19.79 MWh for domestic hot water and 5.01 MWh for lighting (Figure 32). This highlights the need for alternative cooling solutions to enhance energy efficiency.

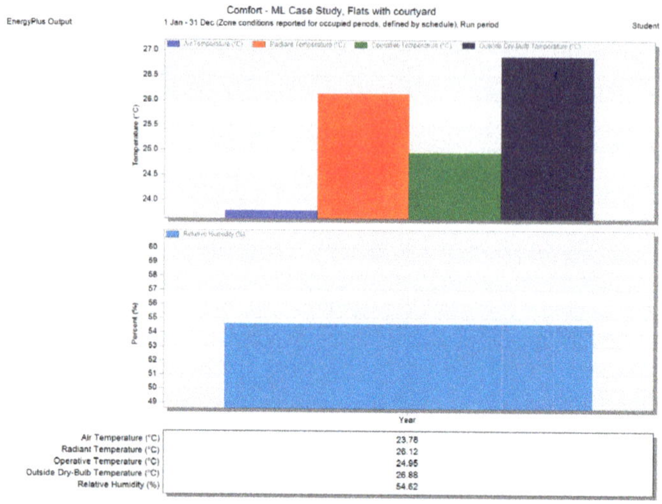

Figure 31. Operative temperature with HVAC integration (source: author via DesignBuilder).

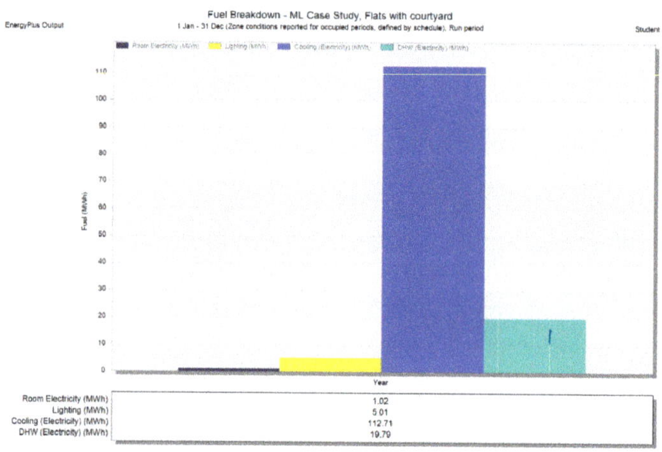

Figure 32. Fuel breakdown with HVAC integration (source: author via DesignBuilder).

6.7. Shading on Thermal Comfort Analysis

Further simulations examined the effects of courtyard roof technologies (CBRT) on indoor thermal comfort, adhering to the methodology of [34]. Models of a courtyard building were created with and without a closed roofing system. The results indicated that implementing a closed CBRT led to a slight decrease in the operative temperature from 24.95 °C to 24.87 °C (Figure 33) during peak conditions, reflecting a reduction of 0.08 °C. This marginal decrease suggests limited cooling benefits, attributable to enhanced shading that reduces solar heat gains. However, the absence of open-sky ventilation and evaporative cooling typical of traditional courtyards may mitigate these advantages [34].

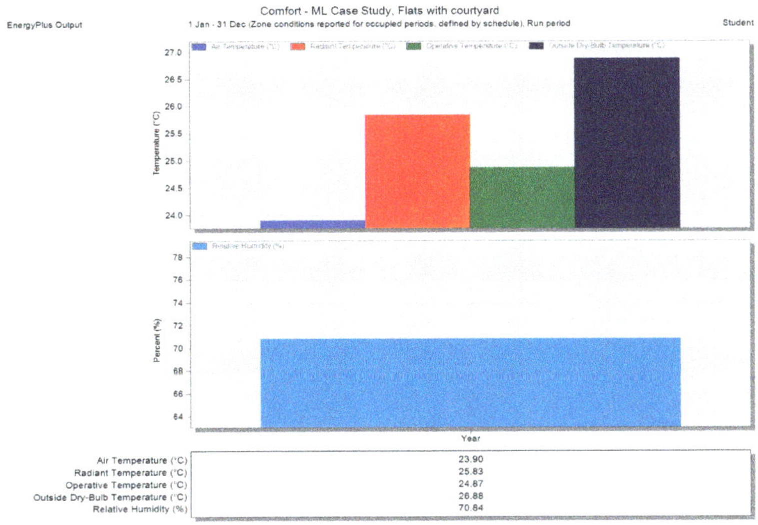

Figure 33. Operative temperature with HVAC and courtyard roofing (source: author via DesignBuilder).

6.8. Thermal Comfort and Energy Efficiency Analysis—Courtyard Size Increment

Further simulation was conducted to evaluate the effects of courtyard size on indoor thermal comfort, with no shading implemented (Figure 34) and no HVAC integration. The baseline model, featuring a 62-square-metre courtyard, yielded an operative temperature of 28.11 °C. Expanding the courtyard to 97 square meters resulted in a 0.29 °C reduction in operative temperature, decreasing it to 27.82 °C, as shown in Figure 35, without increasing energy consumption (Figure 36). This improvement is attributed to enhanced airflow and ventilation, as well as increased thermal mass, which facilitates greater heat dissipation through radiative cooling at night.

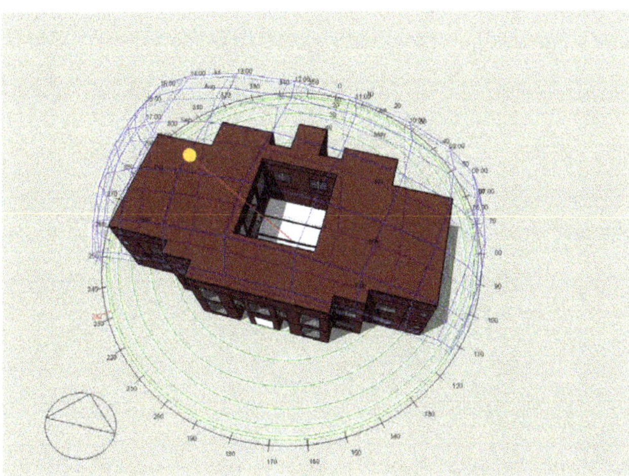

Figure 34. Extended courtyard size (source: author via DesignBuilder).

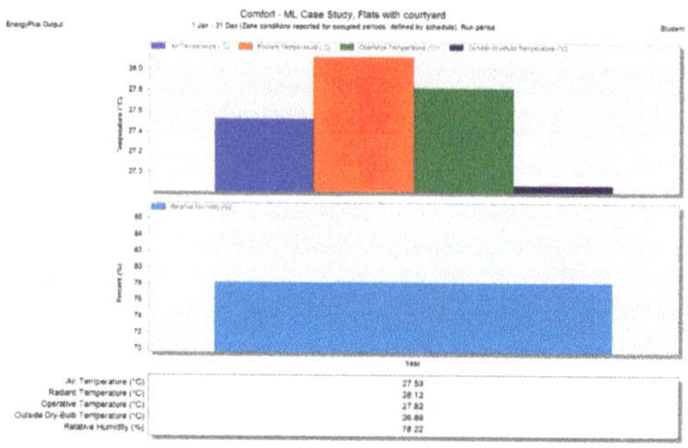

Figure 35. Temperature after courtyard size increment (source: author via DesignBuilder).

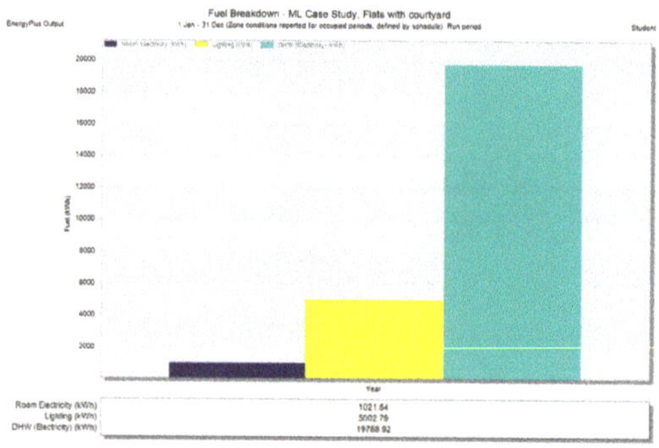

Figure 36. Fuel breakdown after courtyard size increment (source: author via DesignBuilder).

6.9. Daylighting Performance Analysis—Courtyard Size Increment

The multilevel case study building with an enlarged courtyard exhibited a cumulative natural lighting level of 53.3% (Table 3, row 3), with the ground floor (Figure 37) and the first floor (Figure 38) achieving insufficient natural lighting. Key interior spaces on the first floor, including the kitchen, dining area, anteroom, and living room, received abundant natural light, while similar spaces on the ground floor were inadequately lit. These findings underscore the significant impact of courtyard size on natural daylighting. However, the current design fails to meet minimum BREEAM daylight standards for average daylight factors in some occupied areas, particularly on the lower floors.

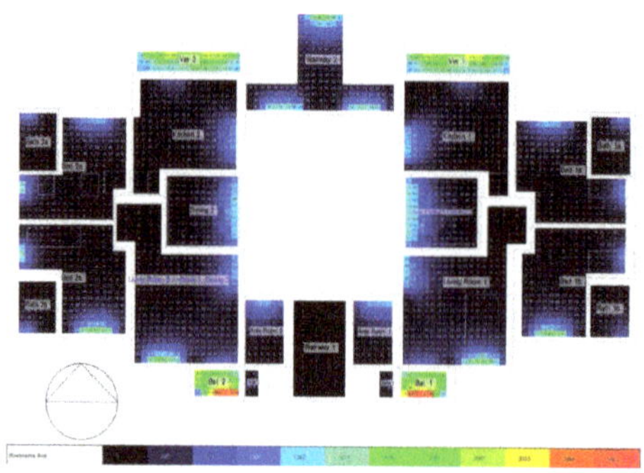

Figure 37. Ground floor—ADF with courtyard size increment (source: author via DesignBuilder).

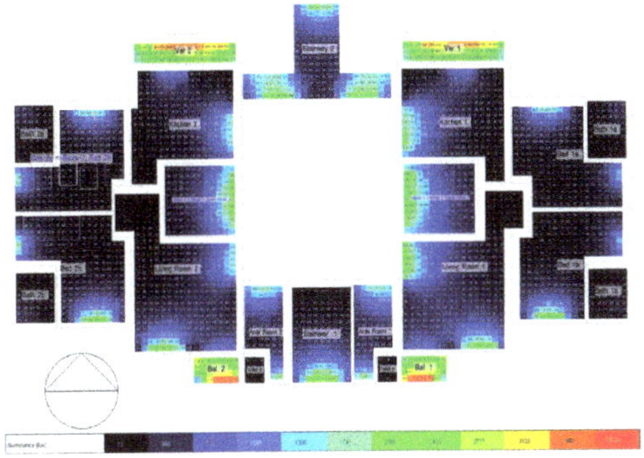

Figure 38. First floor—ADF with courtyard size increment (source: author via DesignBuilder).

6.10. Shading Analysis: Courtyard Size Increment

The enlarged multilevel courtyard building underwent daylighting simulations (Figures 39 and 40) to assess the effects of a polycarbonate shading system. The baseline model, featuring an open courtyard, exhibited an average daylight factor of 53.5% in surrounding areas. However, the integration of the CBRT shading system led to a decrease of 7.6%, resulting in an average daylight factor of 45.7% (Table 3, row 4). Consequently, energy usage for lighting rose from 5002.79 kWh per year to 15,683.88 kWh (Figure 41), accounting for 42.9% of total energy consumption and reflecting a 23.9% increase. This decline in natural daylight necessitates a reliance on artificial lighting, highlighting the need to balance the CBRT system's weather protection and architectural benefits with its adverse impact on daylight access.

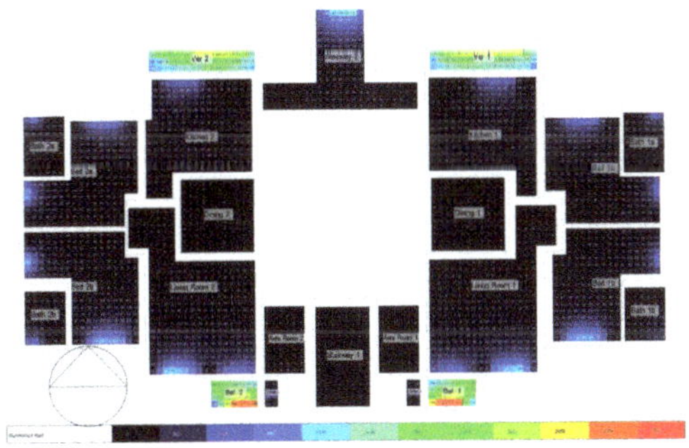

Figure 39. Ground floor ADF with CBRT and size increment (source: author via DesignBuilder).

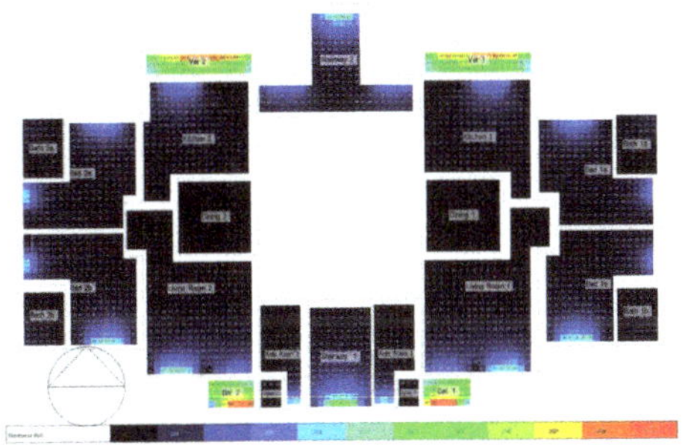

Figure 40. First floor ADF with CBRT and size increment (source: author via DesignBuilder).

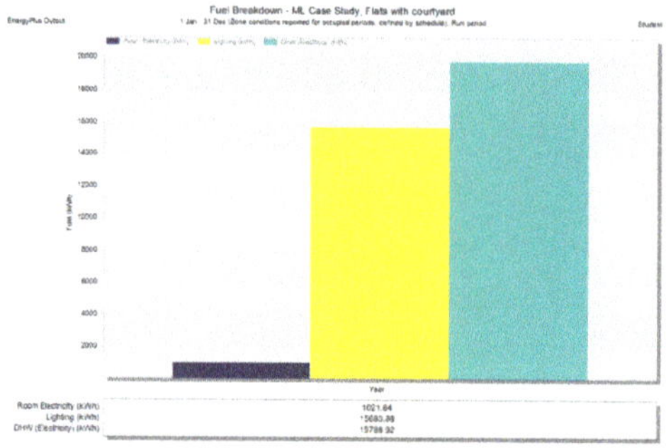

Figure 41. Fuel breakdown with CBRT and size increment (source: author via DesignBuilder).

7. Interviews

Semi-structured interviews with a researcher and a professional in architecture and the built environment were conducted during the fieldwork. These interviews were designed to investigate professional views towards the introduction of the multilevel courtyard typology in Abuja, Nigeria.

7.1. Participant 1: Interview with Architecture Researcher

An interview was conducted with an architecture researcher specialising in energy efficiency, materials, and inclusive design strategies to investigate multilevel courtyards within the Nigerian context. The expert rated the concept's receptivity as 4 out of 5 and identified it as an aesthetically appealing method to enhance functionality through multi-plane spaces. Key recommendations for energy efficiency included utilising compressed earth bricks instead of concrete, incorporating mud brick floors for high thermal mass, and implementing integrated water spraying systems as passive cooling strategies. The expert emphasised inclusive design, advocating for safe height differences of 200–500 mm between levels to ensure accessibility for the elderly and disabled. Zoning these levels for diverse functions was highlighted as a primary advantage. Market feasibility in Abuja was deemed promising with appropriate design, materials, and marketing strategies. The expert preferred alternating level courtyards with a centrally recessed space over descending tiers. Major barriers included perceptions of mud bricks as outdated and concerns about their quality compared to concrete. Nevertheless, the researcher posited that scientifically tested and controlled production methods could validate compressed earth bricks as a sustainable, locally available alternative.

7.2. Participant 2: Interview with an Architect

In an interview with an experienced architect specialising in residential design, the potential of multilevel courtyards was positively evaluated, particularly in tropical climates. These courtyards were deemed effective for apartment buildings, with recommendations for incorporating solar panels and evaporative cooling to enhance energy efficiency. The architect highlighted the significance of optimising geometry, orientation, and zoning to maximise ventilation, sunlight access, and shading while also emphasising the need for accessibility and inclusive design. The acceptance of multilevel courtyards in Abuja was viewed favourably, as they align with local cultural values and resonate with younger populations when designed thoughtfully. However, challenges regarding accessibility were noted. The architect rated market receptivity to multilevel courtyards at 4 out of 5, favouring an alternate-level typology with a central recessed space. These insights provide a foundational understanding for proposing the integration of multilevel courtyards within Abuja's architectural framework.

7.3. Thematic Summary of Interviews on Multilevel Courtyards

By organising interview data into themes, summaries, and key points, Table 4 acts as an analytical tool that supports evidence-based design decisions. This systematic documentation method allows designers and researchers to derive actionable insights from practitioners' experiences, ultimately leading to more responsive and sustainable courtyard environments in modern architecture.

Table 4. Thematic analysis of interviews (source: author).

Interviewee	Themes	Summary	Key Points
Participant 1	Courtyard Building Material	Compressed earth bricks are advised over hollow concrete blocks due to their reputation for energy reduction. Changing their building's walls from hollow concrete to compressed earth bricks resulted in some energy reduction.	• Versatility • Modern • Energy-efficient
	Courtyard Flooring Material	Mud brick flooring is advised, due to its high thermal mass (can help regulate temperature). The use of cement paving in hot regions can lead to the urban heat island effect.	• Sand/gravel • Mud brick paving • Grass
	Cooling	Use sprinklers or sensors to detect high temperatures and automatically sprinkle water to lower the temperature. It also proposes incorporating water pouring or a water fountain in the courtyard, particularly in points with mud flooring.	• Temperature sensors • Sprinklers • Water fountains
Participant 2	Typology Design	The benefits of zoning levels with activities and designing different sections for different functions are emphasized; different levels in the building make it more pleasing to the eye. Zoning different textures and designs for different levels can give an extra edge. The high and low levels can also be utilized practically, with the high level serving as a sitting area.	• Zoning by texture
	Health and Safety	When building a structure with levels, it is essential to consider the height of the levels for safety, particularly for elderly, visually impaired people, and those living with disabilities. Inclusive use of the place should also be considered to prevent tripping, and the height of the levels should match the rise in the staircase. The recommended height is 200 mm, while 300 mm may be too high unless there is another way to access that level.	• Moderate height • Inclusive design

8. Discussion

We provide a comprehensive examination of outcomes and recommendations derived from various research methodologies, including literature reviews, case studies, questionnaires, interviews, and simulations. The structure aligns with the study's primary objectives, clearly demonstrating how these objectives were addressed. Findings from each research approaches have been meticulously analysed and synthesised, contributing significantly to the field. Furthermore, based on the results, recommendations for future research are proposed.

This study evaluates the existing literature on the impact of courtyard typologies in hot-dry and humid climates. A comprehensive review of studies and case analyses highlights the significance of courtyards as traditional bioclimatic strategies that enhance thermal performance and outdoor comfort. Key findings indicate that courtyards can reduce indoor temperatures by over 20 °C compared to the outdoors and diminish cooling demands by 8–18%, influenced by their geometry, orientation, and floor level. Efficient use of open spaces necessitates the optimisation of stack ventilation, night flushing, and cooling effects through strategic placement of vegetation and water features. The incorporation of advanced elements such as algae ponds or pools can enhance microclimatic conditions and lower emissions. Additionally, sustainable materials like ceramic bricks can bolster thermal mass and cooling. Computer models have shown that courtyards are more efficient than atriums for low-rise buildings in hot regions, while the reverse holds for high-rise structures. Beyond their environmental benefits, courtyards play a vital socio-cultural role, facilitating social interactions and promoting natural healing. Contemporary designs should thus focus on preserving cultural heritage while enhancing sustainability with features such as shading, seating, and green spaces. Despite the established benefits of courtyards, there is a gap in knowledge regarding the adaptation of traditional passive design principles for multilevel courtyard configurations suitable for modern densities and climate change challenges. Further research is essential to explore the integration of traditional strategies with innovative designs, such as multi-plane courtyards, coupled with the application of optimisation tools for energy performance modelling. It is recommended that interdisciplinary research involving architecture, engineering, and social sciences be pursued to develop guidelines for effective multilevel courtyard designs.

To evaluate user perception and the feasibility of multilevel courtyards as a sustainable building model for dense urban areas in hot-dry and humid contexts, qualitative interviews and surveys were conducted. The analysis reveals that courtyards are highly favoured among Abuja residents, with all respondents currently living in such buildings. Key functions identified include natural ventilation, social interaction, and daylighting, emphasising the bioclimatic advantages of courtyards. Survey findings indicate a strong interest in multilevel courtyards, with 72.6% of participants desiring to live in one and 38.4% expressing significant enthusiasm. This suggests a notable market potential, especially among young professionals aged 26–35, who represent an important demographic for courtyard innovation. Expert interviews recommend several passive cooling strategies, including the use of compressed earth bricks for thermal mass, mud brick flooring for heat absorption, and integrated water sprinklers for evaporative cooling. Combining these traditional materials with modern technologies could enhance energy efficiency in courtyard spaces. Zoning activities across different levels are proposed to customise space usage, though accessibility considerations, such as ramps, must be addressed to ensure inclusivity. The cultural and bioclimatic significance of courtyards is clear, but there is a demand for innovation to improve their performance. Engaging in rigorous testing of passive cooling strategies, including thermal mass and evaporative techniques, is essential. While zoning may provide functional differentiation, it risks disrupting the open, communal essence of courtyards; therefore, collaborative design involving residents is crucial to balance specialisation and cohesion. Integrating lessons from vernacular architecture with contemporary approaches will facilitate the successful modernisation of courtyards through multilevel typologies. Further research and user engagement will support the implementation of these innovative housing solutions.

To analyse the impact of multilevel courtyard configuration variables (e.g., shading, proportions, height, and building material) on daylighting, energy efficiency, and thermal comfort in hot-dry/humid climates compared to single-level courtyards, DesignBuilder software was employed for quantitative simulations to assess the thermal performance, daylighting, shading, and energy efficiency of multilevel courtyards in these climates. The findings are summarised as follows:

Thermal Comfort:

Compressed earth blocks demonstrated superior thermal mass compared to concrete blocks, yielding a reduction in operative temperatures by 1.19 °C. Altered courtyard levels produced a temperature decrease of 2.15 °C through enhanced stack ventilation and mud brick thermal mass. While the HVAC system achieved a 3.16 °C reduction in temperature, it increased energy consumption, underscoring the need for hybrid ventilation strategies. Expanding the courtyard area by 35 m^2 led to a 0.29 °C temperature reduction by promoting airflow and night cooling. Key strategies like thermal mass, natural ventilation, and courtyard proportions reveal significant potential for passive cooling, with sustainable materials and multilevel designs offering distinct thermal advantages.

Daylighting:

The multilevel courtyard achieved 49.1% daylighting, below the recommended minimum of 80%, especially on lower floors. Ground floor daylighting improved from 29.9% to 56.2% through height reduction but still fell short of targets; additional courtyard enlargement may be necessary. The introduction of a CBRT enclosure reduced daylighting levels by 7.5%, bringing it to 41.6%, due to obstructed sky access necessitating increased artificial lighting. Optimising daylighting is critical for multilevel designs, with open central courtyards showing promise for meeting desired daylight objectives.

Shading:

CBRT shading did not significantly increase energy consumption for cooling or other needs. It contributed to a temperature decrease of 0.08 °C compared to an open courtyard, attributed to enhanced shading. However, it lowered average daylight factors by 7.5%, leading to a dependency on artificial lighting. The implementation of CBRT resulted in a marked increase in lighting energy usage, escalating from 5002.79 kWh to 15,683.88 kWh per year—accounting for 42.9% of total energy consumption, with a 23.9% increase overall. Additionally, residential spaces post-CBRT installation did not meet BREEAM standards due to adverse effects on daylighting, minimal thermal comfort improvement, and elevated energy consumption levels. The reduction in daylighting presents significant implications that merit further consideration.

Energy Efficiency:

The implementation of a multilevel courtyard design has led to a significant reduction in lighting electricity consumption, accounting for only 19% of the total annual electricity usage. This optimisation facilitates enhanced daylight penetration, thereby minimising reliance on artificial lighting during daytime hours. The design also allows adjacent areas to maintain reduced lighting requirements, yielding substantial energy savings. However, the introduction of HVAC systems has markedly increased cooling energy demands, totalling 112.71 MWh annually. It is recommended to prioritise natural ventilation and passive cooling via multilevel courtyards, reserving HVAC for peak conditions to achieve a hybrid approach. This integrated approach prescribed by [56] balances energy efficiency with thermal comfort. Increasing the size of the multilevel courtyard enhances heat dissipation and airflow, resulting in a decrease in operative temperatures by 0.29 °C without additional energy consumption. This underscores the potential for energy optimisation through careful courtyard sizing and geometry. Furthermore, the multilevel courtyard's tiered design promotes vertical airflow and ventilation through the chimney effect, leading to a passive temperature reduction of 2.15 °C compared to a single-level courtyard.

Overall, the multilevel typology offers significant advantages in energy efficiency for lighting and cooling by leveraging effective daylighting, natural ventilation, and passive thermal modulation, as evidenced by the reduced energy consumption for these needs.

9. Conclusions and Recommendations

This research investigated the viability of multilevel courtyards as a contemporary bioclimatic housing model for urban areas characterised by hot-dry and humid climates. One of the objectives was to enhance traditional single-level courtyard designs by optimising environmental performance factors such as ventilation, cooling, daylighting, and energy efficiency. The research question raised was that strategically incorporating varied-level and stepped courtyards could facilitate improved airflow and thermal mass, thus achieving passive climate control and energy savings compared to conventional typologies. A literature review established the historical significance of courtyards in creating favourable microclimates; however, a sustainability-focused re-evaluation is required to address modern urban densities and climate change. Case studies from countries like Vietnam, Morocco, and Nigeria highlighted the cultural relevance and potential for enhanced bioclimatic functionality in contemporary courtyard architecture. Moreover, the examination revealed opportunities for performance improvement through sustainable materials, green features, and innovative configurations. A survey of Abuja residents indicated a strong preference for courtyards due to their functional and aesthetic appeal, particularly in fostering ventilation and communal spaces. The interest in multilevel courtyard designs underscores a shift towards innovative housing solutions. Interviews with experts identified critical considerations for integration, including passive cooling, energy efficiency, and inclusivity. Simulations demonstrated that design elements such as thermal mass walls and tiered levels significantly benefit daylighting and airflow. The study concluded that while multilevel courtyards offer promising bioclimatic interventions, thoughtful design is essential. Findings indicated a cooling effect of 2–3 °C from improved air circulation and shading, yet lower floors may face daylight deficiencies requiring adequate dimensions. Enclosing courtyards was ineffective for lighting and cooling, emphasising the need for climate-responsive designs. Striking a balance between customisation and community connection is vital in multilevel courtyard designs. Different materials and textures can enhance visual differentiation across levels while promoting interaction. Overall, incorporating traditional bioclimatic principles alongside innovative strategies can yield energy-efficient designs suitable for Abuja's context. The multilevel courtyard model presents a climate-responsive, culturally attuned housing strategy for hot-dry and humid environments. Continued pilot studies, community engagement, and expert collaboration are essential for mainstream implementation, potentially establishing multilevel courtyards as a sustainable urban housing paradigm that integrates vernacular knowledge with technological advancements to address contemporary challenges.

To establish climate-specific multilevel courtyard design guidelines that harmonise environmental performance with cultural preferences, the following recommendations are proposed. Further research should investigate the potential of retractable shading systems and water-soaked fabric coverings, with prototype testing to generate comparative performance data. Additionally, exploring methods to reconcile cultural preservation with sustainability is essential. Strategic orientation of courtyards is recommended to enhance solar control, prioritising passive cooling through vegetation and water features. The use of thermal mass materials, such as mud bricks, is advised, alongside minimising glazing areas and incorporating light colours to reflect heat. Zoning multilevel spaces according to cultural, religious, or activity preferences is encouraged, with distinctions made through texture. Safety can be enhanced by maintaining courtyard heights between 400 and 600 mm and using intermediate steps, and enclosing courtyards in high-rises while keeping them open in low-rises. Aesthetic considerations and opportunities for social interaction should not be overlooked in the design process. From an engineering standpoint, accessibility should be facilitated through ramps and railings, ensuring safety. The development of adaptive comfort models and hybrid ventilation systems tailored for multilevel typologies is recommended. On a policy level, providing incentives for integrating multilevel courtyards and mandating inclusive accessibility standards is advisable. Municipal authorities may consider offering property tax rebates to developers who incorporate multilevel courtyards and comply with sustainability standards. This approach encourages the adoption of courtyard designs aimed at improving thermal comfort and energy efficiency; it offers zoning relaxations, such as increased floor area ratios for projects incorporating multilevel courtyards and accessibility features as seen in [57], which allows bonus gross floor area for sustainable buildings, which could be adapted to reward courtyard designs and provide grants for the retrofitting of existing buildings or the construction of new buildings that feature multilevel courtyards. Building codes should permit sustainable materials like stabilised compressed earth blocks, cardboard, and other emerging bio-based building materials, while planning policies should aim to protect socio-cultural spaces such as courtyards. Updating regional building codes to require the incorporation of multilevel courtyards in specific building types, such as residential complexes and public buildings, is an important consideration in hot-dry and humid climates like Abuja's. Ref. [58] asserts that architects in Abuja have shown interest in adopting passive design strategies, indicating the potential for certification programmes that incentivise the integration of multilevel courtyards. Lastly, designated zones should be established in Abuja to serve as testbeds for implementing multilevel courtyard policies. Achieving a balance between environmental sustainability, accessibility, and heritage preservation necessitates interdisciplinary collaboration. Successful pilot projects showcasing multilevel courtyards as a climate-responsive housing strategy could serve as a model for broader implementation.

Author Contributions: Conceptualization, E.C. and A.T.; methodology, E.C. and A.T.; software, E.C. and A.T.; validation, E.C. and A.T.; formal analysis, E.C. and A.T.; investigation, E.C. and A.T.; resources, E.C. and A.T.; data curation, E.C.; writing—original draft preparation, E.C.; writing—review and editing, E.C. and A.T.; visualization, E.C.; supervision, A.T.; project administration, A.T. All authors have read and agreed to the published version of the manuscript.

Funding: This research received no external funding.

Data Availability Statement: Data are contained within the article.

Conflicts of Interest: The authors declare no conflicts of interest.

Abbreviations

The following abbreviations are used in this manuscript:

ADF	Annual daylighting factor
BREEAM	Building Research Establishment Environmental Assessment Methodology
CBRT	Courtyard being roofed technologies
CEB	Compressed earth block
CFD	Computational fluid dynamics
CO2	Carbon dioxide emissions
CSEB	Compressed stabilized earth blocks
HCB	Hollow concrete block
HVAC	Heating, ventilation, and air conditioning
IES	Integrated environmental solution
k av	Average coefficient of thermal conductivity, W/mK

NBRRI	Nigeran Building and Road Research Institute
NT-CBRT	Non-light-transmitting CBRT
NTER	Night-time effectiveness ratio
PET	Physiological equivalent temperature
T-CBRT	Light-transmitting CBRT
U-value	Overall thermal transmittance, W/m^2K

References

1. Krishan, A. (Ed.) *Climate Responsive Architecture: A Design Handbook for Energy Efficient Buildings*; Tata McGraw-Hill Pub. Co.: New Delhi, India; New York, NY, USA, 2001; 409p.
2. Moosavi, L.; Mahyuddin, N.; Ab Ghafar, N.; Azzam Ismail, M. Thermal performance of atria: An overview of natural ventilation effective designs. *Renew Sustain. Energy Rev.* **2014**, *34*, 654–670. [CrossRef]
3. Edwards, B.; Sibley, M.; Hakmi, M.; Land, P. (Eds.) *Courtyard Housing: Past, Present and Future*, 1st ed.; Taylor & Francis: Abingdon, UK, 2004. Available online: https://www.taylorfrancis.com/books/9781134504183 (accessed on 7 April 2025).
4. Gupta, R.; Joshi, M. Courtyard: A Look at the Relevance of Courtyard Space in Contemporary Houses. *Civ. Eng. Archit.* **2021**, *9*, 2261–2272. [CrossRef]
5. Markus, B. Review of Courtyard House in Nigeria: Definitions, history, evolution, typology, and functions. *AFRREV STECH Int. J. Sci. Technol.* **2016**, *5*, 103–117. [CrossRef]
6. Aldawoud, A. Thermal performance of courtyard buildings. *Energy Build.* **2008**, *40*, 906–910. [CrossRef]
7. Ragette, F. *Traditional Domestic Architecture of the Arab Region*, 2nd ed.; Edition Axel Menges GmbH: Stuttgart, Germany, 2006.
8. Foruzanmehr, A. *Thermal Comfort in Hot Dry Climates: Traditional Dwellings in Iran*; Routledge: London, UK; New York, NY, USA, 2018; 1p.
9. Taleghani, M.; Tenpierik, M.; Van Den Dobbelsteen, A. Environmental impact of courtyards—A review and comparison of residential courtyard buildings in different climates. *J. Green Build.* **2012**, *7*, 113–136. [CrossRef]
10. Mohamed, M.A.A. Traditional ways of dealing with climate in Egypt. In *Amman, Jordan: The Center for the Study of Architecture in Arab Region*; CSAAR Press: Amman, Jordan, 2010; pp. 247–266.
11. Zakaria, M.A.; Kubota, T. Environmental Design Consideration for Courtyards in Residential Buildings in Hot-humid Climates: A Review. *Int. J. Built. Env. Sustain.* **2014**, *1*, 50. Available online: https://ijbes.utm.my/index.php/ijbes/article/view/7 (accessed on 7 April 2025). [CrossRef]
12. Khan, H.M.; Lane-Serff, G.F.; Dewsbury, J. Optimizing the design of courtyard houses for passive cooling in hot, dry regions. *J. Build. Perform. Simul.* **2023**, *16*, 231–247. [CrossRef]
13. Almhafdy, A.; Ibrahim, N.; Ahmad, S.S.; Yahya, J. Courtyard Design Variants and Microclimate Performance. *Procedia-Soc. Behav. Sci.* **2013**, *101*, 170–180. [CrossRef]
14. Zamani, Z.; Heidari, S.; Hanachi, P. Reviewing the thermal and microclimatic function of courtyards. *Renew. Sustain. Energy Rev.* **2018**, *93*, 580–595. [CrossRef]
15. Mundra, S.; Kannamma, D. Effect of Courtyard on Comfort—Study of thermal performance characteristics of courtyards in hot and humid climate. In Proceedings of the Architectural Science Association (ANZAScA), Roorkee, India, 28–30 November 2019; pp. 11–20.
16. Fasiku, A.M. The place of courtyard (Akodi) in the cultural upbringing of a child in yorubaland: A caseof omuo in Ekiti state. *Afro Asian J. Soc. Sci.* **2019**, *10*, 1–21.
17. Zhang, D. Courtyard Houses around the world: A cross-cultural analysis and contemporary relevance. In *New Approaches in Contemporary Architecture and Urbanism*; Ahmad Nia, H., Ed.; Cinius Yayinlari: Alanya, Turkey, 2020; pp. 23–44. Available online: http://iccaua.com/page/book-chapter-2020 (accessed on 7 April 2025).
18. Zarghami, E.; Fatourehchi, D.; Karamloo, M. Impact of Daylighting Design Strategies on Social Sustainability Through the Built Environment. *Sustain. Dev.* **2017**, *25*, 504–527. [CrossRef]
19. Al-Jokhadar, A.; Jabi, W. Applying the Vernacular Model to High-Rise Residential Development in The Middle East and North Africa. *Int. J. Arch. Res. ArchNet-IJAR* **2017**, *11*, 175. [CrossRef]
20. Belakehal, A.; Tabet Aoul, K.; Bennadji, A. The Impact of Sunlight and Daylight on the Dwelling's Occupancy: Case of the Hot Dry Regions of Algeria. In Proceedings of the 3rd Conference of the Environmental Psychology in the UK (EPUK) Network, Aberdeen, UK, 23–25 June 2003; Craig, T., Ed.; Robert Gordon University: Aberdeen, UK, 2003; pp. 56–63.
21. Akande, O.K. Passive design strategies for residential buildings in a hot dry climate in Nigeria. In Proceedings of the 3rd International Conference on Harmonisation Between Architecture and Nature, La Coruna, Spain, 12–14 April 2010; pp. 61–71. Available online: http://library.witpress.com/viewpaper.asp?pcode=ARC10-006-1 (accessed on 7 April 2025).

22. Pastore, L.; Corrao, R.; Heiselberg, P.K. The effects of vegetation on indoor thermal comfort: The application of a multi-scale simulation methodology on a residential neighborhood renovation case study. *Energy Build.* **2017**, *146*, 1–11. [CrossRef]
23. Halwatura, R.U.; Jayasinghe, M.T.R. Strategies for improved micro-climates in high-density residential developments in tropical climates. *Energy Sustain. Dev.* **2007**, *11*, 54–65. [CrossRef]
24. Herath, H.M.P.I.K.; Halwatura, R.U.; Jayasinghe, G.Y. Evaluation of green infrastructure effects on tropical Sri Lankan urban context as an urban heat island adaptation strategy. *Urban For. Urban Green.* **2018**, *29*, 212–222. [CrossRef]
25. Kim, J.J.; Rigdon, B. *Sustainable Architecture Module: Qualities, Use, and Examples of Sustainable Building Materials*; National Pollution Prevention Center for Higher Education, University of Michigan: Ann Arbor, MI, USA, 1998.
26. Al-Zubaidi, M.S.S. Sustainability Potential in Traditional Architecture in the Arab World. 2007. Available online: http://eprints.hud.ac.uk/965/ (accessed on 27 December 2023).
27. Fortea, S.; Mohamed, E. An Investigation of Appropriateness Relative to Indigenous and Modern Housing in Libya. Ph.D. Thesis, The University of Edinburgh, Edinburgh, UK, 1989. Available online: http://hdl.handle.net/1842/32203 (accessed on 12 January 2024).
28. Boros, A.; Tőzsér, D. The Emerging Role of Plant-Based Building Materials in the Construction Industry—A Bibliometric Analysis. *Resources* **2023**, *12*, 124. [CrossRef]
29. Ikemiyashiro Higa, D.; Taki, A. Examining Cardboard as a Construction Material for Sustainable Building Practices in Lima, Peru. *Sustainability* **2024**, *17*, 10. [CrossRef]
30. Argalis, P.P.; Sinka, M.; Andzs, M.; Korjakins, A.; Bajare, D. Development of New Bio-Based Building Materials by Utilising Manufacturing Waste. *Environ. Clim. Technol.* **2024**, *28*, 58–70. [CrossRef]
31. Bakam, V.A.; Mbishida, M.A.; Danjuma, T.; Zingfat, M.J.; Hamidu, L.; Pyendang, Z.S. Determination of Thermal Conductivity of Interlocking Compressed Stabilized Earth Block (CSEB). *Int. J. Recent Eng. Res. Dev.* **2020**, *5*, 1–8.
32. Abbakyari, M.; Abuzeinab, A.; Adefila, A.; Whitehead, T.; Oyinlola, M. Designing Sustainable Housing Using a User-Centred Approach: Paipe Case Study. *Buildings* **2023**, *13*, 2496. [CrossRef]
33. Diz-Mellado, E.; Ruiz-Pardo, Á.; Rivera-Gómez, C.; Sanchez De La Flor, F.J.; Galán-Marín, C. Unravelling the impact of courtyard geometry on cooling energy consumption in buildings. *Build. Environ.* **2023**, *237*, 110349. [CrossRef]
34. Wen, B.; Yang, Q.; Xu, F.; Zhou, J.; Zhang, R. Phenomenon of courtyards being roofed and its significance for building energy efficiency. *Energy Build.* **2023**, *295*, 113282. [CrossRef]
35. Aldawoud, A.; Clark, R. Comparative analysis of energy performance between courtyard and atrium in buildings. *Energy Build.* **2008**, *40*, 209–214. [CrossRef]
36. Davila Delgado, J.M.; Oyedele, L.; Ajayi, A.; Akanbi, L.; Akinade, O.; Bilal, M.; Owolabi, H. Robotics and automated systems in construction: Understanding industry-specific challenges for adoption. *J. Build. Eng.* **2019**, *26*, 100868. [CrossRef]
37. Aste, N.; Manfren, M.; Marenzi, G. Building Automation and Control Systems and performance optimization: A framework for analysis. *Renew. Sustain. Energy Rev.* **2017**, *75*, 313–330. [CrossRef]
38. Adewale, B.A.; Ene, V.O.; Ogunbayo, B.F.; Aigbavboa, C.O. A Systematic Review of the Applications of AI in a Sustainable Building's Lifecycle. *Buildings* **2024**, *14*, 2137. [CrossRef]
39. Tabesh, T.; Sertyesilisik, B. An Investigation into Energy Performance with the Integrated Usage of a Courtyard and Atrium. *Buildings* **2016**, *6*, 21. [CrossRef]
40. Adunola, A.O.; Ajibola, K. Factors Significant to Thermal Comfort Within Residential Neighborhoods of Ibadan Metropolis and Preferences in Adult Residents' Use of Spaces. *SAGE Open* **2016**, *6*, 2158244015624949. [CrossRef]
41. Oyedepo, S.O. Towards achieving energy for sustainable development in Nigeria. *Renew. Sustain. Energy Rev.* **2014**, *34*, 255–272. [CrossRef]
42. Weather and Climate. Weather and Climate. Abuja Federal Capital Territory, Nigeria Climate. 2024. Available online: https://weatherandclimate.com/nigeria/abuja-federal-capital-territory (accessed on 3 January 2024).
43. Macrotrends. Abuja, Nigeria Metro Area Population. Abuja, Nigeria Metro Area Population. 2024. Available online: https://www.macrotrends.net/cities/21976/abuja/population (accessed on 3 January 2024).
44. Akubue, J.A.; Adesina, D.O. The Effect of Courtyard Geometry on Airflow Regimes for Ventilation in Tropical Nigerian Environment. *J. Arch. Environ. Struct. Eng. Res.* **2023**, *6*, 1–10. [CrossRef]
45. Nwifa Le-ema, K.; Imaah, N.O. Courtyard enhances natural ventilation and daylighting in buildings in Nigeria. *Int. J. Environ. Sci.* **2019**, *2*, 38–45.
46. Adegun, O.B.; Adedeji, J.M.; Adedeji, Y.M.D. Whither the courtyards? Understanding disappearance and transformation of courtyards in residential buildings in Akure, Nigeria. *IOP Conf. Ser. Mater. Sci. Eng.* **2019**, *640*, 012010. [CrossRef]
47. Saunders, M.N.K.; Lewis, P.; Thornhill, A. *Research Methods for Business Students*, 8th ed.; Pearson: New York, NY, USA, 2019.
48. Rubin, A.; Babbie, E.R. *Research Methods for Social Work*, Student ed.; Cengage Learning: Melbourne, VIC, Australia, 2017; p. 652.
49. Easterby-Smith, M.; Thorpe, R.; Lowe, A. *Management Research*; SAGE: New Delhi, India, 2002.

50. ArchDaily. More than a House. A House in Quang Nam/Lequang-Architects. 2023. Available online: https://www.archdaily.com/1008436/a-house-in-quang-nam-lequang-architects? (accessed on 28 December 2023).
51. ArchDaily. The ReYard House. ReYard House/Team Bosphorus from Turkey. 2020. Available online: https://www.archdaily.com/942491/reyard-house-team-bosphorus-from-turkey (accessed on 16 January 2024).
52. ArchDaily. Villa LL/Mohamed Amine Siana. Villa LL/Mohamed Amine Siana. 2022. Available online: https://www.archdaily.com/994012/villa-ll-mohamed-amine-siana (accessed on 16 January 2024).
53. Field, A. *Discovering Statistics Using IBM SPSS Statistics*, 5th ed.; SAGE Publications: Los Angeles, CA, USA, 2018; p. 1070.
54. Singh, R.K.; Sharma, R.V. Numerical analysis for ground temperature variation. *Geotherm. Energy* **2017**, *5*, 22. [CrossRef]
55. Nguyen, A.T.; Singh, M.K.; Reiter, S. An adaptive thermal comfort model for hot humid South-East Asia. *Build. Environ.* **2012**, *56*, 291–300. [CrossRef]
56. Xi, T.; Sa'ad, S.U.; Liu, X.; Sun, H.; Wang, M.; Guo, F. Optimization of Residential Indoor Thermal Environment by Passive Design and Mechanical Ventilation in Tropical Savanna Climate Zone in Nigeria, Africa. *Energies* **2025**, *18*, 450. [CrossRef]
57. Building and Construction Authority (BCA). Sustainability. Green Mark Incentive Scheme for Existing Buildings 2.0 (GMIS-EB 2.0). 2024. Available online: https://www1.bca.gov.sg/buildsg/sustainability/green-mark-incentive-schemes/green-mark-incentive-scheme-for-existing-buildings-2.0 (accessed on 29 April 2025).
58. Adewale, B.A.; Ogunbayo, B.F.; Aigbavboa, C.O.; Ene, V.O. Evaluation of Green Design Strategies Adopted by Architects for Public Buildings in Nigeria. *Eng. Proc.* **2024**, *76*, 24. [CrossRef]

www.ingramcontent.com/pod-product-compliance
Lightning Source LLC
Chambersburg PA
CBHW061848210926
51850CB00006B/1005

MDPI AG
Grosspeteranlage 5
4052 Basel
Switzerland
Tel.: +41 61 683 77 34

Energies Editorial Office
E-mail: energies@mdpi.com
www.mdpi.com/journal/energies